MANUFACTORIES
Nº 821 CHERRY STREET.
AND
Fifth St: near Columbia Avenue.
PHILADELPHIA.
821 CHERRY ST.
STORE 710 CHESTNUT ST.

P S Duval, Son & Co.
Designers
&
Lithographer
PHILADELPHIA
Portraits & all subjects of fine arts executed in
CHROMO.

PHILADELPHIA

AND ITS

MANUFACTURES;

A HAND-BOOK

OF THE

GREAT MANUFACTORIES AND REPRESENTATIVE MERCANTILE HOUSES OF PHILADELPHIA,

IN 1867.

BY

EDWIN T. FREEDLEY,

AUTHOR OF A "PRACTICAL TREATISE ON BUSINESS," "THE LEGAL ADVISER," "OPPORTUNITIES FOR INDUSTRY," ETC.

PHILADELPHIA:
EDWARD YOUNG & CO.,
144 SOUTH SIXTH STREET.

RESPECTFULLY DEDICATED

TO

THE MERCHANTS OF THE INTERIOR,

In the hope that this exhibit of the first-class MANUFACTORIES and MERCANTILE HOUSES of the largest Manufacturing city on this continent will benefit as well as interest them, and in the belief that they are directly and personally concerned in knowing that Philadelphia is the leading producer of those Staple Goods which, by their intrinsic value and low price, are specially adapted to the wants of the people of the Middle, Western, and Southern States.

PHILADELPHIA AND ITS MANUFACTURES.

MANUFACTURES—CAUSES OF EMINENCE IN.

THE term *Manufacture*, in its derivative sense, signifies —making by hand. Its modern acceptation, however, is directly the reverse of its original meaning; and it is now applied particularly to those products which are made extensively by machinery, without much aid from manual labor. The word therefore is an extremely flexible one; and as Political Economists do not agree in opinion, whether millers and bakers are properly manufacturers or not, we shall, if need be, take advantage of the uncertainty, and consider as Manufactures what strictly may belong to other classifications of productive industry.

The end of every Manufacture is to increase the utility of objects by modifying their external form or changing their internal constitution. In some instances, substances that would otherwise be utterly worthless, are converted into the most valuable products—as the hoofs of certain animals into Prussiate of Potash; the offal into Goldbeater's Skin; and especially rags into Paper. Thus beneficent in their general object, it is scarcely remarkable that modern Manufactures are principally distinguished for their ameliorating influence upon man's social condition. By cheapening manufactured products they put within the reach of the poorest classes what in former times was accessible only to the wealthy and noble. The servant, the artisan, and the husbandman of England, at the present time have more palatable food, better clothing and better furniture, than were possessed by "the gentilitie" in the "golden days" of Queen Bess. In no other equally extensive districts of the world are the people generally so well off as to physical comforts, or so intellectually progressive, as in England and Massachusetts, and in none have Manufactures as yet attained equal prominence as branches of industry. In 1850 there were employed in *Textile Manufactures* alone, the following:

MILLS IN THE UNITED KINGDOM.

	England and Wales.	Scotland.	Ireland.	Total.
Mills,	3,699	550	91	4,340
Spindles,	22,859,010	2,256,408	532,303	25,647,721
Power Looms,	272,586	28,811	2,517	303,914
Moving Power, Steam, (horses)	91,610	13,857	2,646	108,113
" " Water, "	18,214	6,004	1,886	26,104

The persons employed in these mills numbered 596,082, of whom 40,775 were children under thirteen years of age, and 329,577 were females above thirteen. In the United States, the most important of the Textile Manufactures are those of Cotton and Wool. In 1850 there was employed in the Cotton Manufacture a capital of $74,500,931; consuming 641,240 bales of cotton annually, and producing about 763,000,000 yards of sheetings, shirtings, calicoes, &c., and 27,000,000 lbs. of yarn, valued for the entire product at $61,869,184. The number of persons employed was 92,286, of whom 33,150 were males, and 59,136 were females. Massachusetts contained about one-third of the whole number of spindles in the United States, and about one-half the capital invested in the Cotton Manufacture was owned in Massachusetts. The Woolen Manufacture of the United States employed a capital of about $28,000,000; consuming 71,000,000 lbs. of Wool, worth $25,000,000; and the product was valued at $43,207,545. It is more generally distributed throughout the United States than the Cotton Manufacture, yet Massachusetts employs in it one third of the whole capital and consumes one third of the Wool.

But the future of manufacturing enterprise in the United States, except in its effects upon society, must not be judged from its present development in Massachusetts. In 1810, according to the census, Virginia, the two Carolinas and Georgia manufactured greatly more in quantity and value of Cotton and Woolen fabrics than the whole of New England; and North Carolina produced double as many yards as Massachusetts. We doubt not the superior intellectual energy of the people of Massachusetts has attracted much that, with equality in this particular, combined with superior physical advantages, will again be attracted elsewhere. Manufacturing enterprise in the United States is yet in its experimental stage. The people have but recently recovered from the delusion that Manufactures are injurious to national prosperity. They have not had time to study the conditions upon which success in Manufactures depends, or to comprehend the lines that naturally and properly separate human pursuits. In future times, a manufacturer will no more think of consulting merely his personal inclinations, or one favorable circumstance, in the location of his manufactory, than the agriculturist, for a similar reason, would choose for the field of his operations the Pilot Knob of Missouri, or the gold-seeker the sands of New Jersey. As yet manufacturers are working independently not only of each other, but of the general laws that underlie economical production. Being in doubt as to the proper locality, they have not concentrated or combined their efforts; and the buyers of manufactured goods being in doubt as to the Home

Market, give their confidence to European manufacturers. It is the object of the present volume to submit, with due deference, to the consideration of both these classes, some suggestions based on the experience of the past and of other countries; and to endeavor to aid them—First: *by considering what are the requisites to prosperity or the causes of economical production in Manufactures;* Secondly: *by indicating a locality possessing the advantages for manufacturing in the highest degree of perfection;* Thirdly: *by showing the progress already made in Manufactures in that locality.*

I. Political Economists divide the essential requisites of production into two—Labor, and appropriate natural objects. To these, in Manufactures, we must certainly add Capital. But the productive efficacy of all productive agents, as every one has observed, varies greatly at various times and places, and depends upon a variety and due combination of circumstances, partly *moral* and partly *physical.* Foremost among the moral circumstances conducive and essential to prosperity, especially in Manufactures, are *freedom of industry and security of property.* We need but glance at the history of any European nation, France in particular, to discover that governmental interference with industry is baneful in its effects, and that monopolies and corporation privileges retard progress. "I have frequently seen," says Roland de la Platière, a minister of state during the French Revolution, "manufacturers visited by a band of satellites, who put all in confusion in their establishments, spread terror in their families, cut the stuff from the frames, tore off the warp from the looms, and carried them away as proofs of infringement; the manufacturers were summoned, tried and condemned; their goods confiscated; copies of their judgment of confiscation posted up in every public place; future reputation, credit, all was lost and destroyed. And for what offense? Because they had made of worsted a kind of cloth called *shay*, such as the English used to manufacture, and even sell in France, while the French regulations stated that that kind of cloth should be made with *mohair*. I have seen other manufacturers treated in the same way, because they had made camlets of a particular width, used in England and Germany, for which there was a great demand from Spain, Portugal, and other countries, and from several parts of France, while the French regulations prescribed other widths for camlets. There was no free town where mechanical invention could find a refuge from the tyranny of the monopolists —no trade but what was clearly and explicitly described by the statutes could be exercised—none but what was included in the privileges of some corporation."

In England freedom of industry dates from the abolition of monopolies in 1624; and there can be no question, as McCulloch observes, that "Freedom and security—freedom to engage in every employment, and to pursue our own interest in our own way, coupled with an intimate conviction that acquisitions, when made, might be securely enjoyed or disposed of—have been the most copious sources of our wealth and power. There have been only two countries, Holland and the United States, which have, in these respects, been placed under nearly similar circumstances as England; and notwithstanding the disadvantages of their situation, the Dutch have long been, and still continue to be, the most industrious and opulent people of the Continent—while the Americans, whose situation is more favorable, are rapidly advancing in the career of improvement with a rapidity hitherto unknown."

In the United States, industry, it is true, is generally free, and property in most places adequately protected by public opinion against both legislative and mob violence;

but our advantages in these respects for the development of enterprise in Manufactures have been modified and limited by fluctuating legislation on the subject of foreign competition. Very early in our constitutional history the question was agitated—Shall Government, in adjusting its taxes for revenue, so discriminate as to protect and encourage Home Manufacturers, or in other words, to diminish, if not exclude, foreign competition in our markets? This question was submitted to the people, but proved too vast for popular solution. Their opinions changed with the current of argument, like the judgment of the Dutch Justice; and the decision which they had made promptly in accordance with the wish of the attorneys on the one side, was as promptly reversed upon the suggestion of the attorneys on the other side. Finally, not knowing what to do, the majority seem to have concluded that, as posterity had done nothing for them they were under no obligations to do any thing for posterity. In the mean time legislation upon the question fluctuated with the vacillation in public sentiment; and capitalists being unable to calculate with certainty the risks involved, were timid in embarking in manufacturing enterprises. It would seem therefore that, in addition to security of property and freedom of industry, success in Manufactures implies a certain and stable, if not wise policy in governmental action upon questions affecting manufacturing interests.

2. Another moral cause contributing, and in fact essential to eminence in manufacturing industry, is *the general diffusion of intelligence among the people.* By intelligence, in this connection, we do not mean merely the understanding necessary to enable an individual to become the creator or the lord of a machine. The capacity to contrive and invent seems so much a part of the original

constitution of man, that we believe there is in every civilized community sufficient ingenuity and mental power to have originated all in physical science that has yet been devised by any. The mind is God's machine, with powers seemingly unlimited, and capable of producing any thing from a bad pun to the lever of Archimedes, the flying pigeon of Archytas or the calculating machine of Babbage. But the exercise of this faculty, the application of the best intellect in a community in the direction of practical improvements, depends largely upon the approbation and rewards bestowed upon successful enterprise in invention or mechanical labor. It is in vain to hope that ambition will spur intellect to achieve mechanical triumphs, where an inventor is respected less than a tinseled soldier or a ragged lawyer. It is in vain to expect that mechanics will strive to acquire any extraordinary skill where mechanical labor is degraded to serfdom, or even is not appreciated. In the histories of nations, whose rise and fall are classical studies, we learn that the application of mind to invention as well as handicraft operations, was regarded as unworthy of freemen. "In my time," says Seneca, "there have been inventions of this sort—transparent windows, tubes for diffusing warmth equally through all parts of a building; short-hand, which has been carried to such perfection that a writer can keep pace with the most rapid speaker. *But the inventing of such things is drudgery for the lowest slaves.* Philosophy lies deeper. It is not her office to teach men how to use their hands." Another ancient and eminent teacher, who can boast of a disciple here and there in our country, considered the true object of all education and philosophy to be—to fit men for war. Need we wonder there have been dark ages in the world's history. Need we say that in an atmosphere tainted with such a sterile philosophy, the arts which improve man's material condition cannot flourish. The proud position of New England—a position so enviable that her light reflects lustre on States with which she is allied—is due rather to her sound, intelligent, practical philosophy, than to any physical advantages or original intellectual superiority. A Yankee lad inhales from the surrounding atmosphere, if he do not hear from his father's lips, that it is an important part of his duty to aid in extending man's empire over the material world, and every available addition to human force for accomplishing that end, that he may originate, will be a sure passport to the respect of his neighbors, if not to fortune. The women and children are educated to regard ignorance and idleness as vices; and all, deeming it honorable to add something to the aggregate product of their country's wealth, co-operate and lighten the original curse, for

"All are needed by each one,
Nothing is fair or good alone."

3. A third cause of eminence in Manufacturing, and essential to economical production, *is an abundant supply of the most effective laborers, and of those qualified to direct labor.* In view of the improvements already made, it would be rash to assert that a time will never come when automatic machines will dispense entirely with manual labor in manufacturing. So far, the introduction of machinery has stimulated the pressing demand for educated labor; and if we can at all judge of the future, success will depend more and more upon the quality of the labor employed. Labor is effective according as it is dexterous or as it is skillful. In purely routine processes, dexterity may be the quality of chief value, but laborers differ in dexterity almost as much as in mechanical skill. Englishmen say that a laborer in Essex is cheaper at 2*s.* 6*d.* per

day than a laborer in Tipperary at 5*d.*; and as operatives in cotton factories, our manufacturers assert that one American girl can accomplish as much in a given time as two English girls. "In England," said Mr. Kempton, before the Committee upon Manufactures of the House of Commons, "the girls tend two power-looms. In America our girls tend generally four power-looms; some for years tended five power-looms, and some tended six for some time, and each of those power-looms turned off more cloth than I have found any power-looms turn off in this country." Mr. Cowell, in illustrating the comparative efficiency of operatives, remarks—"At Mulhausen, which is styled the Manchester of France, one adult and two children are requisite for the management of 200 *coarse* threads, and they gain *among them* about 2*s.* (48 cents) at coarse work. At Manchester or Bolton one adult and two children can manage 758 threads, and gain among them 5*s.* 6*d.* per day. Thus, although wages are so much lower in France, the difference of product is so great that the cost, in money, of the commodity produced, is greater than in England. In the former, four men and two children are required to manage 800 threads, for which they receive 8*s.*, while in the latter one man and two children are capable, with the best machinery, of doing the same, and their wages are 5*s.* 6*d.*"

If then there be such difference in the productive efficacy of laborers, in operations calling for mere manual dexterity, it is obvious that the higher we ascend in those departments of mechanics and manufactures, in which the mind has a considerable part, the greater must be the advantage in favor of intelligence and skill. And such is the fact. The only standard by which to estimate the cost of labor, is the *amount of work done for the money paid*—the *per diem* earnings of the workmen being in itself no criterion by which to judge of the cost of labor.

That workman is the cheapest who can produce the most of a given quality for a given sum of money; whether he earn one dollar or five dollars per day, and that manufacturer can produce with the most efficiency, and the least expense, other things being equal, who can *at all times* command the requisite supply of such workmen.

As ingenious mechanics and rapid workmen, the Anglo-Americans have no superiors. As skillful workmen in departments for which they have been specially educated, the English are celebrated. Regular and habitual energy in labor, however, is a characteristic of both. They have no life but in their work—no enjoyment but in the shop. What other races consider amusement, is no amusement to them. But in England and America there is a marked difference between the quality of the labor that can be obtained in the country and in the towns. In fact, in or near large cities only can labor of the first quality be obtained. "As iron sharpeneth iron, so a man sharpeneth the countenance of his friend;" and away from the centres of population and competition, the face loseth its sharpness, and the hand its cunning. Cities are in nothing more remarkable than in their attractive, magnetic influence upon talent of every description. "The man who desires to employ his pen," observes Carey, "and who possesses only the ability to conduct a country newspaper, removes to the interior, while the man of talent leaves his country paper to take charge of one in the city. The dauber of portraits leaves the city to travel the country in search of employment, while the painter removes to Philadelphia, New York or London. The inferior lawyer, physician, surgeon, dentist or merchant removes to the West, while the superior one leaves the West and settles in those places in which population is dense; where the means of production are great; where talent is appreciated and best paid; and where reputation, when acquired, is worth possessing."

Superior mechanics and dexterous workmen manifest a similar preference for cities and an abhorrence of isolation; hence, if for no other reason, extensive mechanical or manufacturing operations must be conducted at a great disadvantage in isolated localities. In a limited experience, I have known of several establishments that have failed apparently from no other cause than the impossibility of filling orders promptly, in consequence of difficulty in procuring and retaining an adequate supply of good mechanics in an unattractive locality; and to the disposition to select such situations because of water power or some other circumstance, we ascribe much of the past embarrassments of our manufacturers. In some of the secluded manufacturing villages of New England, it is the custom of the proprietors to fasten such superior workmen as they may have seduced thither, by aiding them to invest their earnings in a house and lot, which they cannot afterward dispose of except at a great sacrifice; but the practice, it would seem, is rather to be commended for its shrewdness than its wisdom. A dependent or dissatisfied workman can hardly be an efficient one.

As respects those who are well *qualified to direct labor*, the supply is, in all places, especially in isolated localities, far short of the demand. Foremost in this class it can be no disparagement to place *scientific men.* As agents of economical production, none are more effective. The progress of Manufactures, in many of its departments, is intimately connected with and dependent upon the progress made in the exact sciences; and to the experiments and investigations of scientific men—the men who peer into the secrets of Nature, whether concealed in plants, in animals or minerals, and who

> "Find tongues in trees, books in running brooks,
> Sermons in stones, and good in every thing,"—

that our Manufactures are largely indebted for their present development, and upon such men, we must rely principally, as we may do with confidence, for the discovery of new sources of wealth, that at a future day will give employment and wealth to millions of human beings. But scientific men are not abundant even in the centres where Libraries, Galleries and Academies are numerous; those best qualified to direct labor prefer the theatres offering the widest scope for the exhibition of their abilities; and even inventors have discovered, that in isolated localities, they may exhaust their efforts in attempting what has been better executed before.

II. Passing to the *physical causes* of eminence in manufacturing industry, we remark they are more obvious than the moral causes, but not more important. To produce manufactured goods of a given quality with the least expense being the great desideratum, it follows that whatever contributes to economy in production, whatever saves labor, or transportation, or raw materials, cannot safely be overlooked or despised. But to investigate carefully all the circumstances that have an influence upon economical production, would require a considerable volume, and be foreign to our main inquiry. Desiring merely to discover a locality within our extended country, that, by the use of the proper means, will certainly become the centre and chief seat of American manufactures, it is necessary to know what *circumstances have more influence than any others in facilitating manufacturing enterprise,* and thus sooner or later lead to superiority; but it is not necessary to exhaust the subject.

England, it is acknowledged, is pre-eminent in Manufactures over all other countries—but why? Her colonial system, her shrewd legislation, the simplicity of other nations and other accidental circumstances, have no doubt widened the market for her manufactured goods to an extraordinary extent, but her superiority nevertheless is the result of solid, substantial, not accidental circumstances. The physical advantages which have contributed more than any others to her eminence, as we think all must agree, are epitomized by the Edinburgh Review, in the following summary: 1st. Possession of supplies of the raw materials used in Manufactures; 2d. The command of the natural means and agents best fitted to produce power; 3d. The position of the country as respects others; and 4th. The nature of the soil and climate.

"1. As respects the first of these circumstances," the writer says, "every one who reflects on the nature, value, and importance of our manufactures of Wool, of the useful Metals,—such as Iron, Lead, Tin, Copper,—and of Leather, Flax, and so forth, must at once admit, that our success in them has been materially promoted by our having abundant supplies of the raw material. It is of less consequence whence the material of a manufacture possessing great value in small bulk is derived, whether it be furnished from native sources, or imported from abroad, though even in that case the advantage of possessing an internal supply, of which it is impossible to be deprived by the jealousy or hostility of foreigners, must not be overlooked. But no nation can make any considerable progress in the manufacture of bulky and heavy articles, the conveyance of which to a distance unavoidably occasions a large expense, unless she have supplies of the raw material within herself. Our superiority in manufactures depends more at this moment on our *superior machines* than on any thing else; and had we been obliged to import the iron, brass, and steel, of which they are principally made, it is exceedingly doubtful whether we should have succeeded in bringing them to any thing like their present pitch of improvement.

"2. But of all the physical circumstances that have contributed to our wonderful progress in manufacturing industry, none has had nearly so much influence as our possession of the most valuable coal mines. These have conferred advantages on us not enjoyed in an equal degree by any other people. Even though we had possessed the most abundant supply of the ores of iron and other useful metals, they would have been of little or no use, but for our almost inexhaustible coal mines. Our country is of too limited extent to produce wood sufficient to smelt and prepare any considerable quantity of iron, or other metal; and though no duty were laid on timber when imported, its cost abroad, and the heavy expense attending the conveyance of so bulky an article, would have been insuperable obstacles to our making any considerable progress in the working of metals, had we been forced to depend on home or foreign timber. We, therefore, are disposed to regard Lord Dudley's discovery of the mode of smelting and manufacturing iron by means of coal only, without the aid of wood, as one of the most important ever made in the arts. We do not know that it is surpassed even by the steam engine or spinning-frame. At all events, we are quite sure that *we* owe as much to it as to either of these great inventions. But for it, we should have always been importers of iron; in other words, of the materials of machinery. The elements, if we may so speak, out of which steam-engines and spinning-mills are made, would have been dearer here than in most other other countries. The fair presumption consequently is, that the machines themselves would have been dearer; and such a circumstance would have counteracted, to a certain extent, even if it did not neutralize or overbalance, the other circumstances favorable to our ascendancy. But now we have the ores and the means of working them in greater abundance than any other people; so that our superiority in the most important of all departments —that of machine-making—seems to rest on a pretty sure foundation.

"It is further clear, that without a cheap and abundant supply of fuel, the steam-engine, as now constructed, would be of comparatively little use. It is, as it were, the hands; but coal is the muscles by which they are set in motion, and without which their dexterity cannot be called into action, and they would be idle and powerless. Our coal mines may be regarded as vast magazines of *hoarded* or *warehoused* power; and unless some such radical change be made on the steam-engine as should very decidedly lessen the quantity of fuel required to keep it in motion, or some equally powerful machine, but moved by different means, be introduced, it is not at all likely that any nation should come into successful competition with us, in those departments in which steam-engines, or machinery moved by steam, may be most advantageously employed.

"Since the introduction of steam-engines, Water-falls, unless under very peculiar circumstances, have lost almost all their value. Steam may be supplied with greater regularity, and being more under command than water, is therefore a more desirable agent. This, however, is but a small part of its superiority. Any number of steam-engines may be constructed in the immediate vicinity of each other, so that all the departments of manufacturing industry may be brought together and carried on in the same town, and almost in the same factory. A combination and adaptation of employments to each other, and a consequent saving of labor, is thus effected, that would have been quite impracticable, had it been necessary to construct factories in different parts of the country, and often in inconvenient situations, merely for the sake of waterfalls.

"It may be supposed, perhaps, that a difficulty of this sort might have been obviated by the employment of horse-power instead of steam; but the following statement, which we extract from Dr. Ure's work, shows conclusively that this would not have been the case:—

"'The value of steam-impelled labor may be inferred from the following facts, communicated to me by an eminent engineer, educated in the school of Boulton and Watt:—A manufacturer in Manchester works a sixty-horse Boulton and Watt's steam-engine, at a power of one hundred and twenty horses during the day, and sixty horses during the night; thus extorting from it an impelling force three times greater than he contracted or paid for. One *steam* horse-power is equivalent to 33,000 pounds avoirdupois, raised one foot high per minute; but an *animal* horse-power is equivalent to only 22,000 pounds raised one foot high per minute, or, in other terms, to drag a canal boat two hundred and twenty feet per minute, with a force of one hundred pounds acting on a spring; therefore, a steam-horse power is equivalent in working efficiency to one living horse, and one-half the labor of another. But a horse can work at its full efficiency only eight hours out of the twenty-four, whereas a steam-engine needs no period of repose; and, therefore, to make the animal power equal to the physical power, a relay of one and a half fresh horses must be found three times in the twenty-four hours, which amounts to four and a half horses daily. Hence, a common sixty-horse steam-engine does the work of four and a half times sixty horses, or of two hundred and seventy horses. But the above sixty-horse steam-engine does one-half more work in twenty-four hours, or that of *four hundred and five* living horses! The keep of a horse cannot be estimated at less than 1*s.* 2*d.* per day; and, therefore, that of four hundred and five horses would be 24*l.* daily, or 7,500*l.* sterling, in a year of three hundred and thirteen days. As eighty pounds of coals, or one bushel, will produce steam equivalent to the power of one horse in a steam-engine during eight hours' work, sixty bushels, worth about 30s. at Manchester, will maintain a sixty-horse engine in fuel during eight effective hours,—and two hundred bushels, worth 100*s.*, the above hard-worked engine during twenty-four hours. Hence, the expense per annum is 1,565*l.* sterling, being little more than one-fifth of that of living horses. As to prime cost and superintendence, the animal power would be greatly more expensive than the steam power. There are many engines made by Boulton and Watt, forty years ago, which have continued in constant work all that time with very slight repairs. What a multitude of valuable horses would have been worn out in doing the service of these machines! and what a vast quantity of grain would they have consumed! Had British industry not been aided by Watt's invention, it must have gone on with a retarding pace, in consequence of the increasing cost of locomotive power, and would, long ere now, have experienced, in the price of horses and scarcity of water-falls, an insurmountable barrier to further advancement: could horses, even at the low prices to which their rival, steam, has kept them, be employed to drive a cotton-mill at the present day, they would devour all the profits of the manufacturer.' "

Water power has heretofore been considered cheaper, especially for small manufacturing establishments, than steam power; but eminent engineers have carefully investigated the subject, and are of opinion that in any position where coal can be had "at ten cents per bushel," steam is as cheap as water power at its minimum cost. Even for cotton factories, the manufacturers of New England, according to Montgomery, consider the *advantages of a good location as fully equal to the extra expense of steam power, even when coal must be transported from Pennsylvania to Massachusetts*, and the largest mills now being erected are to have steam as a motive power. Steam, therefore, until superceded by some more effective agent, will be the power principally relied upon to propel Machinery; and as wood for the generation of steam upon an extensive scale is out of the question, we may safely conclude that at no very distant day, the centre of our Manufactures will certainly be in or near a district possessing inexhaustible supplies of cheap coal.

The importance of coal as a useful agent in the Arts, is not, however, limited to its capacity to produce power. It lies at the base of all manufacturing and mining operations, and surpasses all other natural products in the power of attracting to the vicinity where it can be obtained abundantly and cheaply—industry and population. In England, the Woolen Manufacturers were once scattered over Sussex, Kent, and other southern counties, but they have been attracted, principally by the wonderful magnetism of coal, to the North. In the coal districts of England we find all her great manufacturing cities and towns; Birmingham, with its population of perhaps 300,000; Leeds, with a population of 200,000; Sheffield, whose hardware manufactures are known all over the world, are located in districts abounding with coal, and its usual accompaniment—Iron. Manchester, the great seat of the Cotton Manufactures of Great Britain, whose population now exceeds 600,000, is situated on the edge of an immense and seemingly inexhaustible coal-bed. A like proximity may be noticed in the location of Bolton, Bradford, Carlisle, Huddersfield, Oldham and Wolverhampton in England; Merthyr Tydvil in Wales; Glasgow in Scotland; and Charleroy in Belgium.

The principal manufacturing cities of Europe, in this respect, present a striking contrast to those of the United States. In New England, the sites of the chief manufacturing towns seem to have been chosen solely with reference to abundant water power; and herein we have one reason for believing that their present pre-eminence is destined soon to be overshadowed, and finally obscured by that of other cities possessing all their other advantages, and having, in addition, a convenient proximity to our immense coal-beds. In spite of our warm regard for New England, and sincere wishes for her continued prosperity in Manufactures, we think the sceptre will eventually, and ere long, depart from Judah. But New England will be New England still. The virtues which make a great people are indigenous to her soil, and will continue to animate and ennoble the population when her capitalists and ingenious men have sought other localities, possessing greater physical advantages for the fulfillment of their "manifest destiny."

3. With regard to the third point, viz. *favorable situation as respects commerce with other countries*, its importance is second only to that which we have just considered. It is in the nature of Manufactures to be regardful of distant and foreign markets. The accelerated production which results from the application of machinery, enables one manufacturer to supply the wants of many hundreds of consumers, and a country or part of a country possessing superior facilities for Manufactures, can supply other countries with manufactured goods cheaper than they can produce them. Great Britain, it is well known, exports the bulk of her manufactured commodities. The writer whom we previously quoted, remarks:

> "Owing to the facilities afforded by our insular situation for maintaining an intercourse with all parts of the world, our manufacturers have been able to obtain supplies of the raw materials on the easiest terms, and to forward their own products wherever there was a demand for them. Had we occupied a central situation, in any quarter of the world, our facilities for dealing with foreigners being so much the less, our progress, though our condition had been otherwise in all respects the same, would have been comparatively slow. But being surrounded on all sides by the sea, that is, by the great highway of nations, we have been able to deal with the most distant as well as with the nearest people, and to profit by all the peculiar capacities of production enjoyed by each."

In the United States, the consumption of manufactured goods is so vast, that we are apt to regard any foreign demand as unimportant. But for the year ending June 30, 1855, we exported manufactured commodities to the amount of $30,609,518. The list of articles exported embraced nearly all our prominent Manufactures—Cotton piece Goods being the most valuable item, amounting to

$5,857,181; Manufactures of Iron the next, $3,753,472; and Artificial Flowers and Billiard Tables the smallest, of which, however, the exports amounted to about $8000. The Canadas, the West Indies, the South American Republics, Spain and her dependencies, Russia, China, are all ready and willing to exchange their natural products for our manufactured goods, if we can compete with other manufacturing countries in their markets. Even English consumers have no objection to take our Manufactures, not excepting Cotton goods, if the price can be arranged satisfactorily. As early as 1826 we exported $664 cotton goods to England; in 1837, $11,889; and ever since, we believe, there have been small shipments annually. Hence, though it be true that, in the United States, the Home market is the one at present of chief importance, and though the consumption of manufactured goods is so immense that there is undoubtedly room for the establishment of many important local manufactories, if such can exist, at a variety of points; yet to supply a foreign demand, as well as to obtain the raw materials on the easiest terms, a situation on or near the sea-coast is desirable; and as large establishments can produce more cheaply than small ones, as we shall subsequently show, it is highly important for such to choose a locality possessing, in addition to the other moral and physical advantages, a complete communication, by railroads and canals, with all parts of our own country, and an established commerce or facilities for commerce with foreign countries.

4. A suitable *Climate* is also a consideration of very great importance. The influence of climate upon the productiveness of industry, especially in Manufactures, is very marked. A warm climate not only enervates the body, but enfeebles the mind. It diminishes the utility of money; and by rendering houses and clothing less

necessary to existence, relieves the inhabitants of one great spur to industry and invention. In very cold climates, on the other hand, the powers of Nature are benumbed, and the difficulty of preserving life overrides all considerations for making existence comfortable. The climate which seems most favorable to the development of manufacturing industry, is that which is also most conducive to health and longevity, imparting vigor to the frame and force to the intellect, and if we may judge from the past, it is found especially, if not exclusively, in that part of the Eastern Hemisphere which lies between the parallels of 45° and 55°, and in the Western between 39° and 45° North Latitude. Climate has also a direct influence upon the durability of buildings, the working of machinery, and the dyeing of fabrics—points that we may subsequently consider—and thus becomes an element of important consideration in many kind of Manufactures.

The *Soil* of a country or district well adapted for Manufactures, need not be naturally very fertile. In fact a soil naturally so rich that Agriculture is an easy art, will not afford sustenance to many kinds of Manufactures. In Southern Europe, for instance, where, according to one authority, the only art which the farmers know is to leave their ground fallow for a year, so soon as it is exhausted, and the warmth of the sun alone and temperature of the climate enrich it and restore its fertility, we look in vain for those enterprises which are the product of qualities and virtues that are nourished by difficulties, not facilities. In England, the soil is naturally coarse and stubborn, but capable of being made highly productive by labor, expense, and good husbandry; and such a soil, with the habits of careful cultivation induced thereby, is the safest reliance for supplying the markets of a manufacturing district with the necessaries of life, at moderate prices.

III. But the one thing essential for the cheap production of manufactured commodities, and without which all the other moral and physical advantages are ineffectual, remains to be noticed. It is ASSOCIATION OR COMBINATION OF LABOR. It is unnecessary to show that man, unaided by his fellow men, is a helpless being. If it were, we might refer to the savages of New Holland, who, they say, never help each other even in the most simple operations; and their condition, as may be supposed, is hardly superior, in some respects it is inferior, to that of the wild animals which they now and then catch. The first step in social improvement, is association for mutual security and mutual assistance; and every advance in civilization is directly the result of some new combination of efforts. All the marvels of past times, produced by human agency —the Temples, Pyramids and Catacombs—and all the wonders of the present—its Railroads, Telegraphs, Mines and Manufactures—have a common origin in association of numbers for a common purpose. All industrial pursuits depend more or less upon this principle for development, but in none are its advantages more strikingly manifest than in manufacturing operations.

To combine Labor effectually, it is necessary first to *separate employments into parts—that is, to assign to each co-worker a special occupation.* The Division of Labor, as Wakefield, it is said, was the first to point out, is only a single department of a more comprehensive Law, which he denominated Co-operation, or combined action of numbers. Its efficiency, however, as an aid to production, is none the less important, and has been abundantly illustrated by all who have written on Political Economy. Adam Smith illustrated it from pin-making; and mentioned that ten men, in a small manufactory, but indifferently accommodated with the necessary machinery, could make, by confining themselves as much as possible to distinct operations, upward of 48,000 pins in a day, or 4,800 for each individual, whereas if they all wrought separately and independently, they certainly could not, each of them, make twenty, perhaps not one pin in a day.

M. Say illustrates the principle by reference to the manufacture of playing-cards, and says that each card, before being ready for sale, undergoes no fewer

"Than seventy operations, and if there are not seventy clasess of work-people in each card manufactory, it is because the division of labor is not carried so far as it might be; because the same workman is charged with two, three, or four distinct operations. The influence of this distribution is immense. I have seen a card manufactory where thirty workmen produced daily 15,500 cards, being above 500 cards for each laborer; and it may be presumed that if each of these workmen were obliged to perform all the operations himself, even supposing him a practiced hand, he would not perhaps complete two cards in a day, and the thirty workmen, instead of 15,500 cards, would make only sixty."

Henry C. Carey refers to weaving in India, and says:

"In India each weaver works by himself. He purchases at a high price, on credit, the materials with which he is to work, and the provisions required for his support, and he sells the product at a price not exceeding one-third of its market value. Here is no combination of action—no division of labor. The whole work is to be performed by the single individual; and the time that might be employed in finishing the finest muslins, is wasted upon various processes requiring inferior ability, from the purchase of the cotton to its final sale."

Further illustrations are therefore superfluous. The principle is settled: quantity and economy of production are immeasurably aided by the division of employments into parts for the sake of combination of Labor.

Secondly, to combine Labor to the best advantage, it is essential to conduct operations *on a sufficientlg large scale to have a separate workman, or a separate machine, for each process into which it is convenient to subdivide the manufacture, and to afford each workman or machine full employ-*

ment in that special occupation. This we regard to be the natural limit of a manufacturing establishment. Any extension beyond this may be said to comprise two establishments in one; and any establishment of less size cannot realize the full benefits of a Division of Labor, and consequently cannot produce with the utmost efficiency and economy. The application of the principle, however, would, in most kinds of Manufactures, lead to moderately large establishments; and that such establishments can produce more economically, or in other words afford to work for a less percentage of profit, is simply a well-established fact. A Philadelphia miller is content with the bran alone as his toll for grinding his customer's corn; but a country miller, in a sparsedly populated district, must take a considerable portion of the grain for converting the balance into flour. The *expenses* of a business do not by any means increase proportionally to the quantity of business. A merchant, for instance, who, by advertising, has attracted trade to the amount of $1,000,000 per annum, is not required to pay ten times as much rent, nor does he need ten times more clerks, fuel, lights, &c., than the man who "never advertises," and perchance, does a business of $100,000 a year. In a large manufacturing establishment, the expenses of superintendence, repairs, etc., form but a trifling percentage on the aggregate product, while the time consumed in making a large purchase is very little more than in making a small one. Producers on a large scale can also afford to procure the best and most expensive machinery; and in some kinds of Manufactures, those who produce largely are content with "savings" as their profit, and are enabled to save what would be "waste" in a small establishment. Mr. Whitney, at his car-wheel establishment in Philadelphia, can save from the cinders, we are informed, enough iron to content a gentleman of his moderate views as to profit, but a manufacturer of car-wheels on a small scale, would not find it profitable to provide the machinery requisite for that purpose. From these and other considerations, which want of space forbids us to allude to, we infer that in future the manufacture of leading articles of consumption will be more and more conducted by large establishments, in a locality possessing in the highest degree of perfection, the moral and physical advantages that are essential to manufacturing prosperity. But it does not follow that large establishments will swallow up all smaller ones, unless it be those of a precisely similar kind, situated outside of the centres of combination. The economy which results from producing on a large scale, induces an increased demand for the manufactured goods; and an increased demand leads to a more minute subdivision of a manufacture into parts. When thousands of machines composed of Iron and Wood are required, we find establishments springing up, devoted exclusively to making parts—one, the nuts and washers ; another the screws ; another the bolts ; another the nails ; and others tools and machines to facilitate making parts, and so on, each extensive in its way, and thus large establishments in the leading branches of Manufactures are the parents of other extensive concerns in minor branches. A man who has not the requisite capital to conduct a leading Manufacture where large establishments abound, permit us to suggest, will not benefit himself by moving away from them. His policy is, we submit—to remain at all events, in their immediate vicinity, and then to accommodate his business to their operations and to his capital—that is, he will find it more profitable to be an extensive manufacturer of eyes for children's dolls in the centre of Manufactures, than a small manufacturer of machinery anywhere.

Lastly, to produce with the utmost efficiency and economy, *manufacturing establishments must be together.* The area of England and Wales is only about one-fourth more than that of Pennsylvania. In England all the large manufacturing establishments are situated, as we have stated, in close proximity to the coal beds. Manufacturers—one after another—have abandoned their factories in the Agricultural counties and moved their machinery to the district of which Manchester may be called the central point. Babbage has referred to one of the advantages resulting from this aggregation:

"The accumulation of many large manufacturing establishments in one district," he says, "has a tendency to bring together purchasers or their agents from great distances, and thus to cause the institution of a public mart or exchange. This contributes to increase the information relative to the supply of raw material and the state of demand for their produce, with which it is necessary manufacturers should be well acquainted. The very circumstance of collecting periodically, at one place, as large a number as possible, both of those who supply the market and those who require its produce, tends strongly to check those accidental fluctuations to which a small market is ever subject, as well as to render the average of the prices paid much more uniform in its course."

The accumulation of many large and excellent manufacturing establishments in one district, also gives a character and stamp to the Manufactures, which others who centre there receive the benefit of. There is also a mutuality of interest between manufacturers of even essentially different products, that renders aggregation highly desirable. The finished products of one class of manufacturers are often the raw materials of another. The power-looms of Mr. Jenks are but the instruments of production for the Manufacturer of Cotton and Woolen goods; and the finished commodities of the latter, are the raw materials of those who manufacture ready-made Clothing. Pig iron—the finished commodity of the smelter, is the raw material of him who rolls the bar; and the bar is again the raw material of sheet iron; which, in its turn, is the raw material of the nail and the spike. A sugar-refiner consumes the hogsheads, boxes and barrels of the cooper, paper of the paper-maker, and the finished products of coppersmiths, nail-manufacturers, twine-spinners, printers and various others. In fact, the largest and in many instances the sole consumers of certain manufactured articles, are the Manufacturers of other products; and finished commodities being, as a general rule, cheapest at the place of their production, without commissions or charges for transportation, it is certainly for the interest of those who buy to produce, and those who produce to sell, to be together. Aggregation, in fact, is the only effectual means of accumulating and combining all economies.

In Combination there is mystery like that of the Oak in the Acorn. Like the philosopher's stone, it turns all to gold—like the lever or the screw, in adds to man's power many hundred fold. Protective tariffs are useful as swaddling clothes to the infant; banks facilitate exchanges; but the perfection of combination cannot be attained except by aggregation in a suitable locality. If the Manufacturers of the United States ever hope to attain an independent position—independent of Foreign competition and of Home legislation, independent of commission merchants and of each other—they must centralize, so far as centralization is at all practicable. They must come out from sylvan retreats, deny themselves the advantages of mill-races and the harmonies of frog-ponds. They must tear down the miserable shingles "No admittance on any pretext whatever"—abandon their petty jealousies, enlarge their views, and co-operate like men and brethren. Blacksmiths, Cobblers and Wheelwrights may eke out an existence "in the neighborhood of the plow and the harrow," but in a Democratic country, whose people believe in buying where they can buy the cheapest, whether wisely or not we do not say, Manufacturers, in the true sense of the term, who attempt isolation, will inevitably find themselves, sooner or later, undersold, first, by those who operate in the centres of Combination, and finally, undersold by the Sheriff.

From *all* these considerations, which in substance we believe to be thoroughly sound, and to which we invite the closest scrutiny, we are led irresistibly to the conviction—that but few countries in the world, and but few places in any country, are well adapted for general Manufactures. Secondly: That the *best possible locality in the United States for general manufacturing is an attractive and suitable centre of Wealth, Population and Intelligence, situated in a populous district, abounding in well developed mines of Coal and Iron, and possessing established and superior facilities of intercommunication with all parts of our own country, and for commerce with foreign countries.* And Thirdly: *If there be two or more such localities, the one possessing desirable, in addition to the essential advantages in the highest degree of perfection, and the one already having the greatest number of large and well-managed manufacturing establishments, must be the best market in which to buy the commodities manufactured there, and eventually will be the chief seat of Manufactures in the United States.*

Now, have we such a locality? The centres of Wealth, Population and Intelligence in the United States are not numerous. Suitable centres for manufacturing, situated in close proximity to well-developed mines of Coal and Iron, and possessing established facilities for procuring raw materials on the easiest terms, and sending away manufactured produce, are very few; and of centres of Wealth, Population and Intelligence, we know of *but one* that possesses all the essential and most of the desirable advantages for manufacturing every variety of products, and which already contains many large and well-managed manufacturing establishments. To that one we invite the attention of all who produce, and deal in or consume manufactured commodities. The subject is one in which all these have a deep interest. If it be true that the highest degree of economy in production depends upon a *combination* of certain circumstances, rarely found, but which exist in the highest degree of perfection in a certain place, all who desire to produce cheaply, and all who desire to buy cheaply, have a direct pecuniary interest in knowing the facts, and in aiding to develop its capabilities. The place to which we invite earnest and sagacious attention, as the best manufacturing centre at present in the United States, is PHILADELPHIA, in the State of Pennsylvania.

PHILADELPHIA AS A MANUFACTURING CENTRE.

PHILADELPHIA is a scriptural name, composed of two Greek words, which signify, as usually interpreted, brotherly love. St. John, as we are informed in the Revelations, was instructed to indite a consolatory epistle to "the church in Philadelphia," a city of Asia Minor, about seventy-two miles from Smyrna. The Philadelphia of which we write is a namesake of the biblical city; and though not very ancient, is yet a cotemporary with most of the important events in American history. It was founded in 1682–3, by William Penn, who with a colony of English Friends or Quakers, had come to America to settle a province or tract of land granted to him by Charles II., in payment of a debt due by the government to his father. Before attempting any overt acts of sovereignty, however, Penn was wisely "moved" to acknowledge and purchase the rights of the aborigines, and thus, as Raynal has remarked, signalized his arrival by an act of equity, which made his person and his principles equally beloved. He also promulgated a series of laws, in which *Liberty of Conscience* was the first in order and importance. "A plantation reared on such a seed-plot," says Chalmers, "could not fail to grow with rapidity, to advance to maturity, to attract notice of the world."

The site chosen for the proposed city was a nearly level

plain between the Delaware and Schuylkill rivers, about six miles above their junction, and sixty miles from the ocean, by a direct line, though nearly a hundred miles by the course of the river. The influences that determined Penn in his choice of the spot are said to have been "the approach of the two rivers; the short distance above the mouth of the Schuylkill; the depth of the Delaware; the land heavily timbered; the existence of a stratum of brick clay on the spot, and immense quarries of building stone in the vicinity." In drafting the plan of his American city, Penn is supposed to have had in view the celebrated city of Babylon, which he certainly imitated in the regularity of the streets, and which he seemed desirous to emulate in size, for he gave orders to his commissioners to lay out a town that would have covered an area of 8000 acres. It was found, however, that "hundred-acre lots," which some of the squatter-sovereigns secured, would never answer the end of a city in a new country, and the plan was subsequently reduced. In 1701 it was again contracted, when the city was declared to be bounded by the "two rivers Delaware and Schuylkill, and Vine and Cedar streets as north and south boundaries." These continued to be the corporate limits of the city until 1854—the suburbs, as population extended, being divided into districts, as Spring Garden, Northern Liberties, Kensington, Southwark, Moyamensing and West Philadelphia, which in 1850 contained nearly twice as many inhabitants as the city proper.

The events in the early history of the town, prior to the Revolution, are not very striking. We subjoin a summary of the most important, as far as possible, in their chronological order. In 1687 a printing-press, the second in America, was set up; in 1689 Penn established a public High School with a charter. In 1742 Franklin projected an Academy and Free School, which became presently a College, and finally the "University of Pennsylvania." In 1765, the merchants of Philadelphia, in consequence of various restrictive and ill-advised Acts, particularly the Stamp Act, passed by the Parliament of Great Britain, pledged their word of honor not to order nor sell on commission any goods from Great Britain, except certain articles, more particularly those necessary for carrying on Manufactures, "unless the Stamp Act be repealed." In 1774 the first Congress in America assembled in Carpenters' Hall, (a building still standing in a court back of Chestnut street, between Third and Fourth streets,) to take into consideration the state of our relations with the mother country. In this city was adopted the Declaration of Independence, which was read from a stand in the State House yard, by Captain John Hopkins, July 4, 1776. From September, 1777 to June, 1778, in consequence of the disastrous battles of Brandywine and Germantown, the British army had possession of the city. The Convention that framed the present Constitution of the United States, met in Philadelphia, May, 1787. Here George Washington, when President of the United States, resided, in a building on the south side of Market street, between Fifth and Sixth, the lot being now occupied by a palatial business edifice, widely known as "Bennett's Tower Hall Clothing Store."

The first bank established in the United States was the Bank of Pennsylvania, opened at Philadelphia on the 17th of July, 1780, with a capital of £300,000, its special object being to supply the American army with provisions. In 1782 the Bank of North America went into operation; and in 1791 the United States Bank. In 1792 Congress passed an act establishing "a Mint for the purpose of a National coinage," to be situate and carried on at the Seat of Government of the United States for the time being, which was then at Philadelphia. In 1793, coinage was commenced in a building on Seventh street, opposite Zane, still known as the "Old Mint," and continued there until 1833, when the present noble edifice at the north-west corner of Chestnut and Juniper streets was completed.*

In the autumn of 1793 the yellow fever visited Philadelphia, and carried off more than 4000 persons, out of a population of a little over 40,000, of whom half, it was thought, had fled the city. The pestilence visited the city again in 1798, but was not so fatal as in 1793. The wars commenced by France in 1792 with other European powers, and which were continued until the abdication of Napoleon in 1814, had an immense influence in developing American Commerce, and Pennsylvania shared largely in this prosperity. Large importations were made from China and India into Philadelphia, for re-exportation to European markets. Our ships then enjoyed the carrying trade of the world, and numbers of our citizens accumulated large fortunes.

In January, 1801, Philadelphia was supplied for the first time with water from Water Works erected according to a plan proposed by Mr. Latrobe, viz. "to make a reservoir upon the banks of the Schuylkill, to throw up a sufficient quantity of water into a tunnel, and to carry it thence to a reservoir in Centre Square; and after being raised there, to distribute it throughout the city by pipes." These works were superceded by the

* Since its establishment in 1793, to the close of the year 1866, the Mint at Philadelphia coined 935,430,611 pieces, of the value of $532,429,174,86; the gold coinage being $427,869,536,61, the silver coinage $99,024,014,42, and the copper coinage $5,535,623,55. The entire coinage of the United States to the same period was $987,424,038,30.

The present officers of the Mint at Philadelphia, are:—*Director,* H. R. Linderman; *Treasurer,* Chambers McKibbin; *Chief Coiner,* A. L. Snowden; *Melter and Refiner,* James C. Booth; *Engraver,* James B. Longacre; *Assayer,* Jacob R. Eckfeldt; *Assistant Assayer,* William L. Dubois.

present works erected at Fairmount, which we will subsequently notice.

In 1811, Dr. James Mease published a book which he entitled "A Picture of Philadelphia." At that time Philadelphia was the most populous city in the Union. From an enumeration made the previous year, it appears that the number of dwelling-houses in the city and districts, was 15,814, and the population of the city and county amounted to 111,210. The population of the whole of Manhattan Island, at the same period, embracing the city of New York, was 96,372. Philadelphia then, as now, was the most healthy city in the Union. The average of deaths per day, in Philadelphia, was 5⅔, whereas in New York, with a smaller population, it was 6⅓. We subjoin Dr. Mease's remarks on the Manufactures, from which it will be perceived that Philadelphia was already celebrated in various departments of Manufacturing industry.

"The various coarser metallic articles, which enter so largely into the wants and business of mankind, are manufactured to a great extent, in a variety of forms, and in a substantial manner. All the various edged tools for mechanics are extensively made: and it may be mentioned as a fact calculated to excite surprise, that our common screw auger, an old and extensively used instrument, has been recently announced in the British publications, as a capital improvement in mechanics, as it certainly is, and that all attempts by foreign artists to make this instrument durable, have failed.

"The finer kinds of metals are wrought with neatness and taste. The numerous varieties of tin ware in particular, may be mentioned as worthy of attention. But above all, the working of the precious metals has reached a degree of perfection highly creditable to the artists. Silver plate fully equal to sterling, as to quality and execution, is now made, and the plated wares are superior to those commonly imported in the way of trade. Floor-cloths of great variety of patterns, without seams, and the colors bright, hard and durable; various printed cotton stuffs, warranted fast colors; earthenware, yellow and red, and stone ware are extensively made; experiments show that ware equal to that of Staffordshire might be manufactured, if workmen could be procured.

"The supply of excellent patent shot is greater than the demand. All the chemical drugs, and mineral acids of superior quality, are made by several persons: also cards, carding and spinning-machines for Cotton, Flax, and Wool. Woolen, worsted, and thread hosiery have long given employment to our German citizens: and recently, cotton stockings have been extensively made.

"Paints of twenty-two different colors, brilliant and durable, are in common use, from native materials; the supply of which is inexhaustible. The chromate of lead, that superb yellow color, is scarcely equaled by any foreign paint. There are fifteen rope-walks in our vicinity. We no longer depend upon Europe for excellent and handsome paper hangings, or pasteboard, or paper of any kind. The innumerable articles into which leather enters are neatly and substantially made: the article saddlery forms an immense item in the list. The leather has greatly improved in quality; the exportation of boots and shoes to the Southern States is great; and to the West Indies, before the interruption to trade, was immense. Morocco leather is extensively manufactured. The superiority of the carriages, either as respects excellence of workmanship, fashion, or finish, has long been acknowledged. The type-foundry of Binney & Ronaldson supplies nearly all the numerous printing-offices in the United States. There are one hundred and two hatters in the City and Liberties. Tobacco, in every form, gives employ to an immense capital. The refined sugar of Philadelphia has long been celebrated: ten refineries are constantly at work. Excellent japanned and pewter ware: muskets, rifles, fowling-pieces and pistols are made with great neatness. The cabinet-ware is elegant, and with the manufacture of wood generally, is very extensive. The houses are ornamented with marbles of various hues and qualities, from the quarries near Philadelphia.

"Mars Works, at the corner of Ninth and Vine streets, and on the Ridge road, the property of Oliver Evans, consists of an iron foundry, mould-maker's shop, steam-engine manufactory, blacksmith's shop, and mill-stone manufactory, and a steam-engine used for grinding sundry materials for the use of the works, and for turning and boring heavy cast and wrought iron work. The buildings occupy one hundred and eighty-eight feet front, and about thirty-five workmen are daily employed. They manufacture all cast or wrought-iron work for machinery for mills, for grinding grain or sawing timber; for forges, rolling and slitting-mills, sugar-mills, apple-mills, bark-mills, &c. Pans of all dimensions used by sugar-boilers, soap-boilers, &c. Screws of all sizes or cotton-presses, tobacco-presses, paper-presses, cast iron gudgeons, and boxes for mills and wagons, carriage-boxes, &c., and all kinds of small wheels and machinery for Cotton and Wool spinning, &c. Mr. Evans also makes steam-engines on improved principles, invented and patented by the proprietor, which are more powerful and less complicated, and cheaper than others; requiring less fuel, and not more than one-fiftieth part of the coals commonly used. The small one in use at the works is on this improved principle, and is of great use in facilitating the manufacture of others. The proprietor has erected one of his improved steam-engines in the town of Pittsburg, and employed to drive three pair of large millstones with all the machinery for cleaning the grain, elevating, spreading, and stirring and cooling the meal. gathering and bolting, &c., &c. The power is equal to twenty-four horses, and will do as much work as seventy-two horses in twenty-four hours: it would drive five pair of six-feet millstones, and grind five hundred bushels of wheat in twenty-four hours.

"All kinds of castings are also made at the Eagle Works, on Schuylkill, belonging to S. & W. Richards."

In 1812, *Steam* Works for supplying the city with water were commenced at Fairmount, and in 1815 the use of the Centre Square Works was discontinued. In 1819, Councils resolved to erect the present *Water-power* Works, which for a long time were the only works of the kind in the United States, and which are yet unsurpassed by any in the whole country. The water from the Schuylkill is turned into a forebay 419 feet long and 90 feet wide; whence it falls upon and turns eight wheels, from sixteen to eighteen feet in diameter, and four turbine wheels, three having two pumps each, and which elevate the water ninety-two feet to the top of a partly natural elevation, immediately at the works, and which give them their name. The reservoirs of the entire Water Works of the city furnish storage to the amount of 85,000,000 gallons. This is about equal to two days' supply in July and August. The total cost, including laying pipes, &c., to the present time, is about $3,500,000.*

* In addition to the Works at Fairmount, Philadelphia has, at the

In 1829 the Pennsylvania Canal was completed, which, with the Schuylkill and Union Canals, previously constructed, formed a connection with the Ohio River, *via* Reading and Middletown.

In December, 1831, Stephen Girard, "Mariner and Merchant," died worth nearly $10,000,000, and bequeathed by his Will large sums to public uses, among others the sum of $2,000,000 for the erection of a College, now known as the Girard College. In 1835 Philadelphia was first supplied with gas from Works erected on Market street, near the Schuylkill.* In the same year a part of the Reading Railroad, to connect Philadelphia with the Schuylkill coal region, was put under contract, and in 1842 the first train passed over the whole line between Pottsville and Philadelphia. In 1837 the Philadelphia, Wilmington and Baltimore Railroad was completed. In 1838 the city was disgraced for the first time by a mob

present time, three other Water Works, viz.: Schuylkill Works, Delaware Works, and Twenty-fourth Ward Works. The total amount of water supplied by all these Works in 1866, was 10,614,344,464 wine gallons. The Duplicates of the Water rents for the same year amounted to $626,156 75.

* Since that period the Northern Districts and Germantown erected Gas Works; and in 1854 the city completed new and additional works near Gray's Ferry Bridge, having the largest gasholder, it is believed, in the United States, being 160 feet in diameter and 90 feet high, and capable of holding 1,800,000 cubic feet of gas. The cost of the whole, now belonging to the City Gas Trust, is about $3,834,241 75. We obtain from the Annual Report for 1866, of the Trustees of the Philadelphia Gas Works, the following statistics:

Street mains laid to January 1, 1866, - - -	486½ miles.
Service pipes, " " - - - - -	52,408 "
No. of Services and Meters in use, - - - - -	52,293
" Lights in use, (private.) - - - - -	650,996
" " " (public,) - - - - - -	7,622
Gas made in 1866, - - - - -	915,956,000 cubic feet.
Total, made in 31 years, - - -	9,460,399,000 " "

Present manufacturing capacity, 3 million feet per diem.

that burned the Pennsylvania Hall, fired the Shelter for Colored Orphans, and attacked the negro quarters. In 1844 the city was again disquieted by riots incited by the presumed interference of Catholics with the elective franchise, and several Catholic churches were burned. In 1847, the Pennsylvania Railroad, to connect Philadelphia with the Ohio River, was commenced, and finally completed in 1854. In 1850 a census was taken, which showed that Philadelphia contained 23,601 more dwellings than the city of New York, and a population of 408,762, being an increase of 58½ per cent. in the ten years preceding the census of 1850, and 953½ per cent. in the sixty years since the first National census. Of the population of 1850, 17,500 were born in England; 72,312 in Ireland; 22,750 in Germany; 3,291 in Scotland; and 1,981 in France. Total foreign, 121,699.

In 1854 the corporate limits of the city were made coextensive with those of the county of Philadelphia, covering an area of 120 square miles, and placing the villages and towns of Bridesburg, Frankford, Holmesburg, Byberry, Nicetown, Andalusia, Bustleton, Rising Sun, Milestown, Germantown, Chestnut Hill, Falls, Manayunk, Roxborough, West Philadelphia, Mantua, Haddington, and Hamilton, under the wise guardianship of a Metropolitan Mayor and City Councils, *sans peur et sans reproche.*

These, we believe, may be called the most important events in the Annals of Philadelphia. In the history of a place whose "birth and spring-time" carry us back nearly a century anterior to the American Revolution, there are necessarily many events of greater or less importance that deserve to be commemorated. No city of equal age can present a fairer or more interesting record of the past than Philadelphia; none has been more prolific in men who have been eminent in their day and generation; and not one has been so fortunate in inspiring that species of affection which manifests itself in culling and preserving, as a labor of love, the features and memorials of a time gone by. John F. Watson, in his "Annals," has done all that can be desired to preserve the lineaments and characteristics of what may be called the "olden time" of Philadelphia; and our Historical and Philosophical Societies have accumulated papers and disquisitions upon every conceivable subject pertaining thereto. Truly, if the prosperity of a city be promoted in proportion to the affectionate attachment of its inhabitants—a feeling, as Everett has observed, entitled to respect, and productive of good, even if it may sometimes seem to strangers over-partial in its manifestations—the citizens of Philadelphia may repeat, with confidence, the poetical prediction of Taylor, the astrological Hague of the eighteenth century:—

> "A city built 'neath such propitious rays
> Will stand to see old walls and happy days."

The Past of this city, therefore, has been well cared for; its historical incidents are preserved in its own and in the records of our country; the fame of its great men will survive "fresh in eternal youth"; and neophytes in Archæology may well despair unless they devote attention to its Present, which, with its material progress, its advance, especially in Manufactures, its Railroads and its Fire and Police Telegraphs, would at any time form a theme sufficiently comprehensive in itself to exclude any minute reference to the events of the past.

I. Philadelphia as it is.

PHILADELPHIA is usually described as the second city in the United States; and, if we except Paris, nearly equals the largest capitals on the continent of Europe in population. No census has been taken since 1860; but assuming that the increase has been in the same ratio as that which distinguished the ten years preceding the last national census, its present population cannot be far short of 700,000. Its entire length, as per Ellet's Survey, is twenty-three miles, and average breadth five and a half miles; area, one hundred and twenty-nine and one eighth square miles, or 82,700 acres. The densely inhabited portion of Philadelphia extends about four miles on the Delaware, from Southwark north to Richmond, formerly Port Richmond, and two and a half miles on the Schuylkill, having a breadth between the two rivers, assuming South street formerly the Southern boundary of the city to be the standard, of 12,098 feet 3 inches. The plan of regularity in the streets,—originally adopted by Penn, and which, though condemned by some travelers accustomed to the crooked and narrow streets of European capitals, has been unqualifiedly approved by mathematical and scientific minds,—is adhered to; and in the northern as well as the central parts of the city, there are avenues and streets which, for spaciousness and elegance, are unsurpassed by any. The elegance of the *public* buildings has long been a subject of remark, even in primary geographies; but, within the last few years, the architectural beauties of the city have been vastly enhanced by the erection of numerous costly *private* buildings: banks, stores, churches, dwellings—of granite, iron, sandstone, and marble; and its upward growth, by the addition of stories upon stories, is not less remarkable. Beyond the compact or densely built-up portions, in the northerly direction, there is a wide expanding district between the two rivers, occupied in part by beautiful suburban residences, and by numerous Manufactories. surrounded by the habitations of industrious and contented artisans. The vicinity of Germantown is especially noted for the number of elegant cottages and villas, surrounded by handsomely laid out grounds, delightfully shaded; while the beauties of the Wissahickon, have they not inspired poets? But the citizens of Philadelphia, though appreciating her elegance in architecture, and scenes of natural beauty, cherish them less fondly, and point to them with less pride, than to the number and superiority of her charitable institutions, the excellence of her schools, the refinements of her society, her eminence in the Fine and the Mechanical Arts, the multiplied conveniences of life, promoting domestic comfort, and the celebrity of her Forum and Medical Schools, which, like the works of the Athenian orators, are regarded with veneration and respect by every polished nation.

Upon the minds of strangers and tourists, however, the external aspect of a city seems to leave the most permanent impressions; and if we may judge from their written opinions, that of Philadelphia has charmed those who charm the world. The learned and philosophical author of Mademoiselle Rachel's tour in America, was sagacious enough to remark—and in one so courteous, a trifling geographical inaccuracy can readily be pardoned—that "the capital of Pennsylvania, the Quaker city as it is called, is one of the richest, handsomest, and most flourishing cities in the United States of America." This is much from a gentleman who thanked God that he had visited North America, "because it is a duty disposed of," and he would never have to return there; but he proceeds to add: "Fortunately, it is superb weather here, and we can see this elegant capital at our ease. All the houses have a flaunting, coquettish air, which is pleasant to see The streets are broad and clean. The shops are generally very large and very rich. There are superb goods in them. In fact, this city has a happy physiognomy, which is very agreeable." The ladies, especially the

Fannies, it is consoling to reflect, have also found much to delight them. Fanny Kemble was enraptured, we believe enchanted by the appearance of Fairmount, by moonlight; and Fanny Fern went off like an alarum clock at the beauties, and particularly the butter of Philadelphia. None, however, have expressed their admiration more gravely, deliberately, and ornately, than the writer of the following:

"Few great cities present such attractions for the stranger, as the city of 'Brotherly Love.' The American is proud that here the Declaration of Independence was signed; and his patriotic heart swells with a nobler emotion, while he looks upon the bell that pealed forth the joy of a nation's deliverance; and his heroic spirit will be stirred within him as he sits on the chair on which once sat the Father of his Country, yet, with many a relic of the past, preserved in Independence Hall. The philanthropist feels his heart throb with pleasure as he views the many noble institutions that a munificent charity has erected to ameliorate the condition of suffering humanity, supply the wants of the poor, minister to minds diseased, and alleviate the sufferings of the sick and wounded. The lover of science rejoices to see the city of Franklin abounding in Institutes whose object is the cultivation of all the arts that adorn, and all the sciences that tend to the progress of mankind. The philosopher will find kindred spirits in the great centre from which the rays of intellect emanate, whose brightness appears as a star of glory to the nation and the world. Medical students resort to Philadelphia for their professional training; the young aspirant to forensic honors seeks her classic shades; and while the admirer of the beautiful in architecture, and the architect, may exult in the stately proportions of her solemn temples, her gorgeous palaces, and the genius that adorned her with edifices whose beauty might vie with the Grecian models, the true Christian will find that the piety that erected the ancient church of Gloria Dei in the city's infancy, has diffused itself, and kept pace with its rapid increase. The merchant from other cities may look with wonder upon the commercial facilities of Philadelphia, her double port, the rich mineral treasures poured into her lap from the exhaustless resources of the Commonwealth, and the resources that put the numerous wheels of manufacturing industry in motion, and send the products of her skill, the results of her commerce, and the proceeds of her inland trade, to the furthest regions of the West, and almost all points of the

compass. Her great Railway system, the most complete in the country, makes her pre-eminent for all the facilities of business, giving her a great advantage over all other cities in the Union. The exceeding beauty of her location, and the lovely scenery of the surrounding country, make her the resort of many who delight in beholding the fair face of Nature, seldom so full of beauty as in some portions of her enchanting rural scenery."

Such is Philadelphia as it appears to the optics of intelligent strangers. Such may it ever appear. If, however, a statistical description were wanted to convey a clearer idea of the magnitude of the city, we might say that Philadelphia is a collection of nearly 100,000 Dwellings, Shops, and Manufactories, 8,000 Stores, 391 Churches, 374 School-houses, 32 Banks, 11 Market-houses, 8 Medical Schools, 1 High School, 1 Girard College, 1 Polytechnic College, 1 State House, 1 Custom House, 1 Exchange, 1 Mint, 1 Navy Yard, 1 Naval Asylum, 3 Arsenals, 1 Blockley Almshouse, 2 Insane Asylums, 1 Pennsylvania Institute for Deaf and Dumb, 1 Blind Asylum, 1 Pennsylvania Hospital, 1 Academy of Music, 1 Academy of Fine Arts, 1 Academy of Natural Sciences, 1 Athenæum, 1 Club House, numerous Libraries, 5 Theatres, 1 Masonic Hall, 15 Public Halls, 7 Gas Works, 5 Water Works, 1 County Prison, 2 Houses of Refuge, 1 Penitentiary, 14 Cemeteries, 9 Railroad Depots, 90 Fire Engine-houses, 17 Station Houses, and 3 Race Courses. The officers of the city government consist of 1 Mayor, 1 City Solicitor, 1 City Controller, 1 Receiver of Taxes, 3 City Commissioners, 1 City Treasurer, 1 Chief Engineer of Water Department, 1 Chief Engineer of Fire Department, 1 Chief Engineer of Gas Works, 1 Chief Commissioner of Highways, 1 Commissioner of City Property, 1 Commissioner of Market-houses, 1 Chief Surveyor, and 12 Regulators; 27 Select Councilmen, with 3 officers, and 52 Common Councilmen, with 4 officers; 12 Members of a Board of Health, with 7 officers, and 10 Executive officers,

12 Guardians of the Poor, with 6 officers, and 12 Out-door Visitors; numerous Assistants and Clerks in each Department; with a Police force, consisting of

1 Mayor,	whose salary is		$5,000	per annum.
2 Mayor's Clerks,	- - - - - -		1,500	" "
1 Chief of Police,	- - - - - -		2,000	" "
7 High Constables,	each	- -	1,200	" "
8 Detective Officers,	- - - - -		1,200	" "
1 Supt. of Fire and Police Alarm Telegraph,			1,500	" "
1 Assistant,	- - - - - - -		1,100	" "
18 Lieutenants,	each	- - -	1,155	" "
32 Sergeants,	"	- - -	1,082 40	"
1 Chief Detective,	- - - - - -		1,500	"
1 Fire Marshal,	- - - - - -		1,700	"
776 Policemen, $2 50 per day,				

Who made, in 1866, 43,226 arrests, and restored 3,081 lost children.

It is thus evident that Philadelphia, regarded from every point of view, is a centre of Wealth and Population; and, if the social characteristics of its inhabitants correspond with its external allurements, it must be an *attractive* centre. *What, then, are their characteristics, particularly with reference to the social position of the Mechanic and the Artisan?* What facilities are provided for their physical comfort and intellectual advancement? In the first place, the citizens of Philadelphia, who now give tone and direction to its popular sentiment, it may be relied upon, are far too clear-headed and practical in their views to do any thing tending to degrade labor and check useful enterprise. Even among the numerous sets of exclusives into which the descendants of great people sometimes divide themselves, there are none that I have heard of in this city who make idleness the "open sesame" to the enjoyments of their society. Nearly every citizen has some regular occupation; and prides himself upon diligence in the transaction of business and punctuality in fulfilling his engagements. The circle of those, at least among the male population, who aspire to distinction because of their uselessness, is like a wart on a man's nose, more looked at than important. The mass of the inhabitants believe in the Baconian philosophy, and illustrate its wisdom and beneficence by multiplying human enjoyments and mitigating human sufferings. The Press is emphatically a People's Press. The Quakers, whose influence, though diluted of late, continues to be felt in modifying the characteristics of our society, are true Benthamites in their views on individual and general happiness. They hold that the greatest happiness of the individual is, in the long run, to be obtained by pursuing the greatest happiness of the aggregate. They excel especially in the substantials of character, are fruitful in good works, zealous in education, and liberal in encouraging and rewarding decided mechanical and artistic triumphs. Constitutionally deliberate and prudent, the want of cordiality in their manners, which some strangers complain of, may be, and probably is, an unfortunate manifestation of these excellent qualities: or, in other words, of thinking twice before speaking once. Their city has been so prolific in great men, that the arrival of another does not create a sensation; and being quite inexperienced in the art of giving entertainments at the *subsequent expense of their guests,* they prefer to conciliate mercantile visitors by giving them mercantile advantages. With respect to the want of enterprise—a standing accusation, which our fellow-citizens are accustomed to make against each other in tempestuous weather—we acknowledge the charge is seemingly reasonable and well founded, especially if it mean a total inability to comprehend the morality, or realize the pecuniary value of claptrappery, slap-dashery, or eclat. Adverse to puffing, they even refrain from scattering broadcast, as they ought to do, information relative to the mercantile and manufacturing advantages of their city; practical in their views, they sometimes forget that man does not live by bread alone; and straightforward in their own dealings, and governed exclusively in their own transactions by economical or commercial reasons, they do not suppose it possible that such trifles as "ancient and fish-like smells" in market-houses, can keep one customer away from where he ought to go; or that such vanities as popular preachers, big hotels, capacious theatres, palaces of mirrors, can possibly attract one customer where it is not his interest to go. The late panic, however, has dispelled many illusions; and if, moreover,—disabusing every mind of the feeling of entire security, and of the conviction that perfection is already attained,—it awaken a more active spirit, the anniversary of its advent may hereafter be celebrated as a civic holiday; and this beautiful city, having taken a new lease of Prosperity, will perpetuate the glory, as well as the memory of its Founders.

Secondly, the social and practical characteristics of the citizens of Philadelphia are in nothing more clearly and favorably manifested than in their zealous support of *free education.* According to the Controller's Report of 1866, there were 374 Public Schools in the city, viz.: 1 High School, 1 Normal School, 60 Grammar Schools, 69 Secondaries, 187 Primaries, and 56 unclassified Schools. The whole number of teachers was 1,314, of whom 79 were males, 1,235 females; the expense $864,276.26, and the number of scholars who enjoyed the benefits of gratuitous tuition was 77,164. But Public Schools are only a moiety of the educational establishments of Philadelphia. The city abounds in private schools and institutions of a semi-public character. Yet the quantity of the instruction given in the schools is, perhaps, less noteworthy than its quality. Public teachers must compete with private teachers; while the latter are incited to emulation by the example of numerous eminent professors. From a mechanical point of view, however, the crowning distinction in this respect is the abundance of facilities provided for those who desire to increase their stock of *practical* and *scientific knowledge.* Books are at the command of such, rare in character and unlimited in quantity. The *Philadelphia Library,* one of the largest and best in the country, containing some eighty-five thousand volumes, is open to all, and access is thus given to works that probably are inaccessible to mechanics elsewhere. The work on British Patents, recently donated to the library, is valued at $3,000; the binding of the volumes alone having cost, we are informed, seven hundred dollars. For five dollars a year, any respectable person may enjoy the advantages of the *Mercantile Library,* whose members now number, we believe, 5,766. In various parts of the city there are Institutes with Reading-rooms and Libraries attached, where gratuitous lectures are given, especially adapted to the wants of mechanics. At the *Wagner Free Institute of Science,* twelve lectures are delivered weekly, during the Winter season, on Geology, Mineralogy, Mining, Astronomy, Botany, Anatomy, Physiology, Natural Philosophy, Chemistry, Chemical Agriculture, Ethnology, Comparative Anatomy, Zoology, Meteorology, and Civil Engineering. The apparatus is superior, and the lectures are well attended. The Spring Garden Institute gives instruction in the Mechanic Arts and Architecture, and has lectures on Literary and Scientific subjects. The Mechanics' Institute of Southwark, the Moyamensing Literary Institute, the Philadelphia City Institute, have reading-rooms and lectures, and the last has a School of Design. The Kensington Literary Institute, and the West Philadelphia Institute, are of the same character as the others; the latter having a School of Design. The Board of Trustees, in their report to contributors for 1856, state that the results of these Institutes show "that there is an aggregate of more than 11,000 volumes in the libraries; that during the past year more than 32,000 volumes have been loaned for home-reading; that more than 48,000 visits were paid to the reading-rooms by parties who partook of the intellectual food there dispensed: that one hundred pupils availed themselves of the valuable privileges afforded, for the culture of the eye and the hand in designing and drawing, by the schools of the Institutes; that sixty-seven lectures on literary, scientific, and artistic subjects, many of them replete with useful information, were listened to by thousands; and that, stimulated by your own generous contribution of more than $30,000, more than $50,000 additional have been contributed by our fellow-citizens to help onward the noble work commenced by you."

The *Franklin Institute* provides lectures at cheap rates every Winter, on Mechanical, Literary, and Scientific subjects, publishes a Scientific Journal, the oldest of its kind in the country, possesses a valuable Cabinet of Models and Minerals, and gives an Annual Exhibition that does much to promote progress in the Useful Arts. The *Academy of Natural Sciences* has a fine collection of objects in Natural History, embracing 25,000 specimens in Ornithology, and 30,000 in Botany; a library of over 26,000 volumes; and Mineralogical and Geological Cabinets, noted for their completeness. Professor Agassiz pronounced this institution the best out of Europe for its collections in the department of Natural History. At the *Polytechnic College*, opposite Penn Square, an engineer may obtain instruction in Physics that, before its establishment, he could not have obtained on this side of the Atlantic. In addition to the regular course, which embraces instruction in Civil Engineering, Mechanical Drawing, Mining, &c., the Managers have recently established a department designed to give instruction in "certain branches of knowledge that are demanded in common by every business pursuit, and are alike indispensable to the merchant, the farmer, the manufacturer, mechanic, and the manager of mining and other property." At the Girard College, drawing is taught from models of geometrical solids, and also in the High Schools, by competent teachers. The science of Accounts, Book-keeping, Penmanship, and Commercial Law, are taught at Commercial Colleges, such as Crittenden's, Bryant, Stratton & Kimberly's, and the Quaker City College, presided over by competent professors; and for the instruction of females in many departments of design, as applicable to manufactures, there is a school known as the "The Philadelphia School of Design for Women," established a few years ago, by Mrs. Peter, the wife of the late British Consul at Philadelphia.

Among the educators of the people, too, the *newspapers* of this city are fairly entitled to rank. There are now thirteen newspapers published daily—nine in the morning, and four in the afternoon; numerous weeklies, and nearly fifty publications properly designated as periodicals. The aggregate of those distinguished as newspapers, does not embrace any of a strictly scientific description; but the deficiency is in great part compensated for by many of the dailies, which never fail to advise their readers of whatever is important in the progress of the Mechanic Arts. The complement, also, lacks one or more of a metropolitan character, or those which can be said to possess universal interest; but as a faithfnl local Press, the newspapers of this city are models for those of the Union. The working-man here, for two cents, may enjoy a better morning newspaper than he can, for the same trifling sum, in any other place on the globe; while, for a larger expenditure, he may suit his taste from "grave to gay—from lively to severe." The sources then, it will be perceived, for acquiring that sort of knowledge

which makes superior, efficient, intelligent mechanics, are very abundant in Philadelphia. It would be well, indeed, for affluent munificence to endow more completely one or two colleges, and establish an institution resembling, for instance, the British Museum; but in view of her present advantages, this city deserves now to be the resort of students in Art-education from all sections of the Union, as she long has been of students in Medical science. Here, there is an amount of scientific intelligence and professional skill concentrated, in part by the demands of the various institutions, seemingly sufficient to solve any thing in Mechanics but impossibilities; and which, conjoined with favorable physical circumstances, must enable manufacturers located near this city, to triumph over difficulties under which, in less favored localities, they would be compelled to succumb. Here, an educated hand-craftsman, or an inventor, may be said to stand at one of the great centres of intellectual life, with the world of mechanism in its practical forms on exhibition and in operation before him; Mentors on every side to enlighten him as to the recorded failures and triumphs of the ingenious men of all countries; and with the resources of the most scientific men of the present age, possessing the most perfect apparatus, at his command, to aid him in his experiments, or sustain him in his discoveries.

As a *place of residence*, Philadelphia enjoys the rare distinction of being desirable alike to the capitalist and to the artisan. In this respect, it is generally acknowledged, no other American city can compare with it. To the former, it offers all the attractions that can delight a cultivated mind, and all the luxuries that can please a fastidious palate; while an artisan, if industrious and intelligent, may command probably every thing essential to his present comfort, prospective independence, with constant participation in many of the chief pleasures of the capitalist. In the important particulars of general cleanliness, healthfulness, wholesomeness of water, and the excellence of its markets, Philadelphia is unapproached by any of the other great cities; and, as respects domestic accommodations, its superiority, at least over New York, is strikingly revealed by the census of 1850, which showed that, with a smaller population, this city contained about 23,601 more dwelling-houses: there being an average of 13½ persons to a house in the former city, and only 6½ in Philadelphia. The custom, too, that prevails of selling lots on *ground-rent*, gives to the man of small means facilities that he cannot ordinarily obtain in other cities. For instance, if he have but money enough to erect a house, he can procure a lot on an indefinite credit; and so long as he pays the interest of the purchase-money, he will not be disturbed, nor can the principal be called for. By this means, it is quite common for mechanics, small tradesmen, and even laborers, to become owners of homesteads in the suburbs, which, by Passenger Railways that are being introduced, will be brought nearer to the centre than ever before.

A city, then, so attractive as Philadelphia, and possessing such superior educational advantages, can hardly fail, it would seem probable, to command, at all times, one of the first and most important requisites for success in Manufactures, viz.: *an abundant supply of skilled labor, and of those qualified to direct it.* Experience demonstrates, that not only is the supply of labor generally abundant, but the surplus sometimes troublesome. Here is congregated, at all times, an army of artisans from every civilized nationality—the majority employed, others seeking employment; and should the supply at any time fall short, an advertisement would bring a regiment from every place where it had been seen. Men who would not go to "Raw Cheney," in Georgia, for $1,000 a year, nor to Pittsburg for $900, nor to Lowell for $850, eagerly come to Philadelphia for $800. Philadelphia has thus the pick and choice, at less wages, of the mechanics of the Union. Hence, too, the name, PHILADELPHIA MECHANIC, has become synonymous with skill and superiority in workmanship. We simply state a well tested fact, when we assert that a mechanic, traveling with favorable credentials from reputable workshops in this city, will be preferred to fill the first vacancy in any similar establishment, not merely in most places throughout the United States, but in portions of Europe.

So much for Philadelphia as it is. Its *status* establishes the fact, that it possesses the *moral* circumstances that are essential to success in manufacturing operations, and we might proceed immediately to consider those that are properly denominated *physical*. Before doing so, however, it may be proper to glance at the present,

II. Commercial Relations of Philadelphia.

Referring to the official statistics, we find a very considerable increase in the commerce of this port during the year 1866 as compared with the previous year. The number of vessels that arrived, was seven hundred and twenty-two foreign and thirty-five thousand eight hundred and sixty-four coastwise; being an increase of one hundred and eighty-one in the foreign arrivals, and four thousand one hundred and fifty-nine in the vessels engaged in the coasting trade.

The annexed table gives a list of some of the principal articles imported into Philadelphia during the year 1866.

Imports at Philadelphia.		*Exports of Domestic Produce.*	
Brimstone, tons......	2,670	Flour, bbls...........	78,735
Coffee, bags..........	43,686	Corn Meal, bbls......	27,735
Hides, number.......	15,169	Rye Flour, bbls......	1,386
Iron, tons...........	5,300	Ship Bread, lbs.......	1,766,173
" bars............	58,701	Wheat, bushels	26,034
" bdls......	71,319	Corn, bushels........	752,888
Lead, pigs...........	18,292	Cloverseed, bushels...	6,991
Lemons, boxes..... .	32,938	Beef, lbs..............	437,388
Molasses, hhds.	61,903	Pork, lbs............	964,823
" bbls	2,848	Bacon, lbs............	309,989
Oranges, boxes.......	87,160	Lard, lbs............	2,075,570
Rice, pkgs...........	3,900	Candles, lbs..........	544,850
Salt, sacks...........	154,033	Soap, lbs............	55,284
" bushels.........	253,888	Butter, lbs...........	119,639
Sugar, hhds..........	72,265	Cheese, lbs............	19,105
" boxes.........	35,251	Tallow, lbs...........	2,944,341
" bbls..........	3,712	Coal, tons............	41,871
" bags..........	17,408	Bark, hhds...........	699

The receipts of cotton during the year 1866 were 20,517 bales, and of flour and grain as follows:

	1864.	1865,	1866
Flour, bbls...........	903,447	715,398	574,552
Rye Flour, bbls......	5,446	9,100	10,400
Corn Meal, bbls......	17,774	58,290	22,370
Wheat, bushels.......	2,465,790	1,802,040	1,219,670
Corn, bushels........	1,604,415	1,506,305	1,603,394
Oats, bushels.........	1,434,670	1,527,470	1,570,218
Rye, bushels..........	200,900	220,600	279,673

In Foreign Commerce, the Port of Philadelphia, it is undeniably obvious, has not maintained its original supremacy. In 1796, the exports amounted to $17,513,866; and from 1795 to 1826, the aggregate exports more than trebled those of the last thirty years. For a long period of time, nearly a century, Philadelphia was regarded throughout Europe as, commercially and numerically, the great city of the Western Continent.

The names of her principal merchants were known and respected at every Exchange in Europe. A decline, then, from a position so commanding in the world's markets, would naturally cause a pang of regret in the breasts of every one not indifferent to the city's future, were the causes that induced it, less creditable, patriotic, and honorable than they are. It is especially consoling to know that they do not impinge in the slightest the commercial capabilities of the Port. No one, if called upon to correct the assertions of the willfully ignorant, need trouble himself to controvert the assertion, that the inland situation of Philadelphia is an effectual barrier to her commercial supremacy. The position of the chief commercial cities of the Old World,—London on the Thames, Liverpool on the Mersey, and Paris on the Seine,—proves that immediate proximity to the ocean is not essential to constitute a great shipping port. Besides, the channel of the Delaware is known to be abundantly wide and deep to float, as it has floated, the largest vessels in the Naval service. According to the official chart of the Coast Survey, it is seldom less than a quarter of a mile in width, and ranges in depth, *at low water*, from 4 to 9½ fathoms, excepting at the bar below Fort Mifflin, where, for a few rods, it varies from 18 feet to 25 feet 8 inches, according to the state of the tide. Moreover, the fact that the "Cathedral," a vessel of too large tonnage to obtain entrance into the port of her destination, New York, was therefore sent hither, where she was amply accommodated; the length of wharves on the Eastern front, extending as they do, for about three miles; and especially her former eminent success in Foreign Commerce, all establish and fortify the assertion, that Philadelphia possesses all essential, in fact ample facilities for shipping. To ascertain the true and principal causes of the decline in this particular, then, we must direct our attention to other channels that have absorbed capital; and we may possibly discover that the chief sources of the present prosperity of Philadelphia have their origin in a comparative neglect of Foreign Commerce.

About forty years ago, the citizens of Philadelphia may be said to have become thoroughly aware of the immensity of the riches concealed in the mountains and ravines of their native State. They then, for the first time, comprehended the value and the vastness of the deposits of Coal and Iron near their metropolis, and realized, as vividly perhaps as subsequent experience has justified, how great a boon would be conferred upon the whole country by their development. It is true, that for many years previously, it had been known that a peculiar species of coal abounded in the counties of Lehigh and Schuylkill, but it was regarded as worthless. Even as late as 1817, when Col. George Shoemaker forwarded ten wagon loads of a coal which he had discovered about one mile from Pottsville, to Philadelphia for sale, he could with difficulty find purchasers; and some of those who did purchase it, were wholly unsuccessful in their attempts to use it. "Nearly every one considered it a sort of *stone*, and, saving that it was a 'peculiar stone'—a stone-coal—they would as soon have thought of making a fire with any other kind of *stone!* Among all those who examined the coals, but few persons could be prevailed upon to purchase, and they only a small quantity,—'to try it.' But, alas! the trials were unsuccessful. The purchasers denounced Colonel Shoemaker as a vile impostor and an arrant cheat! Their denunciations went forth throughout the city; and Colonel Shoemaker, to escape an arrest for swindling and imposture, with which he was threatened, drove thirty miles out of his way, in *a circuitous route, to avoid the officers of the law!* He returned home, heart-sick with his adventure. But, fortunately, among the few purchasers of his coal, were a firm of iron factors in Delaware County, who, having used it successfully, proclaimed the astounding fact in the newspapers of the day. The current of prejudice thereafter began to waver somewhat; and new experiments were made at iron-works on the Schuylkill, with like success, the result of which was also announced by the Press. From this time, Anthracite began gradually to put down its enemies; and among the more intelligent people, its future value was predicted." But it was not until 1825 that the first successful experiment to *generate steam*, with Anthracite coal, was made at iron-works at Phœnixville. From that year, too, the existence of the Schuylkill trade may be said to date, though some coal had been shipped previously. The speculative mania in the coal regions, however, did not commence until a few years later.

About 1829, the news of fortunes accumulated by piercing the bowels of the earth, and bringing forth from the caverns of mountains "metals which shall give strength to our hands," became generally current, and aroused an enthusiasm less wide-spread than that which fevered Europe upon the discovery of silver in Mexico, or recently America upon the discovery of gold in California, but certainly not less intense. "Capitalists awoke as if from a dream, and wondered that they had never before realized the importance of the anthracite trade. What appeared yesterday but as a fly, now assumed the gigantic proportions of an elephant! The capitalist who, but a few years previously, laughed at the *infatuation* of the daring pioneers of the coal trade, now coolly ransacked his papers, and ciphered out his available means; and whenever met on the street, his hands and pockets would be filled with plans of towns, of surveys of coal lands, and calculations and specifications of railways, canals, and divers other improvements until now unheard of. The land which yesterday would not have commanded the taxes levied upon it, was now looked upon as 'dearer than Plutus's mine, richer than gold.' Sales were made to a large amount; and in an incredibly short space of time, it is estimated that upward of *five millions of dollars* had been invested in lands in the Schuylkill coal-field alone! Laborers and mechanics of all kinds, and from all quarters and nations, flocked to the coal region, and found ready and constant employment at the most exorbitant wages. Capitalists, arm-in-arm with confidential advisers, civil engineers, and grave scientific gentlemen, explored every recess, and solemnly contemplated the present and future value and importance of each particular spot. Houses could not be built fast enough; for where nought but bushes and rubbish were seen one day, a smiling village would be discovered on the morrow. Enterprising carpenters in Philadelphia, and elsewhere along the line of canal, prepared the timber and frame-work of houses, and then placing the *materiel* on board a canal boat, would hasten on to the enchanted spot to dedicate it to its future purposes. Thus *whole towns* were arriving in the returning canal boats; and as 'they were forced to play the owl,' a moonlight night was a godsend to the impatient proprietors, for with the dawning of the morning would be reflected the future glory of the new town, and the restless visages of scores of anxious lessees."*

* The late Joseph C. Neal, who was one of the motley mass, some years afterward wrote the following humorous description of the speculating scenes:

In the memorable year to which we allude, rumors of fortunes made at a blow, and competency secured by a turn of the fingers, came whispering down the Schuylkill and penetrating the city. The ball gathered strength by rolling—young and old were smitten with the desire to march upon the new Peru, rout the aborigines, and sate themselves with wealth. They had merely to go, and play the game boldly, to secure their utmost desire. Rumor declared that Pipkins was worth millions, made in a few months, although he had not a sixpence to begin with, or to keep grim want from dancing in his pocket. Fortune kept her court in the mountains of Schuylkill County, and all who paid their respects to her in person found her as kind as their wildest hopes could imagine.

The Ridge road was well traveled. Reading stared to see the lengthened

A reaction attended this, as all other speculative manias. But disastrous as it was, and involving hundreds in ruin, it did not prevent the continued investment

columns of emigration; and her astonished inhabitants looked with wonder upon the groaning stage-coaches, the hundreds of horsemen, and the thousands of footmen, who streamed through that ancient and respectable borough; and as for *Ultima Thule*, Orwigsburg, it *has not recovered from its fright to this day!*

Eight miles further brought the army to the land of milk and honey, and then the sport began—the town was far from large enough to accommodate the new accessions; but they did not come for comfort—they did not come to stay. They were to be among the mountains, like Sinbad in the valley of diamonds, just long enough to transform themselves from the likeness of Peter the Moneyless into that of a *Millionaire;* and then they intended to wing their flight to the perfumed saloons of metropolitan wealth and fashion. What though they slept in layers on the sanded floors of Troutman's and Shoemaker's bar-rooms, and learned to regard it as a favor that they were allowed the accommodation of a roof by paying roundly for it, a few months would pass, and then Aladdin, with the Genii of the Lamp, could not raise a palace or a banquet with more speed than they!

One branch of the adventurers betook themselves to land speculations, and another to the slower process of mining. With the first, mountains, rocks, and valleys changed hands with astonishing rapidity. That which was worth only hundreds in the morning, sold for thousands in the evening, and would command tens of thousands by sunrise, in paper money of that description known among the facetious as slow notes. Days and nights were consumed in surveys and chaffering. There was not a man who did not speak like a Crœsus—even your ragged rascal could talk of his hundreds of thousands.

The tracts of land, in passing through so many hands, became subdivided, and that brought on another act in the drama of speculation: the manufacture of towns, and the selling of town lots. Every speculator had his town laid out, and many of them had scores of towns. They were, to be sure, located in the pathless forests; but the future Broadways and Pall Malls were marked upon the trees; and it was anticipated that the time was not far distant when the deers, bears, and wild-cats would be obliged to give place, and take the gutter side of the belles and beaux of the new cities. How beautifully the towns yet unborn looked upon paper! the embryo squares, flaunting in pink and yellow, like a tulip show at Amsterdam; and the broad streets intersecting each other at right angles, in imitation of the common parent, Philadelphia. The skill of the artist was exerted to render them attractive; and the more German text, and the more pink and yellow,

of capital in the coal regions. Every year more mines were opened, more iron-works erected, more improvements of a stupendous character planned, more tons of coal

the more valuable became the town! The value of a lot, bedaubed with vermilion, was incalculable; and even a sky-parlor location, one edge of which rested upon the side of a perpendicular mountain, the lot running back into the air a hundred feet or so from the level of the earth, by the aid of the paint-box was no despicable bargain: and the corners of Chestnut and Chatham streets, in the town of Caledonia, situated in the centre of an almost impervious laurel swamp, brought a high price in market, for it was illustrated by a patch of yellow ochre!

The bar-rooms were hung round with these brilliant fancy sketches; every man had a roll of inchoate towns in the side pocket of his fustian jacket. The most populous country in the world is not so thickly studded with settlements as the coal region was to be; but they remain, unluckily, in *statu quo ante bellum.*

At some points a few buildings were erected to give an appearance of realizing promises. There was one town with a fine name, which had a great barn of a frame hotel. The building was let for *nothing;* but after a trial of a few weeks, customers were so scarce at the "Red Cow," that the tenant swore roundly he must have it on better terms, or he would give up the lease.

The other branch of our adventurers lent their attention to mining; and they could show you, by the aid of a pencil and piece of paper, the manner in which they must make fortunes, one and all, in a given space of time—expenses, so much; transportation, so much; will sell for so much; leaving a clear profit of 000,000! There was no mistake about the matter. To it they went, boring the mountains, swamping their money and themselves. The hills swarmed with them; they clustered like bees about a hive; but not a hope was realized. Calculations, like towns, are one thing on paper, and quite another when brought to the test.

At last the members of the expedition began to look haggard and careworn. The justices did a fine business; and Natty M., Blue Breeches, Pewter-Legs, and other worthies of the catchpole profession, toiled at their vocation with ceaseless activity. When the game could not be run down at view, it was taken by ambuscade. Several bold navigators discovered that the county had accommodations at Orwigsburg (at that time the seat of justice, now located at Pottsville) for gentlemen in trouble. Capiases, securities, and bail-pieces, became as familiar as your garter. The play was over, and the farce of "*The Devil to Pay*" was the after-piece. There was but one step from the sublime to the ridiculous, and Pottsville saw it taken!

sent to market. Canals which had been projected but suffered to languish were speedily completed; rail-roads were built, not only above ground but under ground; and in a comparatively brief period of time, it has been estimated by competent authority, *a hundred millions of dollars* were withdrawn from commercial activity, and invested in productive and unproductive improvements and partially abortive schemes. Many of the works, however, constructed in Lehigh and Schuylkill counties, are imperishable monuments of solidity and beauty, and will be objects of admiration in after ages.

At the present time there are, within the borders of Pennsylvania, upward of 800 miles of Canal, and 1,600 miles of Rail-road, of which the revenues are mainly derived from freight on Anthracite and semi-Anthracite coal. Many of those, projected with other views, have become large transporters of coal; and certainly the amount of capital expended in Pennsylvania for one object, viz., *for constructing avenues to convey Anthracite coal to market*, is now at least SEVEN TIMES GREATER than *the whole amount invested in all the manufactories at Lowell.* (See annexed TABLE.)

Gay gallants, who had but a few months before rolled up the turnpike, swelling with hope, and flushed with expectation, now betook themselves, in the gray of the morn, and then the haze of the evening, with bundle on back—the wardrobe of the Honorable Dick Dowles tied up in a little blue-and-white pocket-handkerchief—to the tow-path, making, in court phrase, "mortal escapes"; and, in the end, a general rush was effected—the army was disbanded—*sauve qui peut!*

The following table, which was prepared in 1857, principally from official information, exhibits the

Names, Length, and Cost of the Canals and Rail-roads in and leading to the Anthracite Coal Regions of Pennsylvania.

	Canals. Length.	Railroads Length.	Cost. Jan. 1, 1858.
Philadelphia and Reading R. R. (including City Br.).....		98	19,262,72C
Catawissa, Williamsport and Erie R. R., including the Quakake Branch, with equipments..........		79	4,200,000
Williamsport and Elmira, with real estate and basins at Williamsport and Elmira, and equipments complete...		78	3,850,000
Lebanon Valley R. R., (consolidated with Reading)......		53½	4,000,000
Schuylkill Navigation Co..........	108		10,950,000
Lehigh Coal and Navigation Co., viz.:			
Canal and Improvements..........	72		4,455,00C
Lehigh and Susquehanna R. R.		20	1,380,000
Summit and Branch Rail-roads..........		25	1,400,000
Delaware Division of Pennsylvania (State) Canal	60		2,200,000
Eastern Division of Pennsylvania (State) Canal..........	46		1,737,236
Susquehanna Canal..........	41		897,160
Lower North Branch Canal..........	73		1,598,379
Upper North Branch Canal..........	94		4,500,000
Union Canal..........	99		5,000,000
North Pennsylvania R. R..........		68	5,773,925
Lehigh Valley R. R..........		46	3,276,523
Philadelphia and Sunbury R. R., (unfinished)..........		33	1,348,812
Sunbury and Erie R. R., whole amount expended on finished and unfinished..........		40	3,693,492
Dauphin and Susquehanna R. R..........		54	4,000,000
Lackawana and Bloomsburg R. R..........		57	1,650,000
Delaware, Lackawana and Western R. R..........		113	8,701,888
Lackawana R. R., (unfinished)		9	300,000
Little Schuylkill Railroad, exclusive of land..........		33	1,402,651
Mine Hill Railroad, Extension and Branches..........		120	2,750,000
Schuylkill Valley Rail-road and Branches..........		28	568,000
Mill Creek Rail-road and Branches..........		13⅓	311,000
Mount Carbon and Port Carbon R. R., including land....		2½	282,000
Mount Carbon R. R..........		7	200,000
Beaver Meadow Rail-road and Branches..........		20	1,000,000
Hazleton Coal Co.'s R. R..........		14½	400,000
Lehigh and Luzerne R. R., (unfinished)..........		8	200,000
Buck Mountain Coal Co.'s R. R., (exclusive of land)......		7	430,000
Big Mountain Coal Co.'s R. R..........		2½	35,000
Trevorton R. R.		14	700,000
Tioga R. R..........		29¾	1,093,263
Barclay R. R. and Coal Co.'s R. R..........		16¼	438,000
New York and Middle Coal Field Co.'s R. R., (unfin'd)..		5	150,000
Columbia Coal and Iron Co.'s R. R..........		7	150,000
Carbon Run Coal Co.'s R. R..........		3½	60,000
Lykens Valley R. R..........		16	443,000
Union Canal Co.'s R. R..........		4	150,000
Swatara R. R..........		6	100,000
Wisconisco Canal..........	12		381,836
Lorberry Creek R. R..........		2½	25,000
Sundry Coal roads, private and underground, to the Mines, say..........		300	6,000,000
	605	1,433⅓	111,444,885
Morris Canal, (present total cost)..........	102		5,612,000
Pennsylvania Coal Co.'s R. R..........		44	1,994,819
Delaware and Hudson Canal and R. R., (estimated)	108	24	3,250,000
Central Rail-road of New Jersey..........		63	5,048,340
Total,	**815**	**1,564⅓**	**$127,350,044**

But the development of the mineral regions of Eastern Pennsylvania was not the only scheme that abstracted the attention and capital of the citizens of Philadelphia from the prosecution of Foreign Commerce. The West was becoming known as "The Great West." Regiment after regiment of hardy pioneers, armed with axes and plowshares, had entered the wilderness to subdue it: each successive year the frontiers of civilization were carried further westward; production outran consumption; and the people of Pennsylvania were called upon to furnish superior avenues and outlets for the produce of the West to the best markets on the Atlantic coast. A grand system of internal improvements was therefore resolved upon, and undertaken, to connect the metropolis of Pennsylvania with the Ohio River and the Lakes. The Erie Canal in New York was then near its completion, and herculean and partially successful efforts were being made to divert the trade of the West away from its natural and geographical channels by a circuitous route to New York. But the means adopted by Pennsylvania to establish superior connections with the West were less successful in execution than praiseworthy in conception. The Alleghanies defied the skill of the engineers, broke up the chain of communication into disjointed links; and the attempts made to unite them—constructing part rail-road, and part canal—instead of affording to shippers and producers the promised benefits, only fully succeeded in arousing the fears of foreign creditors, and provoking the sarcasm of the witty Dean of St. Paül's at the "the drab-coated gentry." No one acquainted with the physical characteristics of this State —its magnificent scenery, its rugged acclivities and im penetrable fastnesses—need be told that to construct railroads and canals within its limits, is and must be a serious and costly undertaking. The cost of the Commercial

Marine of many recognized commercial nations is a mere bagatelle in comparison with the vast sums expended in Pennsylvania for internal improvements alone.

On the first of January, 1858, Pennsylvania had 2773¼ miles of rail-road, costing $135,166,609; or, estimating the population of the State at three millions, the amount expended was at the rate of $45 for each man, woman, and child in the Commonwealth. The cost of constructing the canals within its borders, exceeding as they do 1200 miles in length, has been stated at thirty millions of dollars. To these immense sums, if we add the amounts expended in seeking for minerals, sinking shafts, opening mines, disinterring iron ore, and erecting works to manufacture it, the vastness of expenditure incurred for the development of internal wealth may well astonish and appal even those to whom the theme has become familiar by daily contemplation. In all these enterprises, the capital and credit of Philadelphia are conspicuous. Owning property equal to one third the assessed value of the property in the entire State, the city has contributed more than one half of the cost of public and private improvements. To aid these, her merchants sold their ships: to sustain them, her capitalists declined the profits of Bottomry and Respondentia.

But the prodigies achieved within the limits of Pennsylvania, great as they are, did not exhaust the zeal of the citizens of Philadelphia in behalf of internal improvements. Their brethren in neighboring States, in the South and the West, have drawn largely for contributions to such projects; and, to the extent of our ability, their drafts have not been dishonored. The portfolios of our merchants are now plethoric with such obligations and bonds; and when presently available, will build an Armada of merchant ships. If it were practicable to ascertain how many thousands of merchants are now thriving, how many tens of thousands of farmers in the States of Ohio, Indiana, Illinois, Wisconsin, and the South, are now comparatively wealthy, because of their present facilities for reaching good markets—facilities encouraged and perfected through aid from Philadelphia—the revelation would so interweave the ties of friendship with those of mutual mercantile interests, as to form a bond indissoluble by any assaults.

The citizens of Philadelphia, it is then safe to aver, are eminently patriotic, even in their business predilections. They have withdrawn their capital largely from prosperous commerce, to invest it in Mines, Rail-roads, Iron-works, and Manufactories, preferring to aid the development of the resources of the interior even at the expense of commercial importance and reputation abroad. Without giving assent to the doctrines of Chinese economists, who hold that Foreign Commerce is generally prejudicial to a State, because, by diminishing the quantity of desirable products, it must raise their price to the home consumer, they nevertheless believe that a prosperous, active interchange of products between citizens of the respective States is more conducive to the permanence and well-being of the Republic than even a more profitable commerce with foreigners. Cherishing, then, as they have done, and as they do, what they presume to be the best interests of our whole country, and having proved, by abandoning their share in the rich commerce of the Indies, the sincerity of their desire to accelerate its industrial development, the Merchants and Capitalists of Philadelphia would seem to be entitled to praises rather than taunts for the decline of their city in direct Foreign Commerce; and certainly they have established a claim to the high place which they hold in the friendly regard of their intelligent fellow-merchants throughout the Union.

But while acknowledging a decline in the Foreign

Commerce of Philadelphia, it is but justice to state that the decline is more apparent than real. The number of foreign arrivals, and the amount of duties paid at the Custom House here, are no index to the imports of the merchants of this city. Many of the most extensive importing-houses—and there are some, we were about to say, quite too extensive for the country's welfare—import nearly all their goods *via* New York. The largesses given by Government to steamers connecting with that port, and the peculiar facilities and inducements said to be held out to shippers, not to mention the rumor recently current that duties are sometimes lower there than elsewhere, influence our merchants in directing their foreign correspondents to ship goods to Philadelphia *via* New York. The advantage that the New York importer has over his Philadelphia competitor is simply a saving in freight between the two cities—an item perhaps not exceeding $2 per ton, or at least so unimportant on imported light and costly fabrics as to add no appreciable per-centage to the cost. That this advantage is overbalanced by other circumstances—lower rents, less extravagant expenditures for personal gratification, etc.—is evidenced by the fact that scores of New York jobbers visit Philadelphia every season, to replenish their stocks from the shelves of the importers, knowing that they can do so, besides paying fare, freight both ways, and all other expenses, at a cheaper rate than they can purchase the same goods from any of their neighbors. One Fancy Goods importing-house in particular, whose operations came within the range of my personal observation, attracts New York and Boston jobbers as regularly and more extensively than Cincinnati and St. Louis buyers. This is explained in part by the fact that the house has more favorable connections in Europe than their competitors in other cities, and partly by their ability to sell at a lower per-centage of profit in consequence of diminished expenses. These two circumstances, and especially favorable connections with the foreign manufacturers, would seem to be of more importance, in a regular importing business, than any other; and these, Philadelphia merchants,—whose honorable character and mercantile probity have ever been understood and appreciated in Europe,—enjoy peculiar facilities for obtaining. But in all probability I would not misrepresent popular feeling if I were to say that Philadelphia does not covet the distinction of being a great importing mart. She would be content if other cities monopolized the doubtful honor of importing hither French gimcracks and German cloths in exchange for gold and silver—our commercial life-blood—provided her merchants were encouraged to devote their energies successfully and uninterruptedly, to building up Home Industry and American Manufactures.

III. Commercial Relations with the South and West.

Pennsylvania, it has been frequently observed, is the only State in the Union that has a navigable outlet to the ocean, a footing upon the Lakes, and a command of the Ohio and the Mississippi. This position necessarily gives the metropolis of the Commonwealth points of superiority over all the other great cities on the Atlantic coast, for the purpose of receiving and distributing merchandise to and from a great portion of the South and West. With the ocean, and the principal cities of the Southern seaboard, Philadelphia has regular and direct communication by way of the Delaware River; and in consequence of improvements in locomotion, the distance is now less than at any previous time. With the gate of the West, Philadelphia is connected by canal and a magnificent railway; and at Pittsburg, with all the cities and towns on the navigable waters east of the Rocky Mountains, by thousands of miles of river navigation, and also by rail-roads joining Cleveland and Chicago on the one side, Wheeling and Cincinnati on the other, continuing through Kentucky to Nashville, and prolonged with a continuous, unbroken gauge westwardly, beyond St. Louis, on the Mississippi. Philadelphia has also an advantage over New York and Boston, in being considerably nearer to all the prominent foci of the products of the Great West. The principal rail-road lines from New York—the Erie and Central—it has been aptly remarked, lie on the circumference line to the West; while the great rail-road of Pennsylvania—the Pennsylvania Central—is on a diameter line. Their direction is to the Lakes—ours to the West. But to exhibit more clearly the relative position of Philadelphia, New York, and Boston, with reference to proximity to the chief centres of trade in the West, we have prepared the following Table, from data furnished in "Dinsmore's Railway Guide," published in New York:

	Clevel'd, Ohio. Miles.	Cincinnati, Ohio. Miles.	Chicago, Ills. Miles.	Indianapolis, Ind. Miles.	St. Louis, Mo. Miles.
From Philadelphia, *via* Pennsylvania Rail-road, to Pittsburg; thence by shortest Rail-road route to..................................	501	703	851	746	1000
New York, *via* Hudson River to Piermont, and the Erie Rail-road to Dunkirk, 468 miles; thence by shortest Rail-road route..............	612	867	954	893	1154
New York, *via* Hudson River Rail-road to Albany; thence by Rail-road to Buffalo, 442 miles; thence as above................................	625	880	967	906	1167
Boston, *via* Western Rail-road to Albany and Buffalo, 498 miles; thence as above...........	681	936	1023	962	1223

Hence, it is manifest that Philadelphia has considerable advantage over New York and Boston, in nearness to the principal centres of trade in the West. The saving in distance will be regarded as an important one by the weary traveler, while its effects in reducing the cost of transportation will be shown hereafter. It is true, New York has a shorter route to the places named than by the above-mentioned rail-roads; but that is, *via Philadelphia and Pittsburg*. Pennsylvania is truly the Key-stone State; and those who would pass and repass from the

West to the East, may congratulate themselves that their most direct route carries them over a railroad so well managed as the Pennsylvania Central, abounding in scenery of the most diversified description, and through a city so beautiful as Philadelphia.

Now, it may be said that New York and Boston have the advantage of Lake navigation to many prominent points in the West. We assert—and appeal to the managers of the New York and Erie and Boston and Worcester rail-roads, who receipt through both "all the way by rail-road or by steamer on the Lakes," as shippers prefer—that this is no advantage. The Lake freights are the regulators of the rail-road charges, which barely exceed them by the cost of insurance necessary to cover the great risks attending navigation on the Lakes. But Philadelphia, on the contrary, has a very important advantage, in addition to that stated in the Tables, by communicating at Pittsburg with thousands of miles of safe river navigation, extending southwardly to New Orleans and the ocean, and westwardly to St. Paul, on the Mississippi; and, in fact, to all the cities and towns on navigable waters east of the Rocky Mountains. The advantage in shipping from Philadelphia to Pittsburg, and thence by the Ohio River to Cincinnati and Louisville, over shipping to those points by the Northern rail-road lines, amounts, in addition to the saving stated above, to about $5 per ton on first-class goods, $4 on second, $3 on third-class, and $2 on very heavy goods; while to Nashville, Memphis, Cairo, St. Louis, and all points south of New Albany, Ind., the additional saving is nearly double this amount—that is, about $10 per ton on first-class goods, $8 on second, $6 on third, and about $3 per ton on fourth-class. It is thus evident, as experienced shippers know, that freight from the West, bound for European markets, can be brought to Philadelphia, and shipped hence, landing it at its destined port abroad, at cheaper paying rates than by way of New York. Indeed, the leading products of the West—for instance, flour, the products of the hog, whisky, etc.—can be shipped to Philadelphia, and hence at least *half the distance to Liverpool*, for the cost of transporting them to New York. Further, in view of the facts stated, it is also obvious that a Western merchant, purchasing goods in Philadelphia, may have his preference rewarded by a saving in the cost of transporting them home. The only practical question, then, for him to consider is, *whether it is probable he can make his purchases in the Philadelphia market as cheaply as in any other;* for, supposing the terms to be the same, he will nevertheless, by doing so, obtain an advantage. We beg permission to offer a suggestion or two upon this probability, for the consideration of those who study and appreciate economy.

We may remark, at the outset, that any one who has taken time to examine, compare, and reflect upon the characteristics of the respective markets, the development of Manufactures, and the comparative facilities for manufacturing, will not need any arguments to convince him that the probability of an advantage in price must be altogether in favor of the Philadelphia market. Let those, however, who have not already done so, examine the subject in its details, and they will be astonished to discover how few classes of goods constituting a country trader's usual assortment are not, to greater or less extent, made in or near Philadelphia. For instance, with regard to *Domestic Dry Goods:*—According to the census, Pennsylvania, in 1850, contained within her borders a larger number of factories for the making of cotton and woolen goods than any State in the Union; even more than the great manufacturing State of Massachusetts, and considerably more than New York. The former had 213 cotton and 119 woolen factories, and the latter 86 cotton and 249 for wool; while, in Pennsylvania, there were 588 of these establishments in all, of which 208 were employed in the cotton and 380 in the woolen manufacture. The extent to which Philadelphia is engaged in the production of these goods, will be illustrated in another place (see DRY GOODS MANUFACTURE); but we may state here, that one firm, Messrs. ALFRED JENKS & SON, manufacturers of cotton and woolen machinery, supplied the mills of this city alone, during the past year, with 800 looms for weaving checks, and on which could be woven twenty thousand yards per diem. The *New York Tribune* of May 1, 1857, in an editorial, urging greater attention to manufactures in that locality, remarked, we suppose with truth, "*Philadelphia has at least twenty manufactories of textile fabrics where New York has one; and her superiority in the fabrication of metals, though less decided, is still undeniable.*" Cottonades, checks, carpetings, Germantown hosiery and woolen goods, ribbons, sewing-silks, military goods, &c., are manufactured here to an immense extent; and of these, New York and other jobbers are constant and acceptable customers to the amount of millions annually. But, besides the vast quantities of dry goods manufactured in and near this city, all the principal mills of New England, and elsewhere, consign their fabrics to agencies established here, with authority to sell them, frequently at an abatement from invoice prices. The first agency for the sale of domestic fabrics in the United States, was that of ELIJAH WARING, established in this city about the year 1805; and from that day to this, the domestic Dry Goods Commission-houses of Philadelphia have maintained a position alike honorable to themselves; advantageous to American manufactures; and with one exception, viz., too great liberality in giving credit to strangers,* beneficial to the city.

With respect to *Foreign Dry Goods*, the importing-houses of Philadelphia certainly possess the same facilities for procuring desirable selections on advantageous terms as any others do; and in some instances enjoy unusually favorable connections in Europe, established long since, and by means of them secure perhaps more than their share of bargains. The stocks are generally selected by resident partners, who know the wants and consult the interests of purchasers; and therefore they consist, less than some others, of the unsaleable refuse of London warehouses.

Proposing, as we do, to make a minute and detailed examination of the manufacturing industry of Philadelphia, it would not be proper here to anticipate its results; but, for the benefit of anxious mercantile inquirers, we may state further, that more than four millions of dollars worth of *fine Boots and Shoes* are annually made in this city; while of the common, cheap, pegged-work of New England, Philadelphia is also a large purchaser, consumer, and distributer. The quality of our manufactures in this department is so generally and highly appreciated, that several of the manufacturers in Lynn, Mass., with a view of attracting additional custom, announce on their signs, "Philadelphia Shoes for sale." Of *Educational and Medical Books*, the publishers of Philadelphia are generally recognized as leaders; and for the distribution of books of all kinds, Henry C. Carey, the distinguished political economist, has asserted that Philadelphia has the largest book distributing house in the world. As respects *Iron*, the last census showed that nearly one-half of the pig, cast, and wrought iron, made in the

* The panic of 1857, disclosed the fact, that a prominent dry goods jobbing-house that failed, in New York city, was indebted to a commission-house in Philadelphia, in a sum but little short of $100,000—a line of credit entirely beyond the limits of prudence.

United States, was the product of the furnaces and forges of Pennsylvania; and official statistics show, that of the 782,958 tons of iron produced in the United States, in 1856, Pennsylvania produced 448,515 tons. Of the *Manufactures of Iron*, as stoves, hollow-ware, and those articles, usually denominated *Hardware*,—nails, screws, saws, forks, shovels, enameled-ware, hinges, bolts, nuts and washers, Philadelphia is an immense producer; and, for the sale of their products, the hardware manufacturers of Old and New England have agencies established in this city, authorized to sell at factory prices. In short, the market of Philadelphia differs in many important respects from most others, resembling from one point of view a Leipsic Fair, and from another the Eastern Bazaars. Manufacturers' depots are often situated between a commission-house and a house importing the same class of goods; fabrics, fresh from the loom, may be found close to the gold-tipped embroideries of France, or the crasse dresses of Turkey; factories adjoin stores, and stores are surrounded by manufactories; while, diverging from the city, are numerous roadways, constantly traversed by iron horses, bringing fuel from Nature's vast magazines not far distant; and from the East, caravans of boats, propellers, cars, come laden with the products of distant workshops, seeking here a central point for redistribution throughout the South and the West. Hence it is obvious, that a purchaser of a miscellaneous stock, adapted to the wants of a rural, town or city population, must be, when in Philadelphia, as near the fountain head, where goods are as yet in first hands, as it is possible for him to get; while the merchant, who visits the city to replenish his mind as well as his stock, can hardly fail, in a world of machinery, literature and art, as this is, to note much that is to him novel, and carry back suggestions that will be useful to himself and his neighbors.

Is it not probable, then, that the merchants of Philadelphia, in view of their advantages, with manufactories all around them, consignments from abroad seeking their markets and supplying their auction-houses, with abundance of capital and good credit, can buy and sell on terms as favorable as any of their competitors? We have no doubt they do this; but we go further, and insist that those now doing business have mistaken their vocation, unless, to responsible buyers, they *actually do undersell all others.* One reason that we have for entertaining this opinion is, that expenses for conducting business are less here than in most other large cities. In the city of New York, the leading Dry Goods jobbing-house pays, or did recently pay, as we are informed, an annual rent of $50,000 for their store; and a prominent wholesale clothing-firm pays, or did pay, $40,000; while the greatest amount of rent paid by a leading firm, in a similar business in Philadelphia, that I have heard of, and for which equal, or at least all necessary accommodations are procured, is $8,000. It is true, the "Stewart" of Philadelphia deems $14,000 a moderate compensation for his magnificent store, but his customers are principally the wealthy of the city. A proportionate difference in favor of Philadelphia prevails in rents, generally, for dwelling-houses as well as stores. The room for expansion afforded by the plan and locality of the city multiplies the number of eligible sites, and consequently diminishes speculation and prevents monopoly. The demands of fashion and extravagance, also, though sufficiently exorbitant, are less onerous in Philadelphia; and, from these and other circumstances, it would seem evident, without ocular demonstration, that a merchant in Philadelphia can afford to sell at a per-centage of profit, which, on the same amount of business, would not pay the expenses of his less favorably situated competitor.

These are the deductions of reason and common sense. Their importance entitles them at least to consideration, reflection, and experiment; hence we beg those who are engaged in buying and selling, inasmuch as their mercantile success, and the prosperity of the mercantile class throughout the country, depend upon the wisdom of their action, to test the respective markets fairly,—disregarding "baits," which are quite too common in all, and extending their view beyond exceptional circumstances,—and if there be an atom of truth in that principle of political economy, which demonstrates that the nearer the place of production the cheaper the price, they will discover, as thousands of thriving merchants have already done, that Philadelphia is the CHEAPEST SELLER, and NATURAL DISTRIBUTER OF MERCHANDISE ADAPTED TO THE WANTS OF THE SOUTH AND THE WEST.

Returning from this digression to subjects more immediately connected with our inquiries, and having already adverted to the moral circumstances that have an effect upon economy of production in Manufactures, we now proceed to consider the position of Philadelphia with respect to

IV. Physical Advantages for Manufacturing.

In considering Philadelphia as a Manufacturing centre, it must be obvious, from previous remarks, and still more obvious from minute information respecting the topographical and geological features of Pennsylvania, and the intimacy of connection between the metropolis and the principal mineral sections of the State, that Philadelphia and its vicinity command, in the first place, the *most important raw materials used in Manufactures; and secondly, the agents best fitted to produce power.* But the celebrity of Pennsylvania for its vast deposits of IRON and COAL —those primary sources of England's manufacturing greatness—is so widely extended, that to dilate upon their abundance would hardly convey any additional information to any person of ordinary intelligence. The census of 1850, as we previously stated, showed that nearly one half of the pig, cast, and wrought Iron made in the United States was from her forges and furnaces; while her mines of "black diamonds," it is a proverb, are only equalled in national importance by the gold mines of California. The district in Pennsylvania that produces the most Iron and the cheapest Coal, viz., the Valleys of the Lehigh, the Schuylkill, and a part of the Delaware—is directly tributary to Philadelphia, procuring its supplies from this city, and selling its products here almost exclusively. We therefore record the latest statistics of these important products.

I. IRON. The statistics of the Iron production of Pennsylvania, for 1866, as furnished us by the Secretary of the *American Iron Association* in Philadelphia, are as follows:

	Tons of 2000 lbs.	
Anthracite Iron.		
Valley of the Delaware and Lehigh, - -	230,636	
Valley of the Schuylkill, - - - - -	117,626	
" " " Upper Susquehanna and Juniata,	108,586	
" " " Lower Susquehanna, - - -	103,376	
Total Anthracite iron - - - - - -		566,224
Charcoal Iron.		
Western Pennsylvania, - - - - -	8,800	
Eastern Pennsylvania, - - - - - -	51,378	
Total Charcoal iron - - - - - -		60,178
Raw Coal and Coke.		
In Pennsylvania, - - - - - - - - - -		146,077
Total of all kinds of pig iron made in Pennsylvania -		772,479

This is about 60 per cent of the whole amount of pig iron made in the United States during last year. Of the total production of iron in the state nearly 60 per cent, or 450,000 tons, was made east of the mountains, 350,000 tons being anthracite iron made in the immediate vicinity of Philadelphia. The value of the crude iron made in the State in 1866 may be set down in round numbers at $35,000,000. The following statement exhibiting the growth of the pig iron manufacture in the State we take from the carefully prepared tables of the American Iron and Steel Association before alluded to.

Year.	Quantity of Pig Iron made in tons of 2000 lbs.	Year.	Quantity of Pig Iron made in tons of 2000 lbs.
1854	408,688	1861	437,729
1855	419,888	1862	478,411
1856	494.187	1863	571,103
1857	444.356	1864	695,081
1858	382,839	1865	551,303
1859	473,693	1866	772,479
1860	531,208		

It is thus manifest, that Philadelphia is situated in the district that is entitled to be called the *centre of the Iron production of the United States*. It is further manifest, that the centre of the Iron interest is likely to *remain* in the district tributary to Philadelphia, inasmuch as the production has been an increasing one; and, the establishments situated within its limits have been able to survive disasters that have borne down those in other places, and consequently there must exist circumstances peculiarly favorable to economy of production.

2. Coal. The quantity of Coal sent to market from the district tributary to this city, was as follows:—

Product of the Anthracite Coal Fields of Pennsylvania, for 1866.

	1866. Tons.	1865. Tons.
1. Southern Coal District, comprising the Schuylkill, Pine Grove, and Lyken Valley regions,	4,633,487	3,649,063
2. Middle Coal District, comprising the Lehigh, Mahanoy, and Shamokin regions,	3,029,644	2,388 147
3. Northern Coal District, comprising the Wyoming and Lackawanna regions,	4,736,616	3,960,836
Total of the three fields,	12,399,747	9,998,046

The value of this product at $2.50 per ton, the minimum

price at the mines, would be $30,999,367, while the market value certainly exceeds *seventy millions of dollars.* In addition to Anthracite, the mines of Eastern Pennsylvania produced, last year, 539,281 tons of semi-Anthracite coal; and those West of the Alleghanies, produced three millions of tons, making the total production of a value not less than $80,000,000.

The qualities of different coals have necessarily been made the subject of careful analysis; and their relative value has been tested by frequent experiments. We believe it is conceded by both scientific authority and practical experience, that *Pennsylvania Anthracite is practically the cheapest and best fuel that the United States afford.** It contains about 90 per cent. of carbon, and,

* For the purposes of *steam navigation*, an impression formerly prevailed that the Pennsylvania Anthracite was inferior to the Cumberland coal, which, it is acknowledged, surpasses in strength the foreign bituminous coals of Newcastle, Liverpool, Scotland, Pictou, and Sydney. In January, 1852, a series of experiments were undertaken at the New York Navy Yard with the boilers of the United States Steamer "Fulton," to settle the question of relative value and superiority for this purpose. The result is given in the following extract from the Report of the Engineer-in-Chief, CHARLES B. STUART, to Commodore JOSEPH SMITH, Chief of Bureau of Yards and Docks.

COMPARISON.

The coals used in these experiments were the kinds furnished by the agents of the Government for the use of the United States Navy Yard and Steamers, and was taken indiscriminately from the piles in the yard, without assorting.

The bituminous was from the "Cumberland" mines. The anthracite was the kind known as "White Ash Schuylkill."

From the preceding data, it appears that, in regard to the rapidity of "getting-up" steam, the anthracite exceeds the bituminous thirty-six per cent.

That, in economical evaporation per unit of fuel, the anthracite exceeds the bituminous in the proportion of 7.478 to 4.483, or 66.8 per cent.

It will also be perceived, that the result of the third experiment on the boilers of the pumping-engine at the New York Dry Dock, which experiment was entirely differently made and calculated from the first and second

next to charcoal, gives out more heat than the same weight of any other fuel. So far as at present known, Pennsylvania is the only State where this valuable mineral can

experiments, gave an economical superiority to the anthracite over the bituminous of 62.8 per cent.—a remarkably close approximation to the result obtained by the experiments on the "Fulton's" boilers (66.8 per cent.), particularly when it is stated that the boilers and grates of the pumping-engine were made with a view to burning bituminous coal, which has been used since their completion, while those of the "Fulton" were constructed for the use of anthracite. The general characters of the boilers were similar, both having return drop-flues.

Thus it will be seen, from the experiments, that, without allowing for the difference of weight of coal that can be stowed in the same bulk, the engine using anthracite could steam about two-thirds longer than with bituminous.

These are important considerations in favor of anthracite coal for the uses of the Navy, without taking into account the additional amount of anthracite more than bituminous that can be placed on board a vessel in the same bunkers; or the advantages of being free from *smoke*, which in a *war*-steamer may at times be of the utmost importance in concealing the movements of the vessel; and also the almost, if not altogether, entire freedom from spontaneous combustion.

The results of the experiments made last spring on the United States steamer "Vixen" were so favorable, that I recommended to the Bureau of Construction, &c., the use of anthracite for all naval steamers at that time having, or to be thereafter fitted with, *iron* boilers; particularly the steamers "Fulton," "Princeton," and "Alleghany," the boilers for all of which were designed with a special view to the use of *anthracite*, and with the approval of that Bureau.

The "Fulton's" bunkers are now filled with anthracite; and the consumptions referred to in the engineer's report on that steamer show, during the short time she has been at sea, that the anticipated *economy* has been fully realized.

In view of the results contained in this report, I would respectfully recommend to the Bureau of Yards and Docks, the use of anthracite in the several Navy Yards, and especially for the engine of the Dry Dock at the New York Navy Yard.

In conclusion, I desire the approval of the Bureau to make such investigations as my duties will permit, with regard to the *experience* of the durability of *copper* boilers, when used with bituminous or anthracite coal; which can be done without any specific expenditure.

The inquiry may prove highly important to the Navy Department, as the

be obtained cheaply and in unlimited quantities; but within her borders the supply is seemingly sufficient to satisfy the probable wants of this country for centuries to come.

The rapidity with which Anthracite coal has appreciated in popular estimation, is shown by the increase in the demand for it. In 1820, only 365 tons were sent to tide-water; in 1840, the product amounted to 867,000 tons; in 1852, it had reached five millions of tons: being an increase in 12 years, from 1840 to 1852, of 600 per cent. Supposing this rate of augmentation to continue up to 1870, Gov. Bigler once amused himself by calculating that the production would be forty-five millions of tons, worth, at the present prices of the Philadelphia market, the sum of $180,000,000. No wonder the worthy Governor was moved to pronounce this a gratifying picture, confirming his belief "that, before the close of the present century, Pennsylvania, in point of wealth and real greatness, would stand in advance of all her sister States."

In the cost of fuel, Philadelphia has an admitted advantage over New York of about twenty-five per cent.; over Providence, R. I., from $1.75 to $2.25 per ton; and over Boston, from $2 to $2.50 per ton. The advantage, moreover, which Philadelphia enjoys from controlling the production of the best fuel, in addition to proximity, is too evident to need illustration; and being also the central and chief market of the district producing the

use of anthracite under copper boilers has been heretofore generally considered as more injurious than bituminous coal, and is consequently not used by Government in vessels having copper boilers.

Respectfully submitted, by your obedient servant,

CHARLES B. STUART,
Engineer-in-Chief, U. S. Navy.

Commodore JOSEPH SMITH,
Chief of Bureau of Yards and Docks.

best and cheapest iron, it would seem almost superfluous to inquire further as to her capabilities for Manufactures.

But Iron and Coal, though the most important, are not the only useful mineral products that abound in Eastern Pennsylvania. *Copper* exists extensively in several counties; *Plumbago* is obtained in Bucks County, and *Zinc* in the vicinity of Bethlehem. *Marble*, well adapted and extensively used for building purposes, has long been obtained from quarries in Montgomery County, a few miles above Philadelphia. *Steatite*, or *Soapstone*, is quarried extensively on the Schuylkill, above Manayunk. Roofing and Ciphering *Slates* of the best quality are found in the counties of Lehigh, Monroe, and Northampton; there being in the county of Lehigh alone some thirty quarries open, with a capital of $60,000 invested, employing about 300 men, and producing at least 25,000 squares of roofing-slates per annum, valued at $3 per square on the quarry bank, and at $5 and $6 in the Philadelphia market. Nearly all the best school-slates in this country are from the Pennsylvania quarries; and many of them are manufactured at an establishment in this city. Of *Salt*, the census of 1850 states the produce of Pennsylvania at 184,370 barrels. *Kaolin*, or *Porcelain earth*, is abundant at several points within a radius of thirty miles from Philadelphia. About 2½ miles north of Camden, N. J., there is an extensive bed of Fire Clay, of which specimens have been sent to England, and pronounced by competent judges superior to the German clay, which commands $25 per ton. Besides these, Barium, Chromium, Cobalt, Nickel, Magnesium, Titanium, Lead, Silver, Zirconium, and Fire and Potter's Clay, are scattered over the State, and in some instances of superior quality.

With all the points in Pennsylvania producing mineral and mining products, Philadelphia is directly connected by rail-roads and canals, and thus may be said to be situated in close proximity to the original sources of many of the most important articles that can be enumerated in a list of raw materials of Manufactures. And if we were to pass from the products of the mine to those of the forest and of Agriculture, we would find them equally abundant, cheap, and accessible. *Lumber*, in immense quantities, is obtained on the Susquehanna and the Delaware, and floated down those rivers every Spring and Fall. In 1852, it was estimated that 250,000,000 feet were sent down the former river; while the Lehigh region supplied the Philadelphia market, *via* canal, in the same year with 52,123,751 feet. At the present time, we are informed by persons intimately acquainted with the subject, Philadelphia has a larger stock of *seasoned* lumber than any other mart in the Union.* *Wool*, of the very

* Many of the forest trees most useful in the Arts, Manufactures, and Medicine, are natives of Pennsylvania. We condense from *Trego's Geography of Pennsylvania* the following list, which may be of value to some of our readers:

Oaks. At least twelve varieties. The *White Oak*, the most esteemed of this noble family of trees, is found throughout the State; and in the Southeastern districts the wood is exceedingly compact and tough. The *Black Oak*, which is very abundant, and one of our largest trees, furnishes *Quercitron Bark*, which is exported in large quantities, and used in dyeing wool, silk, &c., a yellow color. When used by tanners, it imparts a yellow tinge to the leather. The *Spanish Oak*, of which the bark commands a high price, is less common in Pennsylvania than further South. The other species, valuable for their bark, which is highly esteemed by tanners, is the *Rock Chestnut Oak*, the *Scarlet Oak*, and the *Red Oak*. In addition to these, there are the *Iron Oak*, confined to the Eastern part of the State, and resembling the White Oak; the *Swamp White Oak*, the *Swamp Chestnut Oak*, *Laurel* or *Shingle Oak*, *Scrub Oak*, and *Pin Oak*.

Walnuts. Two principal kinds, the *Black* and *White Walnut*. The former is much used for cabinet-work, and for the stocks of military muskets; also for the posts of fences, which, it is said, will last from twenty to twenty-five years. The bark of the *White Walnut*, or *Butternut*, yields an excellent cathartic medicine, said to be efficacious in cases of dysentery. It is also used in the country for giving a brown color to wool

best American grades, is grown in the Western counties of the State ; and all, or nearly all, of which, as the woolen manufacturers of Rhode Island and Massachusetts, who

HICKORY. The most common species are the *White Heart Hickory, Pig Nut, Bitter Nut, Shell Bark*, and *Thick Shell Bark*—highly valued for axle-trees, handles, flails, &c., and also as a fuel, affording in the same bulk more combustible matter than any other wood.

MAPLE. The *Red Maple* is the most common, and probably the most valuable species. Its wood is much used by chairmakers, and for bedsteads, saddle-trees, &c. In many of the old trees, the fibres of the wood, instead of following a perpendicular direction, are undulated and waving. This is known as the *Curled Maple*, and when skillfully polished, produces the most beautiful effect of light and shade. The bark of the Red Maple yields a purplish color by boiling, which, by the addition of copperas, becomes dark-blue, approaching to black. It is used in the country for dyeing, and for making ink. The *true Sugar Maple* is abundant, particularly along the elevated range of the Alleghanies, and the *Black Sugar* tree along the Western rivers. Large quantities of maple sugar are made in the Northern and Western counties. The *Striped Maple* grows in the mountainous parts of the State, and the *Ash-leaved Maple*, or *Box Elder*, west of the mountains.

DOGWOOD. The most valuable species grows to the height of twenty or thirty feet. The wood is used for tool handles, and other purposes, and the inner bark has medicinal properties resembling those of the cinchona or Peruvian bark, from which quinine is made, and has been successfully used in intermitting fevers.

The POPLAR or TULIP tree is common in Pennsylvania, and surpasses most of our forest trees in height and the beauty of its flowers and foliage. Its wood is applied to many purposes where lightness and strength are desirable, as trunks, chairs, &c., and the bark is said to possess tonic and antiseptic qualities; and a decoction of it, combined with a few drops of laudanum, has been found efficacious in giving tone and vigor to the stomach after fevers and inflammatory diseases. It has been also used in dyspepsia and cholera infantum.

WHITE and RED BIRCH grow abundantly along the Delaware above Philadelphia, and *Black*, or *Sweet Birch*, in deep, loose, and cool soils. It is said that articles of furniture made from this acquire with time the appearance of mahogany.

Of woods remarkable for their durability, we have the *Locust*, which is abundant in limestone valleys; and the *Red Mulberry*, frequently met with in fertile soils, when seasoned, is nearly equal to the Locust; also, the *Red Cedar*, exceedingly durable, and highly esteemed for making fence posts, is common in most parts of Pennsylvania.

come hither to purchase it, can testify, is secured to the Philadelphia market. About ten millions of pounds are sold annually. But a still wider range of raw materials is open to the manufacturers of Philadelphia. Those which are the product of other States or foreign countries are, by means of direct commerce, brought to her wharves, and concentre in her warehouses. The hides of Buenos Ayres, the woods of Guiana, the marble of Italy, the dye-stuffs of Calcutta, and the cotton of our Southern States, are delivered to the doors of our factories, in many instances as directly from the producers as the Minerals, Lumber, and Wool of Pennsylvania.

The CHESTNUT may also be ranked among very durable woods. It grows most abundantly in the hilly regions, and frequently attains an extraordinary size; one on Mount Etna being 53 feet in diameter, or 160 feet in circumference, but *hollow* to the bark. The wood is much used for posts and rails; and it is largely consumed in the manufacture of charcoal for the supply of the iron-works in the interior of the State. Its fruit is particularly appreciated by the boys.

Of PINES, there is every variety, though the *true Yellow Pine* is not very common in the State. The *Pitch Pine* is abundant, and in some places tar is manufactured from the more resinous parts of it. *White Pine*, so useful. and applied to such a variety of objects, is becoming comparatively scarce. in consequence of the enormous consumption for shingles, lumber, &c.; but nevertheless, it is still found in considerable quantities on the upper streams of the Lehigh, the head waters of the Susquehanna, and some of the tributaries of the Alleghany.

The *Hemlock Spruce*, however, which is more common, growing on the steep banks of streams, and in dark and shaded situations, is being substituted for White Pine, wherever it can well be done.

The other forest trees which are natives of Pennsylvania are the *White* and *Red Ash*, highly esteemed for strength and elasticity, several species of the *Aspen*, *White* and *Red Beech*, *Buttonwood* or *Sycamore*, *Catalpa* or *Bean* tree, *Crab-apple*, *Cucumber* tree, so called because the cones or fruit somewhat resemble a small cucumber, *Chincapin*, *White* and *Red* or *Slippery Elm*, *Sweet* and *Sour Gums*, *Hornbeam*, *June Berry* or *May Cherry*. *Linden*, *Lime* tree or *Basswood*, *Magnolia* or *Beaver* tree, *Papaw*, *Persimmon*. *Sassafras*, *Black* or *Double* Spruce, *Tamarack* or *American Larch*, *Willow*, and *Wild Cherry*, of which the wood is used as a substitute for mahogany, and the bark as a valuable tonic medicine.

But the term raw material, though ordinarily limited to natural or unmanufactured products, is more comprehensive in its scope, embracing *Chemicals, substances used as food, and such substances of vegetable and animal origin as are used in Manufactures.*

3. CHEMICALS. We before remarked that the finished product of one class of manufacturers is often the raw material of another class. If the proposition needed further illustration, we might advert to Chemicals, which are such important reagents in manufacturing operations, that without them it would be difficult, (in fact, by any known processes,) impossible, to produce several articles of daily and essential utility. Without Sulphuric Acid or Oil of Vitriol, for instance, we could not probably produce Alum, Ammonia, Sal-ammoniac; Iodine and Bromine, upon the existence of which the daguerreotype art is dependent; Bleaching powder or Chlorid of Lime; Corrosive Sublimate and Calomel; Bichromate of Potash, and consequently the pigments of chrome-red, chrome-green, and chrome-yellow; Phosphorus, and consequently friction matches; or lastly, Stearic acid candles. By means of this acid, more than 100,000 tons of Soda-ash are extracted from common salt in Great Britain yearly. Without Muriatic and Nitric acids, the art of refining gold and silver, the jeweler's art, the art of electrotyping, and numerous other branches of industry, could not flourish, and some of them could not exist. The useful Arts and Manufactures, it is thus evident, are largely dependent upon Chemicals; and, consequently, a locality possessing those of the best quality in abundance, has necessarily secured an important and undoubted advantage. The chemical factories of Philadelphia, every one acknowledges, rank among the first in extent and celebrity throughout the Union. About seventeen millions of pounds of Sulphuric Acid are made yearly, and other acids and

alkaline salts in proportion. The products of the establishments of POWERS & WEIGHTMAN, ROSENGARTEN & SONS, C. LENNIG, HARRISON BROTHERS, WETHERELL, PHILLIPS and others, are recognized as of standard excellence in the markets of the world; and where such establishments exist, we can hardly err in presuming, that at least, those Manufactures which are dependent upon the Chemical Arts must certainly flourish.

4. AGRICULTURAL PRODUCTS, PROVISIONS, &c. Again, no one need be told that with *substances used as food*, the markets of Philadelphia are always abundantly supplied, at moderate prices. As a wheat-growing State, the census of 1850 shows that Pennsylvania excels all her sister States; the product for that year having been 15,367,691 bushels, which exceeded that of Ohio, and was two millions of bushels more than that of New York. Of Rye, the product was 4,805,160 bushels; of Indian corn, 19,835,214 bushels; of Oats, 21,538,156 bushels; and hay, grass seeds, wool, butter, maple sugar, &c., in proportionate quantities. The counties immediately surrounding Philadelphia vie with each other, and rival the best counties in any other State, both in the quality and quantity of their productions. In 1850, Montgomery produced greater quantities of *hay* and *butter* than any other one county in the State; Lancaster produced more oats than any other county in the United States, more wheat than any, excepting Monroe County, New York, and more corn than any other county in Pennsylvania. In Chester, the quantity of corn produced exceeded that of any other county of the State except Lancaster, and of hay, except Montgomery; while Delaware excels in dairy products, supplying the markets of Philadelphia with butter, cheese, milk, and ice-cream, and the Union with whetstones.

Fifty years ago it was remarked, and the remarks are as

Powers & Weightman, Manufacturing Chemists, Philadelphia.

true now as then, "Much of the land within five or six miles North and South of the city is devoted to the purpose of market-gardens, and is kept in the highest state of cultivation. Two crops are very commonly produced on the same ground in one season. The neighboring State of New Jersey contributes to the abundant supply of those species of fruit and vegetables to which its light soils are particularly adapted: such as the grateful musk-melon, the water-melon, sweet-potato, cucumbers, and peaches, immense quantities of which are brought in boats across the Delaware. The superiority of the butter of Philadelphia, and the great neatness with which it is prepared for market, are generally acknowledged. One fourth of a dollar may be said to be the average price of a pound of butter, throughout the year."*

* The abundance and superior quality of the Agricultural products, for which the markets of Philadelphia are distinguished, are probably the fruition, and certainly the just reward, of the interest that has always been manifested by her citizens in Agricultural improvement. As early as 1785, a number of gentlemen, among others Robert Morris, Dr. Rush, and Richard Peters, met together and established the first Agricultural Society on this continent, under the title of the "Philadelphia Society for Promoting Agriculture," which still survives, surrounded now, however, by almost innumerable sister associations, diffusing information on rural affairs throughout the entire Union. At a later period, in September, 1826, a company of Philadelphians, principally through the instrumentality of the late Dr. James Mease, founded the "Pennsylvania Horticultural Society," which, like its predecessor, has the proud distinction of having led the way in its own particular sphere, and induced the creation of many kindred associations, promoting refinement and kindling a taste for Horticulture, even at the verge of Western settlements. One of the means early adopted by both associations to stimulate improvement was holding Annual Exhibitions, at which live stock, implements, fruits, vegetables, and flowers, were brought into competition. The exhibitions of the Agricultural Society attracted, years before they were held elsewhere, throngs of intelligent observers and practical cultivators from neighboring States, as well as from Pennsylvania, and diffused a most salutary and beneficial influence. The development which it is possible for such societies to attain, was witnessed in October, 1856, when the exhibition of the "United States

Of *Fish*, the markets of Philadelphia are constantly supplied, from the river, the bay, and the sea, with almost every desirable variety. We can imagine the delight with which epicures, a half century ago, read, that "early in the spring large sun-fish are caught in the Bay, and are succeeded by herrings, shad, roach, four kinds of cat-fish, four kinds of perch, rock, lamprey eel, common eel, pike, sucker, sturgeon, gar-fish. These are river fish, and appear in the order mentioned. From the sea, come

Agricultural Society," held in Philadelphia, by invitation of the Philadelphia Society and the officers of the City Government, attracted the most imposing display probably ever witnessed in the United States, on any similar occasion. Upward of *forty thousand dollars* were received, and the entire sum expended in premiums and for the necessary preparations. Competitors from distant States carried off many well-earned and important premiums; but it would be only justice rewarding merit, to record the fact, that to a Philadelphia firm, that of DAVID LANDRETH & SON, was awarded the first and most important premium, viz., that *for the best display of Agricultural Implements manufactured by the exhibiter.*

In the importation of *Live Stock*, Philadelphians were among the first to embark, and they have had the satisfaction of introducing to agriculturalists some of the most valuable foreign breeds that are known. The first "short-horned" cow that probably ever crossed the Atlantic, was landed at the wharf in Philadelphia, in 1807. This importation was in advance of the appreciation of such stock, and the cow was returned to England; but a bull-calf, dropped by her whilst here, was fortunately retained, and impressed his stamp on the cattle of the country.

About the year 1828, Mr. JOHN HARE POWELL, imported *South Down Sheep;* and the same enterprising gentleman, some years subsequently, commenced his importations of Short Horns, (Durhams). Not long afterward, Mr. Whittaker, the noted English breeder, consigned similar animals to his care for sale. Other gentlemen in this vicinity followed the example of Mr. Powell; and shortly afterward further importations were made for Kentucky, and other Western States. Mr. Sarchet, of Philadelphia, has the credit of the first importation of "Alderneys"; afterward, in 1840, the late Mr. Nicholas Biddle, imported specimens of the "Jersey" or "Alderney" cattle. Their descendants are now spread into the adjoining counties, and have produced a sensible improvement in the quality of the cream and butter wherever the strain has been infused. It seems to us proper, that early enterprise in this direction should be recorded.

cod, sea-bass, black-fish, sheep's-head, Spanish mackerel, haddock, pollock, mullet, halibut, flounder, sole, plaice, skait, porgey, tom-cod, and others. Of Shell-fish, there are oysters, (several kinds,) clams, lobster, crab, snapping-turtle, and terrapin—all excellent. Oysters abound throughout the year, and are sold at a low price. The shad caught in the vicinity of Philadelphia are generally esteemed superior in flavor, and more delicate than those caught elsewhere. It is supposed that the situation of the fishing-places influences the size and the flavor of shad." But the abundance, cheapness, and excellence of provisions in Philadelphia are conceded. The *New York Tribune* of May 1, 1857, stated that "*Philadelphia has about twenty-five per cent. the advantage of us in fuel, and perhaps ten per cent. in the average cost of provisions.*"

III. Situation, Rail-road Connections, &c. The third point that we have considered essential to success in Manufactures, is a *favorable situation.* Viewing Philadelphia with respect to situation, we remark, in the first place, that it is far enough from the ocean to be exempt from a salt atmosphere, which has been found decidedly injurious in several Manufacturing and Chemical processes;* yet it is near enough to the great highway of nations to partake of the advantages of a port on the seacoast, in receiving raw materials and sending away manufactured products. Secondly, Philadelphia now possesses unrivaled means of communication with the interior of our country, and directly or indirectly with all foreign countries. Shippers of freight, destined for other seaports, have a choice of routes to the ocean, viz., the Delaware River—the ordinary and natural channel—and the Camden and Amboy Rail-road, the Philadelphia and Trenton Rail-road, and the Delaware and Raritan

* In paints, a pure Carbonate of Lead cannot well be made near to the sea.

Canal. By way of the river and the ocean, merchandise may be forwarded and received cheaply and expeditiously from all parts of the world, regular lines being established to all principal cities of the United States—Boston, New York, Baltimore, Richmond, Savannah, New Orleans, California, and to Liverpool, &c. During the last year, (1866,) as we have stated elsewhere, there were 722 foreign, and 35,864 coastwise arrivals, principally at the Delaware wharves. Since 1845, the vessels annually employed in the coal trade alone, from Port Richmond, largely exceed in number and capacity the whole foreign tonnage of the city of New York. But, though the Delaware River be the natural channel for freight destined to distant sections, it is by no means the only one. Immense quantities of goods are daily sent and received by the Propeller lines, *via* the Delaware and Raritan Canal; and the Camden and Amboy, and Philadelphia and Trenton Rail-roads, which are far-famed thoroughfares. Another and more direct route to the ocean than any of these has been opened within a few years by the completion of the Camden and Atlantic Railroad; and still another by the construction of the Delaware Bay and Raritan railroad extending from Camden to Port Monmouth and which is now a favorite route for the transportation of heavy freights between Philadelphia and New York.

But the highways in which Philadelphia has investea the greatest amount of capital, and which probably will in future be of the most advantage to her industrial interests, are those which communicate with the interior. To the North, and connecting her with the coal regions, there are several canals and two principal railroads—the Reading, and the North Pennsylvania. The latter is a new and promising road, communicating with the populous towns of Lehigh and Northampton counties, and in connection with the Lehigh Valley Rail-road, affording another outlet for the Coal and Iron products of the Lehigh regions. The former was constructed primarily as an avenue for the transportation of coal from Schuylkill County; but by means of connections established with other roads, it now forms part of a great through route to the Falls of Niagara, the Lakes, the Canadas, and the West.

The READING RAIL-ROAD, being unquestionably one of the most magnificent freight roads in the world, is entitled to further notice. It was the first rail-road that revolutionized popular opinion with respect to the adaptation of railways for carrying heavy burdens. Having a slightly descending grade in the direction of the loaded trains, the entire distance from Schuylkill Haven to the Falls of Schuylkill, 84 miles, it is able to transport heavy freight at a cost which is insignificant, even in comparison with the usual tolls on canals. The cost of transporting a ton of coal, per round trip of 190 miles—that is, from the coal region to tide-water and back with empty cars, was, during the last year, only 36.3 cents; whereas, the tolls on a ton of merchandise on the Erie Canal were nearly double that amount. The average load of an engine, during the busy season, is nearly 500 tons of coal; and a single engine has conveyed a train of 166 cars, weighing 797 tons of 2240 lbs. each.

The original charter, passed in 1833, contemplated Reading as the northern terminus of the road—hence its name; but subsequently the charter was extended, and the road constructed to Pottsville. The first locomotive and train passed over the entire line on the first day of January, 1842. The event was celebrated with military display, and "an immense procession of seventy-five passenger cars, 2,225 feet in length, containing 2,150 persons, three bands of music, banners, &c., all drawn by a single engine. In the rear was a train of fifty two burden cars, loaded with 180 tons of coal, part of which was mined the same morning, 412 feet below water level." The road now consists of a double track; rail of the H pattern; whole length 306 miles, of which 127¾ miles have been relaid during the last seven years, at a cost of $796,735.43. The rolling stock includes 267 locomotives, 105 passenger cars, 1340 merchandise cars, and 9,111 iron and wooden coal cars, besides over 600 used by the company, but owned by other parties; and the whole, if placed in a line, *would extend for a distance of thirty miles.* The equipments are ample for the transportation of 5,000,000 tons per annum; the tonnage in 1860 being 4,751,805. The road has nearly ninety stone and iron bridges, and over forty wooden bridges, four tunnels, the largest of which, at Phœnixville, is 1,934 feet, cut through solid rock; numerous depots, wharves, and workshops, (those at Reading furnishing employment to about 400 hands, including boys,) and a vast deal of valuable real estate.* The entire cost of the whole, on

* "At Richmond, the lower terminus of the road, at tide-water on the river Delaware, are constructed the most extensive and commodious wharves, in all probability, in the world, for the reception and shipping, not only of the present, but of the future vast coal tonnage of the railway; forty-nine acres are occupied with the company's wharves and works, extending along twenty-two hundred and seventy-two feet of river front, and accessible to vessels of six or seven hundred tons. The shipping arrangements consist of some twenty wharves or piers, extending from three hundred and forty-two to eleven hundred and thirty-two feet into the river, all built in the most substantial manner, and furnished with chutes at convenient distances, by which the coal flows into the vessel lying alongside, DIRECTLY FROM THE OPENED BOTTOM OF THE COAL CAR IN WHICH IT LEFT THE MINE. As some coal is piled or stacked in winter, or at times when its shipment is not required, the elevation of the tracks, by trestlings, above the solid surface or flooring of the piers, affords sufficient room for stowing upward of two hundred and fifty thousand tons of coal. Capacious docks extend in-shore, between each pair of wharves, thus

November 30, 1866, was $29,929,440.27. The officers are; President CHARLES E. SMITH; Treasurer, SAMUEL BRADFORD; Sec'y, W. H. WEBB, and Gen'l Supt., G. A. NICOLLS.

With Pittsburg, and the "Gate of the West," Philadelphia is connected by a magnificent Railway, to which we have more than once referred, and to which it seems proper to refer again, if for no other purpose than to aid in perpetuating the names of those who have been most active in contributing to the success of so great an undertaking. While the Reading cheapens fuel to the citizens of Philadelphia, the Pennsylvania Central cheapens food, and both are entitled to rank among the most important enterprises of modern times.

The act incorporating the Pennsylvania Central Railroad Company was passed April 13th, 1846. As soon as the news of its passage had reached Philadelphia, a large meeting was held, and a committee appointed to prepare an address inviting the co-operation of the citizens. This committee consisted of THOMAS P. COPE, (since dec'd,) Chairman; DAVID S. BROWN, JOHN GRIGG, THOMAS SPARKS, GEORGE N. BAKER, RICHARD D. WOOD, JAMES MAGEE, and J. R. TYSON. The address issued by these gentlemen met with a warm response, and public and private subscriptions were freely tendered. The city,

making the whole river front available for shipping purposes. Over one hundred vessels can be loading at the *same moment;* and few places present busier or more interesting scenes, than the wharves of the Reading Rail-road, at Richmond. A brig of one hundred and fifty-five tons has been loaded with that number of tons of coal in less *than three hours time,* at these wharves. The whole length of the lateral railways extending over the wharves at Richmond will probably exceed ten miles, affording a shipping capacity for upward of *three millions of tons!* and it will probably not be many years before this amount, extraordinary as it may seem, (as, indeed, it really is,) will be annually transported over this great thoroughfare. The company has laid the *foundation* for a trade as broad as the future destiny of the coal trade itself."

in its corporate capacity, subscribed two and a half millions of dollars, and this gave an impulse to the enterprise that left no longer any doubt of its success. The first Board of Directors consisted of the following gentlemen, most of whom had been active in promoting this great work, viz.: S. V. Merrick, Thomas P. Cope, Robert Toland, David S. Brown, James Magee, Richard D. Wood, Stephen Colwell, George W. Carpenter, Christian E. Spangler, Thomas T. Lea, William C. Patterson, John A. Wright, and Henry C. Corbit. First officers—S. V. Merrick, President; Oliver Fuller, Secretary; George V. Bacon, Treasurer; J. Edgar Thomson, Chief Engineer; William B. Foster, Jr., Associate Engineer, of the Eastern Division; Edward Miller, of the Western.

During the year 1857, this Company made a most important and extensive negotiation, being no less than the purchase from the Commonwealth of 285 miles of Canal, between Philadelphia and Pittsburgh; and 37 miles of Railway, between Johnstown and Hollidaysburg; and 80 miles of double track between Philadelphia and the Susquehanna River, with all the appurtenances, giving their bonds, bearing five per cent. interest, for the sum of $7,500,000, payable $100,000 on July 31st, 1858, and $100,000 annually thereafter, until July 31, 1890, when the payments will be at the rate of $1,000,000 per annum until the whole is paid. Present total cost of roads and canals belonging to Company, $30,896,403. The rolling stock consists of 452 locomotives, 240 passenger cars, 103 baggage cars, and 6,953 freight cars. The aggregate tonnage of the road, for 1866, was 3,452,718. The surplus earnings were $3,792,973,57. The present officers of the road are—President, J. Edgar Thomson; Vice-Presidents, Thomas A. Scott and Herman J. Lombaert; Treasurer, Thomas P. Firth; Secretary, Edmund Smith; General Superintendent, Edward H. Williams; Controller and second Vice-President, Herman

J. LOMBAERT; Auditor, SAMUEL G. LEWIS; General Agent, G. C. FRANCISCUS; Superintendent Philadelphia Division, WILLIAM F. LOCKHARD; General Freight Agent, H. H. HOUSTON; General Passenger and Ticket Agent, H. W. GWINNER.

The last link in the chain is now perfected, connecting Philadelphia and Chicago, *via* Pittsburg, Fort Wayne and Chicago Rail-road; other connections are constantly being made; and the Pennsylvania Central Railway, fortunate in its mode of construction, and fortunate in its officers, will hereafter still further reduce the cost of transportation between Philadelphia and the West, and perpetually prove an increasing source of benefit to both.

The other great trunk line diverging from Philadelphia, and increasing its rail-road connections with the South, is the PHILADELPHIA, WILMINGTON, AND BALTIMORE RAIL-ROAD. This road forms part of the great Southern mail route—and being one of the oldest, is consequently one of the best known rail-roads in the country. The low charges for water carriage between Philadelphia and the prominent points of the South, have heretofore deprived this road of any considerable revenue from freight; but, nevertheless, the Company is now free from floating debt, has paid all the demands that were made upon it, and its regular dividend, without borrowing a dollar. This Company is peculiarly fortunate in its President, ISAAC HINCKLEY Esq., who is regarded as one of the ablest rail-road officers in the country.

Recently, another avenue of communication between Philadelphia and the Great West was opened, by the completion of the PHILADELPHIA AND ERIE RAILROAD. This road was chartered as long ago as April, 1837. But for fifteen years afterward, nothing was done toward its construction, and after it commenced the work progressed so slowly that it was not opened for use until 1864. Since its completion, it has been leased, and is now operated by the

Pennsylvania Railroad Company, and though not as yet profitable to its stockholders, it is daily fulfilling the predictions of its managers in developing the resources of a section of the State that was heretofore almost a wilderness, while it is drawing a legitimate portion of the trade of the Northwest to Philadelphia, which is now nearer the great inland seas than either of her rivals—Baltimore or New York.

The Minor railroads diverging from Philadelphia, are the *Philadelphia, Germantown, and Norristown,* which, in 1866, carried 2,469,354 passengers; and the *Westchester and Philadelphia Railroad,* which has as its tributary the *Philadelphia and Baltimore Central;* and *the Camden and Atlantic and Cape May Railways.*

The following table exhibits the

Names, Length, and cost of the Railroads centering in Philadelphia, with their Receipts and Expenses for the year 1866.

Names.	Length	Cost.	Gross Receipts.	Expenses.
Pennsylvania	354	$21,135,439.82	$16,717,289 20	$10,616,362
Reading	147	26,380,004.18	10,902,818.87	4,886,288
Philada., Wilmington and Balt	95	10,469,300.00	2,470,958.64	1,413,271
Philadelphia and Trenton	26	1,299,120.00	849,445.69	532,692
North Pennsylvania	55	6,420,184.73	902,213.17	519,713
Philada. Germt'n and Norrist'n	17	1,407,567.96	605,345.91	332,619
Westchester and Philadelphia	26	1,492,108.36	357,590.06	213,940
Philadelphia and Erie	288	17,869,732.84	2,541,051.79	2,740,129
Philada. and Baltimore Central	90	1,095,346.40	149,218.35	115,469
Camden and Amboy	88	10,099,001.00	4,312,895.00	3,801,732

The railroad system of Philadelphia, we may remark, in conclusion, adopting the language of one who has made it the subject of careful consideration, extends to all points of the compass, pushes out toward the ocean, pierces the coal regions of the North, reaches eastward to the great seaports of the nation, drains the rich and fertile agricultural counties of our own State, and extends Westward toward the Rocky Mountains and the gold region beyond.

IV. The fourth and last subdivision of *essential* physical advantages is a *suitable climate*—a climate favorable to vigor of mind and health of body, and chemically adapted for manufacturing processes. The climate of Philadelphia, in common with other portions of the State, we may say the country, has undergone important changes within a half century. The winters are less uniformly cold than formerly, and the summers less uniformly warm. Except during the winters of 1855–6, and 1856–7, which were entirely exceptional, ice in the Delaware has not presented any formidable obstruction to navigation for many years, and sleighing has been a sport of short duration. In fact through some seasons lately, no snow, worth mentioning, has fallen up to the middle of February; and the weather during January was as genial as spring. In the summer, the thermometer sometimes rises for a few consecutive days above 93°; but the temperature invariably diminishes sensibly after sunset, and the nights are generally comfortable and refreshing. The most disagreeable feature of the climate in summer is liability to sudden variations, amounting in some rare instances to 30° in twenty-four hours. These variations, however, it would seem, are more unpleasant than permanently injurious in their effects.

The air of Philadelphia, compared with that of New York, has less keenness; and being free from saline impregnation, it is less irritating to weak lungs. It was observed long since, and remarked by physicians, that persons did spit blood in New York who were entirely free from any pulmonic affection in Philadelphia. Compared with New England, generally, the winters in Philadelphia are less severe, and consequently *less fuel is consumed; while the days are of greater average length*, thereby diminishing the consumption of gas. Both of these items have a bearing upon economy of production in Manufactures. But the climate of Philadelphia has further some peculiar and remarkable properties, as is evidenced by its effects upon certain chemical processes. It is conceded, even by Englishmen, that the woven fabrics of southern Europe are superior to those of England in the richness and clearness of their colors; and this superiority is accounted for by ascribing it to atmospheric qualities and peculiarities, for which neither the science of chemists nor the skill of dyers in England, has been able to provide a complete equivalent. So, experience demonstrates, that it is possible in Philadelphia to attain a degree of excellence in dyeing fabrics, unattainable by the same processes anywhere else except in Southern Europe. A celebrated French dyer, whose local partialities are distant from this city, experimented in various localities in France and the United States, and found the climate and water nowhere in either country so well adapted for his purposes as those of Philadelphia. Hence, every year the practice is becoming more common with the merchants of Philadelphia, New York, and elsewhere, to import silks, and woolen goods in an unfinished state, have them dyed in Philadelphia, and then they readily command prices equal to the best French or European finished fabrics.

In addition to these circumstances, which are considered essential to success in Manufactures, there are many others so desirable and important, that they can scarcely be ranked as secondary. Foremost in this class is—

1. Purity of Water. Water, like climate, has a sanitary, and also a chemical bearing. The water principally used in Philadelphia proper is from the Schuylkill; while in Frankford, Bridesburg, and other important manufacturing adjuncts, there are springs possessing some remarkable properties. The Schuylkill water, as we learn from the report of Messrs. Booth and Garrett, who, in 1854, made it the subject of careful analysis, is

distinguished above almost all other waters for its purity and freedom from organic matter. Their very able report concludes with the following opinion:—

"We may further observe, that a comparison of our waters, with waters used elsewhere in the United States and in Europe, highly esteemed for their excellency, may be characterized by its greater purity, its slightly alkaline impregnation, and by being nearly free from organic matter. In conclusion, we infer that the Schuylkill water has deteriorated, in no important respects, from its former excellent quality; is superior to most waters for domestic and manufacturing purposes; and lastly, a comparison of the past and present, leads to the inference, that no plan of improving the water will be required for many years to come."

By analysis, it has been ascertained that the water of the Cochituate, (used in Boston,) contains 1.16 grs. of solid organic substances in one gallon; and the Croton, (used in New York,) contains 4.28 grs., and that, too, after it had passed through forty-one miles of aqueduct; while the Schuylkill water, taken directly from the river, before it had entered into the reservoir, and had time to deposit its solid particles, contained but a trace of organic matter. The chairman of the Philadelphia County Medical Society concludes, that we possess the advantage of a purer quality of water for drinking purposes than any other city in the United States, or perhaps the world over, a prerequisite as essential to the enjoyment of health, as it is necessary for the preservation of life itself.

The sanitary results of the climate and the water are manifested in—

2. The Statistics of Health. The comparative healthfulness of various cities has been made a subject of careful observation by physicians and others, for more

than a half century, and the tables of mortality have uniformly shown that *Philadelphia is the most healthy of the great cities of the United States.* In 1806, when the city contained a larger population than New York, the deaths per day in the former were 5⅔, and in the latter 6⅓. In 1810, the proportion of deaths to population, in Philadelphia, was one to fifty. In 1855, WILSON JEWELL, M. D., as chairman of the Committee on Epidemics of the State Medical Society, presented a report full of valuable suggestions, and containing the following Table and remarks relative to the sanitary condition of our principal cities:—

1855	Population.	Total mortality.	Ratio of deaths to population.	Deaths to every 1000 inhabitants.	Per ct. of deaths under 5 years to total mortality.	Deaths under 5 years to every 1000.	Ratio of still-born to deaths.
New York	650,000	22,728	1 in 28.59	35.	53.40	18.67	1 in 13.70
Philadelphia	500,000	10,458	1 in 47.81	20.91	44.86	9.38	1 in 17.85
Baltimore	215,000	5,465	1 in 39.52	25.41	44.88	11.40	1 in 14.01
Boston	162,748	4,308	1 in 39.36	26.59	46.63	12.40	1 in 19.33

"The averages, deductions, and comparisons drawn in this Table, prove conclusively that the mortality in our own city is much less, compared with the total of deaths, with the deaths to population, or with every thousand, than in the other Atlantic cities.

"While in New York 1 in every 28 of the population dies annually, and in Baltimore and Boston 1 in every 39, in Philadelphia there is only 1 in every 47; more favorable by one half than the death rate of New York; and, by nearly one fourth, more favorable than that of Boston and Baltimore.

"Again, the health of Philadelphia, contrasted with that of the other cities named in the Table, is shown by estimating the deaths to every thousand of the population. While New York contributes 35, Boston 26, and Baltimore 25, Philadelphia gives only 20.

"Nor can it be overlooked, that the infantile population in New York suffers by death to a far greater extent than in either of the other cities. Those under five years of age (exclusive of still-born) make up 53 per cent. of the total mortality; Boston 46 per cent.; while Baltimore and Philadelphia are each 44 per cent.: less by 8 per cent. than the former, and 5 per cent. than those under five years in the latter city.

"The deaths under five years in every thousand of the population presents an equally favorable contrast; New York furnishing 18, Boston 12, Baltimore 11, and Philadelphia only 9 in every thousand.

"It will be seen, too, while the population in New York was but 13 per cent. greater than that of Philadelphia, the deaths for the year 1855 were 35.90 per cent. more than in our own city. The ratio of still-born children to the mortality is less in Philadelphia than in either of the other places.

"The preceding estimates are sufficiently clear to maintain the position, that we are the healthiest of the large Atlantic cities, and that for salubrity, we should have the preference before the others named in the Table.

The aggregate mortality in the four cities, in 1865, and 1866, was as follows:—

	1866.		1867.			
Philadelphia,	17,169	-	16,803	-	Decrease,	366
New York,	24,843	-	26,815	-	Increase,	1,972
Baltimore,	4,695	-	5,623	-	Increase,	928
Boston,	4,541	-	4,379	-	Decrease,	162
Total,	51,248		53,820		Increase,	2,372

The proportion of deaths to population, it will be perceived, is about the same as in 1855, and the result equally favorable to Philadelphia.

3. Protection against Fires, &c. Disastrous fires, it is well known, occur more frequently in American than European cities; and there was a period when Philadelphia enjoyed an unenviable distinction in this respect, even among her sister cities. Fortunately that period has gone by, and we now may proclaim confidently that in no American city is life more secure, or property better protected, than in Philadelphia. One of the causes, it was ascertained, of the former prevalence of fires, and the destruction of property, was the feuds which in course of time had sprung up between the various organizations originally established for the extinguishment of fires. The system of voluntary association for this purpose,—inaugurated, it is said, by Franklin, in 1732,—though manifestly calling forth a great deal of self-sacri-

fice and heroism, was regarded by many as a failure, or in other words, as better adapted for small towns than for large cities. But many of the evils developed from this source have been obviated by the *reorganization of the Fire Department*, recently effected: that is, by disbanding the most disorderly companies, dividing the city into districts, permitting only a prescribed number of companies to go into service except in case of a large fire, when the general alarm rung on the State-House bell calls the whole Department into requisition.*

In 1856, another very important improvement was made by the establishment of a *Police and Fire Alarm Telegraph*, by which information can be communicated, at a moment's notice, to and from any of the sixteen Police-stations that comprise the jurisdiction. During 1857, by this means, 34,207 messages were transmitted, 3,430 lost children restored to their parents, 884 strayed and stolen animals were restored to owners, 392 fire alarms given, the Coroner notified 387 times, and 1,361 Police-officers subpœnaed to testify before the courts. Still more recently, another safeguard was originated by the establishment of the *Fire Detective Police*—a department of the General Police—specially charged with the duty of investigating fires and detecting incendiaries. The inception of this wise measure is due, we believe, to the Mayor, the Hon. Richard Vaux, and its success and efficiency, largely to the signal ability of the chief officer, A. W. Blackburn. But the improvement that has been found the most effective of all, as a protection against serious loss by fire is the introduction of *Steam Fire Engines.*

* The Fire Department consists of 88 companies, 47 Steam Fire Engines, 13 Hand Engines, 107 Hose Carriages, 9 Hook and Ladder Trucks, having 1574 feet of Ladders. The Department now has 93450 feet of Hose and 14035 members. Officers—Chief Engineer David M. Lyle; Secretary, P. West Blake; Assistant Engineers, 1st Division, Terence McCusker; 2d Division, George Hensler; 3d Division, Joseph H. Vanosten; 4th Division, Edmund Wright; 5th Division, James L. Wilson.

There are now nearly fifty Steam Fire Engines in use in the city, including some that, for beauty, are in striking contrast to the original models constructed in the West; and by their efficiency have revolutionized popular opinion as to the practicability of steam for this purpose. One Philadelphia builder (J. B. HAUPT) constructs engines with the cylinders and pumps cast in one, which gives great solidity to the working parts, and of such power that an engine of the third class will throw water, out of a one and a quarter inch nozzle, 270 feet.

There have also been ordinances passed regulating the erection of buildings which, to some extent, will diminish the number of fires. But the most efficient protection against serious loss from this cause, which manufacturers and owners of property in Philadelphia have, consists in the reliable character of numerous Insurance Companies that are always ready to take risks at low rates, and prepared to meet losses with creditable promptness.* It has been ascertained that within a period of seven months, the losses from fire to the owners of property in Philadelphia, that is, over and above insurance, amounted to only $54,780.

3. ABUNDANCE OF CAPITAL. Another matter that has a bearing upon the adaptation of localities for manufactures, is *the quantity of floating or loanable capital—or, in other words, the normal state of the money market.* The success of the English manufacturer, compared with that of the American, is probably due less to the low rate of wages in England, or to any other one circumstance, than to the low charges for the use of capital. In this country the rates of interest, advanced by the competition engendered by the tempting opportunities for profit,

* We may probably insert in the Appendix a list of the Insurance Companies that are of undoubted solvency, as an item of important information to our distant friends, and, in some degree, a protection to our own citizens, who may be deceived by unreliable agencies.

are in most places too high for Manufactures yet in their infancy, and weighed down by an inhospitable political sentiment, to sustain. There is a marked difference, however, in this respect between different localities, and though large fortunes are, it is probable, less numerous now, or at least less prominently conspicuous, than formerly; and though the banking capital* of the city, being about $15,500,000, is much less than that of New York or Boston; nevertheless, that unfailing barometer of money centres—the average rate of interest—has generally indicated in Philadelphia an abundance of loanable capital. If our Manufacturers have not as yet derived their proper share of benefit from this circumstance; or, if bank officers, in distributing their loans, have not exercised a wise discrimination in their favor, we sincerely hope that the mistake originated solely

* The following is a statement of the Capital, Loans, Deposits, and Circulation of the Philadelphia Banks for the week preceding Monday, April 22d, 1867.

BANKS.	CAPITAL.	LOANS.	DEPOSITS.	CIRCUL'N.
Philadelphia, National	$1,500,000	$4,744,000	$2,836,000	$1,000,000
North America	1,000,000	3,833,274	2,515,915	800,000
Farmers' & Mechanics, National	2,000,000	4,890,222	4,042,037	711,430
Commercial, National	810,000	2,178,000	1,126,000	629,000
Mechanics, National	800,000	2,316,000	1,214,000	483,499
Nat. Bank N. Liberties	500,000	2,468,000	2,038,000	465,000
Southwark, National	250,000	1,380,134	1,166,039	218,781
Kensington, National	250,000	1,252,659	1,081,175	232,043
Penn, National	500,000	1,171,092	888,018	175,042
Western, National	400,000	1,322,461	1,250,803	7,025
Manufacturers, National	570,150	1,625,000	1,053.547	450,534
Nat. Bank of Commerce	250,000	1,000,343	707,739	223,055
Girard, National	1,000,000	3,206,000	2,042,000	596,000
Tradesmen's, National	200,000	1,054,020	843,688	184,220
Consolidation, National	300,000	1,026,769	769,024	270,000
City, National	400,000	1,236,584	795,712	364,891
Commonwealth, National	237,000	831,274	705,610	212,220
Corn Exchange, National	500,000	1,790,000	1,189,000	429,000
Union, National	300,000	1,698,000	1,205,000	229,000
First National	1,000,000	4,576,000	2,748,000	795,000
Third National	300,000	957,158	594,767	262,394
Fourth National	225,000	868,450	897,264	130,750
Sixth National	150,000	420,000	233,000	135,000
Seventh National	250,000	672,000	446,000	218,000
Eighth National	275,000	704,000	497,000	234,000
Central National	750,000	2,610,000	1,904,000	598,000
Nat. Bank of Republic	558,000	1,002,000	522,000	417,500
National Exchange	300,000	778,009	509,242	175,750
Total, April 22	$15,575,150	$51,611,449	$35,820,580	$10,647,134

in misconception as to the predominant interest of the city; and that, with the aid of the late panic, in destroying the blinding fascination of "gilt-edged paper,"—and perhaps in some humble degree, with the aid of this volume—Manufacturers and Mechanics will hereafter approximate more closely to that position in the scale of mercantile credit, to which the advantages of the locality, and their own solvency and usefulness, unquestionably entitle them.

5. SUPERIOR MACHINES. The immense productive power of machinery, compared with mere manual operations, can require at this day no illustration. For instance, by the improvements effected in Spinning Machinery, one man can attend to a mule containing 1,088 spindles; each spinning three hanks, or 3,263 hanks a day; so that, as compared with the operations of the most expert spinner in Hindostan, an American operative can perform the work of *three thousand* men. The efficiency of machinery, however, like that of labor, depends upon its quality; and this, it would seem, depends upon the cheapness and abundance of the materials that enter into the composition of machines. In England, it has been supposed, that if Iron, Steel, and Brass, were less abundant, the machines would be in a less degree superior; and in the United States, though the mechanical appliances in use are almost everywhere deserving of admiration—none are probably more remarkable for power and efficiency than those in Philadelphia. A gentleman, who has quite recently made the Manufactures of Iron in this city the subject of investigation, publishes the following observations respecting the machines in use in the Iron establishments:

"In the course of our inquiries into the Manufactures of Iron in this city, the bearing of machinery upon production has been constantly brought to notice, and striking instances of its value have been observed.

In a leading establishment, where foundry work is the principal business, six thousand tons of iron being melted per year, the economical power of machinery in moving all the masses of iron is such that the production of each man exceeds three thousand dollars annually for the average of all the employed. This is three times the production of equally skilled workmen, without machinery. The lowest average for foundry work, as well as for artisans in wrought iron, is below a thousand dollars per annum, and this whether they handle a large weight of iron or not, if the processes are conducted by physical strength alone, and wholly without the use of machinery. In short, the economy of machinery applies alike to all forms of iron working, and to the processes which change its value least, equally with those which increase its value many times.

"The introduction of machinery has revolutionized the simple production of Iron from the ores also. It has been stated to us that the anthracite furnaces now make six thousand tons of iron more easily than six hundred tons were made fifteen or twenty years ago. In every thing that relates to the making or working of iron there is the greatest possible inducement to the employment and perfecting of machinery, intended to economize the force required, and the labor employed. In this direction investment is safe, and capital is certain of satisfactory returns. The leading departments of iron manufacture furnish articles of universal use and universal necessity, in which accumulation of stocks is not to be dreaded so much as the narrow margin between cost and sale prices. Reduce the cost of manufacture fifteen or twenty per cent., and the proprietor may proceed in the face of even a dull market, and indeed, under a total cessation of orders. The direction in which improvement lies is in perfecting and introducing powerful machinery, and every inducement concurs to urge attention to this point.

"It is noticeable that the machinery employed in American manufactures of iron is new and original in almost all cases. The most signal economies of power in the establishments of this city are not by the use of purchased machinery, but they are the creations of the proprietors who use them, suggested in the course of their work, and devised and applied by themselves. In all forms of machinist manufacture those inventive and constructive processes are making rapid progress. The great capital they represent when finished, is capital created by the establishment, and not an investment from the outside. This fact guarantees the permanent efficiency of these manufactures, since such capital is not easily withdrawn, and the establishment is not broken up by temporary depression of a business, or even by the dispersion of workmen for a considerable time.

"It has been recently stated that the machinery invented and applied in American armories, private, as well as those belonging to the government, is much sought in Europe, and will soon be in almost universal use abroad. This fact bears directly on the point we are stating. Machinist machinery is equally advanced here; and at two or three of our great establishments it is confessedly superior to that of the celebrated Lowell Machine Works, while constant improvements are being made. In the appliances for handling iron in heavy foundry work, the world may be challenged for comparison with the machinery of at least one great establishment here, and the most important items in that case are the absolute creation of the proprietors. It is obvious that such machinery differs widely, in its economical importance, from that which is purchased by direct expenditure, and particularly from any form of machinery imported from other quarters.

"The direction in which this city always will excel is in the handling of heavy masses of metals. Power is cheapest here, and necessity first impels to the economy of forge-work, iron rolling, foundry-work, ship building, and costly machine building. In the minor manufactures the application of improvements is more rapidly made at the North; but this is from want of attention here, instead of from want of the requisite field and facilities. No location in the Union can compare with this in natural advantages for the manufacture of arms of every sort, cutlery and tools, implements of every kind, and the multitude of minor manufactures in which inventive talent and machinery decide the whole question of profitable attention to the business. The market is the whole world. At this moment many superior instruments of steel and iron are actually made here for European sale; and the skill which does this now on a small scale, only requires the aid of more perfect machinery, and the capital necessary to work it, to make the business all that the most sanguine might wish."

In the production of MACHINE TOOLS, and fine as well as heavy machinery, very marked success has attended the efforts of our mechanical engineers. The Lathes, Planers, Drills, Borers, and the machinery for working metals generally, made in Philadelphia, are wonderful specimens of workmanship, and celebrated not only throughout the United States, but in portions of Europe. A few years ago, Commissioners were sent to this country to procure tools and machines for the government workshops in Russia. Discharging their duty faithfully, they visited, we believe, all the manufactories of these important articles in New England and the principal cities; and, though they found the prices in some instances nominally cheaper, their order was reserved until they again reached this city. The machines of New England, in consequence of the great cost of iron, are remarkable for their *lightness;* but in substantial excellence and quality of workmanship, none can compare with those of Philadelphia.

In reflecting upon the causes conducive to superiority in this particular, it has occurred to me as probable, that the establishment and continuance of the United States Mint, in this city, have tended in some degree, by creating a demand for a finer and higher class of workmanship, to centre here the best skill in this department of Mechanics. No expense being spared by the able managers of that Institution to procure the most perfect machines; and every reasonable facility being afforded for experiment, we need scarcely wonder at the degree of perfection that has been attained. Our Mint has probably originated a greater number of valuable improvements than any similar establishment in the world; and all persons familiar with its past history and present management, unite with the Committee of the Board of Assay Commissioners, in stating "that the Institution, in their opinion, is conducted and maintained in such a manner as to merit the highest confidence of the Government and the public."*

* The Director of the Mint has favored me with the following letter, in answer to a request for some information respecting the machines, and the curiosities to be seen in that establishment; and, as it will be read with interest, I trust he will pardon its publication.

"MINT OF THE UNITED STATES,
Philadelphia, Jan. 21, 18[illegible]

'DEAR SIR:

"Without being able at present to go minutely into the subjects mentioned in your note of the 5th, I may state, that the establishment and

6. ESTABLISHED REPUTATION. Established reputation, though in its nature etherial, is an object of substantial value—a power in the money market. It is of two kinds—personal and local. The marketable value of

continuance of the Mint, in this city, have undoubtedly had their share in calling forth the various kinds of scientific and mechanical talent, which are requisite for the successful conduct of such an Institution.

"Within a period, now embracing more than sixty years, there has been a large amount of machinery manufactured for, and within the Mint establishment, from the more ordinary workmanship up to the most delicate and elaborate. A number of important mechanisms and processes have had their origin and invention here; and others, borrowed from other places, have been modified and improved. Some instruments, it is true, are still imported; but they are now of comparatively trivial account, being such as are of so limited demand, as not to be an object for the attention of our artists.

"The most important improvements introduced into the Department of the Chief Coiner, have been, the press for cutting out blanks or planchets; the draw-bench for equalizing the strips—afterward adopted in the London Mint; the old self-feeding lever-coining press, and after it the steam press; the milling machine; the counting machine; and the arrangements for cleaning; also, many fine balance beams, large and small; and an assorting machine, not as yet brought into use. The system of hardening dies was originated at the Mint, and is greatly superior to the methods heretofore practiced.

"In the Melter and Refiner's Department, we may specify the parting arrangement, for separating gold and silver; the hydraulic press, for condensing the powdered gold or silver; the sweep machine; and the various arrangements by which the melting has been made a neat and economical operation.

"In the Assayer's Department: the delicate balances; the gas-bath; and generally, the systematic arrangements for the assay of gold, silver, and copper.

"The Cabinet of coins, medals, and ores, which occupies a suite of apartments at the Mint, is an attractive feature in the Institution. The collection is not very large, if compared with similar Cabinets in Europe; but it is sufficiently so to furnish valuable information on the subject of Coinage, and useful monuments of history. Besides, an examination of the collection gratifies popular curiosity, as well as educated taste.

"I may add in conclusion, that the Mint has, within a year or two past, been rendered thoroughly fire-proof in all its departments, and the arrange-

the products of mechanical industry, every one will concede, is affected not merely by the reputation of the maker, but also to a greater or less extent by the general reputation of the place of their manufacture. No illustration of the principle can be necessary; but if it were, we might refer to France, the stamp of whose city, "Paris," on articles of *vertu*, of itself commands a premium; or again, we might refer to New England, whose stamp unfortunately, in many instances, does not tend to elevate the price of articles to which it is attached. The value of a good name is appreciated, perhaps, by none so forcibly as by those who have lost it. The manufacturers and mechanics of New England would no doubt give millions to obliterate from human recollection the impressions produced, in part, by operations in wooden nutmegs, mahogany hams, oak-leaf cigars, and paper-soled shoes. Deceptions of this kind, and trickeries frequently practiced by Yankee operators, though we believe and insist only by a few, and the production of a vast quantity of cheap, fragile fabrics, have so impaired confidence in Yankee contrivances in general, that all, no matter how excellent in themselves, are prejudged unfavorably from the place of their origin. To avoid this prejudice, or to partake of the advantages of an established reputation, New England manufacturers are often tempted to put foreign or fictitious stamps on their best fabrics; and thus our country loses its share of credit for the excellence it has achieved, while it must

ment of the rooms appropriated to the different branches of business greatly improved. It is thus in a condition of great efficiency and security, and is believed to be unsurpassed by any similar institution.

"I am, very respectfully,

"Your ob't servant,

"JAMES ROSS SNOWDEN,

"*Director of the Mint.*

"To E. T. Freedley, Esq."

bear the reproach of its defaults. But mechanics in Philadelphia, fortunately, have none of these difficulties to overcome. The same manufacturers, if located here,—and we welcome them,—would find the way clear before them, the prepossessions of people at a distance in the South and the West in their favor, and their products commanding a readier sale in consequence. Every auctioneer will testify that a Philadelphia made carriage will command more spirited bidding, and most probably a higher price, than a Connecticut carriage of equal quality. The stamp, "Philadelphia," is everywhere regarded as *prima facie* evidence of good materials and superior workmanship. A Philadelphia mechanic is everywhere a title of reputable distinction, and a very acceptable passport to employment in every intelligent master-workman's shop. Hence our Manufacturers reverse the practice of their competitors in New England, and put their names and stamp on their *best* products, leaving the *inferior* in some instances to those who choose to adopt them.*

7. Lastly. OPPORTUNITIES FOR ART-CULTURE. Art, in its relations to Manufactures, has not, until quite recently, been appreciated by any considerable portion of the American, or even the English people, to a degree in anywise approximate to its importance. Both have long known, it is true, that certain goods sell better than others—that English and American prints, for instance, would be less saleable at the same price than those of France; yet, even while claiming superiority in the quality of the cloth, neither has been willing to attach any special importance to beauty and originality of design.

* The principal exception to this rule is, that New York dealers sometimes pay such irresistibly tempting prices to have their names affixed, as makers, to articles actually made in Philadelphia, that our Manufacturers forego the honor for the sake of the money.

This is the more remarkable, inasmuch as it must be evident to the least imaginative, how many articles are valued mainly for their style of ornamentation. We might mention carpetings and floor-cloths, carved wood and furniture, curtains, and other hangings; inlaid floors, ornamental glass, stained glass, metal work, grates and stoves, gas fittings, paper and other hangings, porcelain, pottery, works in the precious metals, works in stone, and a great variety of garment fabrics. The French, in the meanwhile, have unceasingly aimed at perfection in the Ornamental Arts. To improve the national taste, they long ago established Schools of Design and National Collections of Art; and to train up a band of skilled workmen, they more recently erected National Manufactories, employing the best painters, sculptors, and designers, as well as men of the most scientific acquirements in Botany, Mineralogy, and Chemistry. In these establishments the cost of repeated failures is totally disregarded, and every effort made to bring to perfection the fabrics wrought in them, both as to the highest excellence in workmanship and materials, and to their embellishment in ornamental design. The result is, that both English and American Manufacturers must admit, as Cobden did before a Manchester audience, "we do not know what we shall have to print, nor what the ladies will wear, till we find out what the French are preparing for the next spring." But with all their schools, Art collections, and national manufactories, we do not believe that the French would have attained any notable success in decoration, if they had adopted the Yankee system of segregation; and instead of carrying on their manufactures in cities like Paris and Lyons, they had sought cheap lots, gentle water-falls, and the mossy banks of meandering streams. Taste is a thing of culture—it is only in isolated instances, if ever, a gift of

Nature. The ability to judge, and especially to execute what is tasteful in works of Art, is the result of long familiarity with good models and constant observation of the master-pieces in Art. The sight of excellence in the products of skilled workmanship stimulates to exertion, and produces excellence in other fabrics perhaps essentially dissimilar. Hence, the great advantage of carrying on the higher class of Manufactures in or near the cities abounding in the best models, and where the eye, if not the hand, may be educated almost imperceptibly to a high degree of artistic perception.

Now, if we were seeking some one of the various cities in which to apply these principles, where, we would ask, is the principal home of the Arts in America? Which contains the finest models in Architecture, Sculpture, and Painting? There could be but one answer—Philadelphia. No other city in the Union contains so many buildings that are models of classic beauty—so many evidences of a cultivated taste—so many eminent artists—and, we may say, so many devotees of Music, for no other city has been able to sustain the Italian Opera with equal success. A procession of those in this city, who make Art their study, would be imposing from its numbers, as well as the talents of its members. At their head, by common consent, we would find the veterans Sully, Neagle, and Peale; and not far behind them, Lambdin, Waugh, Scheussle, Hamilton, Rothermel, Weber, Van Starkenborg, Moran, Schindler, Conarroe, Boutelle, and Bowers; and among the younger men, George C. Lambdin, George F. Bensell, Edwin Lewis, Haseltine, Richards, Furness, and many others, who are entitled to a niche in the temple of artistic fame; while in the ranks there would be many who, when the leaders fall, can fill their places—many Engravers on wood and steel, and Lithographers, who give to our Government's costly publications their principal value and attractions—many Designers and Artists in bronze, whose chandeliers and lamps, at the World's Fair, extorted admiration from the English and French for "lightness and purity of design,"—some beautiful women, too, whose cultivated fancies, stamped on paper or woven fabrics, gladden the eye in thousands of homes; and sculptors, whose works in stone and marble grace Galleries and Capitols, and whose sarcophagi and mausoleums adorn almost all the Cemeteries in the land. Ornamental Art is without a home in America, if it be not in Philadelphia. Here then is the proper place for the establishment of a Normal School of Design, to supply manufacturing towns throughout the country with competent teachers, who may aid in elevating the Art-products of America to a level with those of the most advanced European countries. We trust some one of our men of fortune will inherit the blessings of future ages by endowing such an Institution, and in connection therewith establish a Museum of Art, which shall contain all the best models—ancient and modern—in every department of Decorative Art, from a coffee-pot to an original Apollo Belvidere.

There are many other advantages that might be noted —the law of limited liability in Partnership for instance —tending to show that Philadelphia ought to attain eminence in Manufactures. We, however, pass them by, for they may all be included in one point, viz., *Philadelphia is already a great Manufacturing city.* I hold it to be eminently safe to infer, that a locality in which manufacturing industry has already taken a deep, permanent root, particularly if it manifest an indigenous growth, possesses a soil adapted therefor, whether by analysis we can perceive the ingredients or not. Moreover, it seems probable, almost certain, that the spot in this country now exhibiting the most varied and extensive development of mechanical industry, in conjunction with enduringly favorable circumstances, will remain for a century to come the central and chief seat of the higher and more artistic Manufactures in America, notwithstanding the growth and promise of other places possessing theoretically marked advantages.

To illustrate the present development of manufacturing industry in Philadelphia, I herewith submit the results, not generally of my own observation or knowledge, but that of others, and principally of reports made to me by gentlemen specially employed to report on certain branches—men far more competent and more experienced in mechanical matters than myself—and not one of whom is a native of this city. Months have been occupied in this investigation; but as comparatively few facts,—especially statistical facts,—after due inquiry, could be precisely and accurately ascertained, and none others were desired, the reports give no indication of the labor involved.*

* Numerous attempts have been made at different times to investigate the manufacturing industry of Philadelphia. Several years ago a Statistical Society was organized, we believe, for the express purpose of ascertaining the capital in trade and manufactures, the number of hands employed and wages paid, and the aggregate of production; but its officers, we understand, have not as yet submitted their report. More recently, a committee of highly respectable and trustworthy gentlemen, appointed by the Board of Trade, undertook the commission; but the most important information that they ascertained and reported was, that "inquiries of this kind are exceedingly impertinent and offensive, and they will not be answered; nor can any authority compel a response to them. They will be either treated with silence; or, if replied to, they will elicit no full and reliable intelligence. We do not make this assertion without ample reason." The Board of Trade consequently recommend, and their advice has been heeded by us, not to extend inquiries beyond what can be precisely and accurately ascertained. If, by this course, a less number of important facts are elicited, many rash or doubtful assertions are avoided. Our conviction with respect to statistics is, that the *mean* of estimates of intelligent

They may also, to a certain extent, be considered the opinions of one or more of the leading men in each branch of industry; for large indebtedness is due to this source, both for original suggestions and confirmation of points otherwise doubtful. The reports submitted are not intended to exhibit the entire manufacturing industry of Philadelphia—to ascertain that would require the purse of Fortunatus, and inquisitorial powers far greater than any possessed by the Pope of Rome, the King of Naples, or the Emperor of all the Russias, or all of them combined—but simply to state the facts that have come within the range of our observation, and submit them in illustration of the position and assertion, that *Philadelphia is already a great Manufacturing city, most probably the greatest in the Union.*

men, familiar with the branch with which they are connected, or with the business of their neighbors, is likely to lead to more reliable aggregate results than any direct personal inquiries of each individual. In the latter case, the small operators who reply at all, are habitually disposed to exaggerate, and the larger ones, who have a mortal aversion to the tax-gatherer and competitors, frequently report a small product and a gloomy state of affairs. It is probable, however, that each succeeding attempt will be attended with more success than the previous ones; and the time will come when it will be possible to exhibit statistically the particulars, as well as the aggregate of the mechanical and manufacturing industry of Philadelphia. At present, the best than can be done is to make a *readable* exhibit.

REMARKS

UPON THE

PROGRESS AND PRESENT DEVELOPMENT

OF THE

LEADING BRANCHES OF PRODUCTIVE INDUSTRY

IN

PHILADELPHIA.

ASSUMING that an Alphabetical arrangement of subjects would be most convenient for reference; but, deeming it advisable to group together those which have practically some points of affinity, whether through identity of raw material or similarity in uses, we come first to—

I.

Agricultural Implements and Garden Seeds.

The manufacture of Agricultural Implements, we are somewhat astonished to learn, is comparatively a new branch of industry in Philadelphia. It seems almost incredible that her citizens, ever foremost, as we have shown them to have been, in enterprises designed to promote agricultural improvements, were, until within a few years, content that the farmers of Pennsylvania and New Jersey should be dependent upon other States for the improved implements with which to till the soil.

In 1854 was founded the first establishment in Eastern Pennsylvania, for the manufacture of Agricultural Implements generally; prior to that, there were shops located for specific objects, as for instance Grain Drills, of which those made by Steacy, and by Pennock, had acquired marked celebrity; but for the manufacture of Farm Implements generally, we believe none of any moment existed. In the year above-mentioned, DAVID LANDRETH & SON, who, with their predecessors in the house, had for many years kept large supplies in Philadelphia obtained from various sources, established their Steam-works at Bristol, not only for the supply of their principal warehouse in Philadelphia, and their branch-houses in Charleston, S. C., and St. Louis, but for the trade in general. Others have since engaged in the business, and at this time, Philadelphia, once dependent upon other cities for tillage implements, is now not only independent, but capable of ministering to the wants of her sister States; and we trust all from distant points, whom business or pleasure may bring among us, will examine the rural machinery manufactured in and near our city.

GARDEN SEEDS.

The Seed trade of Philadelphia, though, in comparison with many other branches, one of very limited extent, is nevertheless entitled to consideration, when discussing the industrial pursuits of our citizens. From its nature, it cannot be expected that we should count the amount of sales, in this department, by millions—a few hundreds of thousands, at the most, complete the aggregate; but the reputation which our city sustains in this especial branch, is more worthy of note than the amount of sales, however large they might be. In no city of the Union, is the sale of Garden Seeds conducted as at Philadelphia. In New York, Boston, and Baltimore, the only other points at which the wholesale trade in Seeds approaches a distinct pursuit, the supplies are mainly obtained from Europe, though it is well known that many kinds of seeds suffer by a sea voyage, and swell so greatly, that the twine on paper parcels, is not unfrequently imbedded, or burst by the expansion.

It may be said truly that in the production of Seeds, Philadelphia stands pre-eminent—if not alone, almost without a rival; and the productions of one establishment, which dates its origin within a few years of the Revolution, are sought for and exported to nearly every country to which American commerce reaches. Tons are annually shipped to the British possessions, to India and South America, the West Indies, and the shores of the Pacific, each of which call for annual supplies. One firm, which is specially alluded to, by reason of its greater prominence, viz., that of DAVID LANDRETH & SON, has Seed Grounds, at Bloomsdale, near Bristol, embracing nearly four hundred acres, cultivated in drill crops, requiring a large force of hands, twenty head of working stock, and a steam-engine for thrashing and cleaning seeds. The estate, in its entirety, exceeds any similar establishment in the world. ROBERT BUIST, JR., HENRY A. DREER, MAUPAY & HACKER, and COLLINS, ALDERSON & CO., are also extensive growers; and we proudly claim for Philadelphia a class of seed merchants, worthy the confidence of all who may have occasion to purchase, whether for personal use or purposes of trade.

ESTABLISHED IN 1828.

ROBERT BUIST, Jr.

SEED GROWER,

AND IMPORTER AND DEALER IN

AGRICULTURAL AND HORTICULTURAL

IMPLEMENTS.

-WAREHOUSE-

Nos. 922 & 924 MARKET STREET, above Ninth,

PHILADELPHIA.

Gardeners, Planters, and all others who purchase Garden Seeds, who wish a pure and reliable article, are invited to try

THEY ARE POPULAR

BECAUSE "RELIABLE."

They are grown by us from the Choicest Stock, and always warranted as represented. They can be obtained from mostly all the responsible Druggists and dealers throughout the Southern and Western States; but should they not be kept by your merchant, send to us for our Price List and Garden Manual, from which you can make out your own order.

FIELD, GARDEN & FLOWER

WILLIAM HACKER.

W. A. MAUPAY.

FARMERS, GARDENERS AND AMATEURS

Are invited to examine the superior assortment of SEEDS, mostly grown by ourselves, or to our order.

WARRANTED FRESH AND TRUE TO NAME.

Private families residing at a distance can have their orders forwarded by mail, at an extra charge of only *Eight Cents* per pound.

Country Merchants, Druggists and Dealers in Seeds, supplied in bulk or package suited for immediate sales, at the lowest wholesale rates.

HYACINTHS,

TULIPS,

CROCUS,

GLADIOLUS,

and every variety of Flowering Bulb. Also, Trees, Plants, and Shrubs, furnished during their seasons.

Wholesale and Retail Descriptive, Seed and Plant Catalogues, furnished on appiication.

MAUPAY & HACKER,

Seed Growers and Importers,

Nos. 803 & 805 MARKET STREET,

PHILADELPHIA.

II.

Books, Magazines, and Newspapers.

The honor of having established the first printing-press in America, must be awarded to Cambridge, Mass. Philadelphia, however, may claim, with laudable pride, that in less than six weeks after the city was founded, a printing-press was established, being the second set up in the North American colonies; and, moreover, that many of the most important works in American literature bear the imprint of her publishing houses.

Prior to the Revolution, and for some years afterward, the most notable issues of the Philadelphia press,—in fact, the American press,—came within Webster's definition of a pamphlet; that is, a small book, consisting of a sheet of paper. The first publication in book or pamphlet form issued from the Philadelphia press was a sheet almanac, for the year 1687, in twelve compartments: the year beginning with March, and ending with February, as was usual before the eighteenth century. A copy of this early specimen of American typography, bearing the imprint of "William Bradford, Printer," is preserved in the Philadelphia Library. His second work was a quarto pamphlet, on the subject of "The New England Churches, by G. Keith," dated in 1689. The name of Bradford continued to be identified with the history of printing in Philadelphia until a very recent period.

In 1699, the press established by Bradford passed into the hands of of Reynier Jansen, evidently a Dutchman by name, who managed it until the year 1712. There are now in the Philadelphia Library two very curious pamphlets, bearing his imprint, and so rare that they are probably the only copies extant. The first was published in 1700, and is entitled "Satan's Harbinger Encountered: his False News of a Trumpet Detected: his Crooked Ways in the Wilderness laid open to the view of the impartial and judicious. Being something by way of answer to Daniel Leeds, his book, entitled, 'News of a Trumpet Sounding in the Wilderness,' &c., C. P., (Caleb Pusey). Printed at Philadelphia, by Reynier Jansen, 1700." The second bears date 1705, and is entitled "The Bomb Searched and Found Stuffed with False Ingredients. Being a just confutation of an abusive printed half sheet called 'BOMB,' originally published against the Quakers by Francis Bugg; but espoused and exposed, and offered to be proved by John Talbot. Printed at Philadelphia, by Reynier Jansen, 1705."

The second printing-office in Philadelphia was established by S. Keimer, in 1723. The first publication, bearing his imprint, of which we have any knowledge, is a very curious and rare one, entitled "The

Craftsman: a Sermon composed by the late Daniel Burgess, and intended to be preached by him in the High Times, (sic.,) but prevented by the burning of his Meeting-house. Philadelphia: Printed by S. Keimer, (*Circa*), 1725."

In 1735, Christopher Sower published a Quarterly Journal, in German, which was the first work of the kind in a foreign language published in the Colony. The same year he published a Newspaper, the first German Almanac, "Extracts from the Laws of the Province, by William Penn," and several other works. At that time all the type used in the Colonies was brought from Europe, and finding this very inconvenient, he commenced a Type Foundry and Manufactory of Printing Ink. This was the first Type Foundry in the country, and the celebrated house of L. Johnson & Co., Philadelphia, claim, through Binney & Ronaldson, to be the legitimate successors of Christopher Sower in the business. In 1743, he printed a quarto edition of the German Bible, Luther's translation, having 1272 pp. This was the largest work which had then been issued from any press in the Colony, and was not equaled for many years after. Copies of this Bible were sold, bound, at fourteen shillings, and are now highly prized by book collectors. About 1744 he resigned his press to his son, and died about 1760. He was a man of large influence among his countrymen, and frequently acted as their representative in their intercourse with the Government.

His son, also named Christopher Sower, continued the business of his father on an enlarged scale, printing many valuable Books, and a Weekly Newspaper. In 1762, he printed a second edition of the German quarto Bible of two thousand copies; and in 1776, completed a third edition of three thousand copies. He had by far the most extensive Book manufactory, then, and for many years afterward, in the country. It employed several binderies, a paper-mill, an ink manufactory, and a foundry for German and English types. He was well educated by his father—was ordained a minister of the German Baptist Society; and, as a man of integrity, was deservedly esteemed. He died at an advanced age, in 1784. He left several children, among whom Christopher, (third,) David, and Samuel, were practical printers and publishers. The name continues to be very popularly represented in the trade, by his descendant, Mr. Charles G. Sower, senior partner of the firm of Sower, Barnes & Co.

In 1782, Robert Aiken published, it is believed, the first American Bible in the English language. But Book Publishing in the United States, even as late as 1786, was yet in its embryo state. It is recorded, that in that year *four* Booksellers held a consultation as to the policy of publishing an edition of the New Testament, deeming the matter a work of great risk, requiring much consultation previous to the determination of the measure; but the change in the state of public affairs soon infused life and vigor into the business. Less than four years afterward one of the prudent gentlemen above referred to, ventured upon the publication of an *Encyclopedia,* in eighteen quarto volumes. When the first half volume was published, in 1790, he had but two hundred and forty-six subscribers, and could only procure two or three engravers. One thousand copies of the first volume were printed; two thousand of the second; and when he had completed the eighth, the subscription extended so far as to render it necessary to reprint the first. He then found difficulty in procuring printers for the work.

In 1804, Mathew Carey set up the Bible in quarto form, and this is believed to have been the first Bible, kept standing in type, of that size in the world—over 200,000 impressions were published.

Within the last few years, the demand for minute and exact information in every department of learning has become so pressing, that the subdivisions which may be remarked in mechanical pursuits are also noticeable with respect to the publication of books. Publishers are no longer divided merely into book, newspaper, and magazine publishers, as formerly; but each of the various classes, Medical, Law, Theological, School, German, Statistical, and Miscellaneous works, has its representatives among the publishing houses. For instance, the publication of Medical books is made a specialty by HENRY C. LEA and LINDSAY & BLAKISTON, and carried on largely by J. B. LIPPINCOTT & CO. These firms, it is said, publish nine-tenths of all the Medical books issued in the United States.

NOTES ON THE LEADING PUBLISHERS.

J. B. LIPPINCOTT & CO. are, it has been said by an old bookseller and political economist, the largest book distributing house in the world. Their general business combines that of *Publishers, Printers, Bookbinders,* and *Wholesale Booksellers* and *Stationers.* As publishers, they have frequently set up in a year twenty thousand solid octavo pages of new standard works, besides printing large editions from the stereotype plates of over two hundred different volumes, now in their vaults. Within the last few years they have issued a number of most costly and valuable books, as for instance, their Gazetteer of the World, at a cost of fifty thousand dollars; Indigenous Races of Mankind, by Nott & Gliddon; and Chambers' Cyclopedia. The character of their leading publications, as well as the enterprise of the publishers, will be inferred from these; or perhaps more distinctly, when we state that the original cost of *four* of their works, including their illustrated edition of the Waverly Novels, and the Comprehensive Commentary, was one hundred and eighty-six thousand

The department of School literature is represented by SOWER, BARNES & POTTS, COWPERTHWAIT & CO., J. B. LIPPINCOTT & CO., E. H. BUTLER & CO., E. C. & J. BIDDLE, CHARLES DESILVER, URIAH HUNT & SON, and others, while HENRY CAREY BAIRD and EDWARD YOUNG & CO.

and three hundred dollars. They have recently incurred an important outlay to secure to Philadelphia the publication of Webster's Dictionary, of which they now publish five different editions.

In 1853, Mr. Lippincott erected on Market street, an immense six-story building with marble front, which is now occupied by the firm for the purposes of their business. It covers the greater part of a lot having a front of forty-one feet on Market street, and extending back a depth of three hundred and fifty-six feet.

This firm is now comprised of Messrs. Lippincott, Claxton, Remsen, Mitchell, and Haffelfinger. They employ a capital in their operations of not less than half a million of dollars.

SOWER, BARNES & POTTS are a leading publishing firm in the department of school book literature. The senior partner is a lineal descendant of the Christopher Sower referred to in the text. The educational series published by this firm, range from the most elementary to the most abstruse, from Raub's Series of Spellers to Fensmith's Grammars, from Sanders' Readers to Brook's Trigonometry.

A large number of their works are adapted alike for use in schools, and for popular information. Of this class are Bouvier's work on Astronomy, Peterson's Familiar Science, Hillside's Primary work on Geology, which probably are better adapted for those who wish to acquire elementary knowledge on the subjects of which they treat, than any other similar works that can be mentioned. Pelton's Series of Outline Maps, published by this firm, have been remarkably successful, and for a time employed two manufactories. Sheppard's admirable text books on the Constitution, and Gilpin's remarkable book on the Gold Regions of Colorado, are among their miscellaneous publications.

COWPERTHWAIT & CO. is the name of another firm in Philadelphia, that for many years has been identified with the publication of School Books. Their list comprises many works of acknowledged merit and wide popularity, such as Warren's Geographies, Greene's Grammars, Colburn's Arithmetics, Berard's History of the United States, Swan's Readers and Spellers, Potter & Hammond's Penmanship and Book-keeping. They are also the publishers of *Barnard's Education in Europe*, and *Blake's Biographical Dictionary.*

URIAH HUNT & SON are an old and highly respectable firm of Booksellers and Stationers, who also publish School Books extensively. Among the works published by them are "Hazen's Speller and Definer," "Comly's Speller," "Bonnycastle's Mensuration," "Gummere's Surveying and Progressive Speller," "Becker's Copy Books and Ornamental Penmanship," "Ainsworth's Latin Dictionary, and Anton's Abridgement."

confine themselves almost exclusively to the publication of Statistical and Industrial works for the benefit of Trades.

Law books are a specialty with T. & J. W. JOHNSON & Co., and KAY & BROTHER, while the Theological publishers are JAMES S. CLAXTON, RICHARD McCAULEY, PERKENPINE & HIGGINS, HOWARD CHALLEN, and

HENRY CAREY BAIRD has selected as his specialty the publication of works of an Industrial and Scientific character. Wherever a trade has found an author to elucidate its theory or practice, it may safely be said his work is in Mr. Baird's collection. He has issued books explanatory of Mechanical Engineering and Architectural Drawing, Cotton Spinning and Carding, Soap Making, Leather Dressing, and Encyclopedias of Chemistry and Metallurgy.

EDWARD YOUNG & Co. have chosen a somewhat similar field, and confine their publications principally to works of a practical character. They have recently issued in three volumes, octavo, and at an expense of over $20,000, *Dr. Bishop's History of American Manufactures*, which was the first attempt ever made to show the gradual and successive steps by which American industry has attained its present wonderful and magnificent development. It is handsomely illustrated with over one hundred steel portraits of prominent American Inventors and Representative Manufacturers.

HOWARD CHALLEN has, for many years, been identified with the publishing interests of Philadelphia, and has issued a number of popular books relating to the East, as for instance, "The City of the Great King," "Palestine, Past and Present," "Hadjii in Syria," "Little Pilgrims in the Holy Land," and "In and Around Stamboul." Recently he has issued some works of a practical character, as "Riddell's Modern Carpenter and Builder," "Carpentry Made Easy," etc. He is also the publisher of the *Uniform Trade List Circular*, which supplies a want that has long been felt both by booksellers and book buyers. It consists of a Trade List giving the title, author, size, binding, and price of all the books published during each month in the country; and semi-annually the monthly issues are revised and bound in a volume of about three hundred pages. It is acknowledged as a valuable aid to all who are interested in the doings of the American publishing world.

JAMES S. CLAXTON is the successor of William & Alfred Martien, who for many years were prominent publishers in Philadelphia of Theological and Juvenile books. His catalogue now contains a list of nearly two hundred distinct works, the greater part of them especially adapted to students in theology, Sunday-school teachers, and others engaged in the moral and intellectual training of the young. His publications are distinguished for neatness of style and the moderate price at which they are sold.

RICHARD McCAULEY, 1314 Chestnut street, succeeds to the works formerly published by Herman Hooker, which are principally of a devotional character, and adapted more particularly to the wants of members of the Episcopal

the various Societies and Boards of Publication organized for the purpose.

Miscellaneous works, including Novels, are published largely by J. B. LIPPINCOTT & Co., and T. B. PETERSON & BROTHERS, and of the one hundred and twenty booksellers in Philadelphia, there are probably fifty

church. Among them we find "A Plain Commentary on the Four Gospels," "Blunt's Sermons," "Bishop Odenheimer's Prayer-Book," and "Adams' Elements of Christian Science." His recent publications are of more miscellaneous character, and his general catalogue is quite extensive.

SAMUEL D. BURLOCK, formerly of Miller & Burlock, has long been prominently identified with the business interests of Philadelphia. For nearly forty years, he and the firm with which he was connected, have been prominent binders, and to a limited extent, publishers of books. Recently he has removed to a large and commodious building, 927 Sansom street, which he has fitted up with improved machinery for manufacturing books, and, in connection with the Quaker City Publishing Company, has engaged extensively in the publication of standard works. His Royal Quarto Family and Pulpit Bible contains the Apocrypha, Concordance, Psalms in Metre, fine Steel Engravings, and Illuminated Family Record. Burlock's Bibles have long been a standard in the market. They are printed on fine white paper, with the marginal references in the centre of the page, handsomely and substantially bound, and are not excelled by any in style or price. He has recently issued new editions of Pocket Bibles and Testaments, also beautifully printed on fine white paper, and substantially bound, that are sold at extremely low prices.

Mr. Burlock is a man of considerable originality of mind, and does not hesitate to depart from the beaten tract of traditional outline. He was one of the first to engage in the manufacture of PHOTOGRAPH ALBUMS, and by the improvements which he made in the method of manufacturing them, contributed to the wonderful popularity which these albums have obtained. The designs for the sides, clasps, and edges, are original and superior, and his panelled and velvet albums are magnificent specimens of bibliopegic beauty. Among his latest publications is Dr. James Moore's *History of the Rebellion*, which contains, in a moderate duodecimo volume, a comprehensive account of the essential facts regarding the late Civil War, and is peculiarly adapted for those who have not time to read the great folios that have been published on the subject.

JOHN E. POTTER & Co. are comparatively a new firm that has rapidly risen to great prominence as publishers, especially of Subscription works. Thirteen years ago Mr. Potter stereotyped his first book, a large octavo, of which he has since sold forty thousand copies. Their catalogue, consisting exclusively of their own publications, now numbers over five hundred works, many of them in handsome and expensive bindings, and their sales reach a quarter of a million annually. They occupy premises on both sides of Sansom street, and bind all their own stock, employing for this purpose alone

who occasionally publish a book when assured of a sufficient sale to justify reasonable expectations of profit.

There is also a class of publishers who prepare works with a view to sale mainly by *subscription*, or *through agents*. Of these the most prominent are JOHN E. POTTER & Co., J. W. BRADLEY & Co., GETZ & Co., P. GARRETT & Co., JONES, BROTHER & Co., and "The Quaker City Publishing Company," with which SAMUEL D. BURLOCK is connected.

Bibles are published in this city largely at prices ranging from sixty-five cents to one hundred and fifty dollars, and in every style of binding, from the plainest to brown morocco, illuminated and with painted edges. The styles are generally distinguished by the name of the publisher, as Harding's Bible, Burlock's Bible, Potter's Bible, and Lippincott's Bible.

about fifty hands. They are the publishers of most of T. S. Arthur's popular books, and manufacture over one hundred different styles of family Bibles, some of which are models of excellence and finish. They have a subscription department connected with their business, and are said to do as large an agency trade as any house in this country.

Their success is due in part to their adherence to sound business principles, discarding speculative ventures, and buying for cash, and also to the admirable judgment they have displayed in the selection of works to publish. So satisfactory has their method of doing business proved to all their business connections, that it is said they have never lost a customer they had once secured.

WILLIAM W. HARDING, who succeeded to the business established by his father, Jesper Harding, has the largest Bible manufactory in the city, and probably in the country, conducted by individual enterprise. He occupies for this purpose the greater part of the old Post Office building, on Dock street, and manufactures Bibles in all styles of binding, and of all sizes of type, at prices ranging from the cheapest to the most expensive. Mr. Harding possesses unusual advantages for the economical manufacture of books, in the fact that all the operations, from the manufacture of the paper to the final binding, are carried on in his own establishments. Mr. Harding is also an extensive manufacturer of *Photograph Albums* of all kinds, including a peculiar kind known as "Harding's Patent Flexible Chain-Back Album." Each of the links in this Album forms a half circle around the edge of the intervening leaf when the book is closed, and is nearly straight when the book is opened; consequently, the links do not cut or break by opening and closing the book. Mr. Harding is also the publisher of the *Philadelphia Inquirer*, which initiated a new era, if it did not effect a revolution, in the journalism of Philadelphia.

ALTEMUS & Co. have, at 401 Race street, one of the largest bookbinderies and Photographic Album manufactories in the United States. This firm has been established since 1842, and now occupy a building sixty by one hundred and twelve feet, in which they employ one hundred and fifty hands. They supply orders from all parts of the country.

The publication of GERMAN BOOKS has become an important branch of the general trade, but is controlled principally by three houses—SCHAEFFER & KORADI, F. W. THOMAS & SON, and IGNATIUS KOHLER.

SCHAEFFER & KORADI are the sole publishers in this country or in Germany of complete editions of the celebrated Heine's works, and they were the first American booksellers, it is said, who engaged regularly and successfully in exporting German books from the United States to Germany. A large German Dictionary published by them, has been sent to Germany to the number of ten thousand copies.

F. W. THOMAS & SON are an old and highly respectable house in this trade, who publish stereotype editions of German classics, such as Goethe, Schiller, Lessing, etc. They are also the publishers of the daily *Freie Presse*, and weekly *Republican Flag*, and other leading German newspapers that are issued in this city.

IGNATIUS KOHLER has for twenty years been established in Philadelphia in the German book trade. Originally a bookbinder—a business in which he is still extensively engaged, he gradually embarked in publishing, until now his list of publications includes some eighty works. His editions of the Bible, and English translations of Schiller's works, are executed in a style creditable alike to the works and the publisher. His publications are distributed throughout the United States and the Canadas, and to some extent exported to Europe.

The publishers of Philadelphia occupying, as they do, a more central position than their brethren in New York and Boston, and having pecu-

Of the various organized Societies for the publication of books, the most important and extensive is the AMERICAN SUNDAY SCHOOL UNION, which was established in 1824. This Society has disbursed, since its organization, over five millions of dollars, principally in the preparation of small works of a moral and religious character. Their catalogue now contains a list of over two thousand distinct publications of which eight hundred and fifty are bound books for children's reading, or for the use of teachers and advanced pupils. They publish four select, cheap libraries of one hundred volumes each, the circulation of which has amounted to more than five millions of bound volumes, involving an actual outlay of at least $600,000. "The Village and Family Library," which is probably the most important of these collections, consists of original works on subjects of universal interest, written by able authors, and is sold at the extremely low rate of sixteen cents for an 18mo volume of two hundred pages.

By the charter of this Society, its officers and managers must be LAYMEN, and all the books it publishes are examined before publication by a committee of twelve men, only three of whom can be of the same denomination. Its members are from the various religious denominations of Christians, and its means for doing good are the free will offerings of benevolent people.

ESTABLISHED IN 1837.

JAS. B. SMITH & CO.,

WHOLESALE

Blank Book Manufacturers,

No. 27 SOUTH SEVENTH STREET, PHILADELPHIA.

A LARGE VARIETY OF

Pass Books,

Memorandums,

Tuck Memorandums,

Composition Books,

Receipt Books,

Note and

Draft Books,

Copy Books,

Day Books,

Journals,

Ledgers,

Cash, Sales, and

Invoice Books,

Records, Dockets,

IN HALF AND FULL BINDINGS.

Bank Books, Time Books, Pencil Books, Pocket Ledgers & Hotel Registers,

CONSTANTLY ON HAND.

☞ Blank Books of all kinds made to order according to pattern, for the Trade. Price Lists sent free per mail on application.

ALTEMUS & CO.,

BLANK BOOKS

AND

PHOTOGRAPH ALBUM MANUFACTORY,

☞ 401 ☜

RACE STREET,

PHILADELPHIA.

HAVE on hand, and constantly manufacturing, ***Patent Hinge-back Blank Books.*** Also, the ***Patent Hinge-back Albums,*** being the sole owners and makers of the above. Albums of all kinds. Also, ***Blank*** and ***Memorandum Books,*** at the ***very lowest cash prices.***

THE AMERICAN

SUNDAY-SCHOOL UNION,

ESTABLISHED IN MAY, 1824.

By the charter, the Officers and Managers of THE AMERICAN SUNDAY-SCHOOL UNION must be LAYMEN. Its members are from the various religious denominations of Christians. The books it publishes are examined before publication, by a Committee of twelve men, only three of whom can be of the same denomination.

Books published by the Society have the words AMERICAN SUNDAY-SCHOOL UNION on the title page. ☞Mark this, as many denominational books are published with the words SUNDAY-SCHOOL UNION on the title page, which are not AMERICAN Sunday-School Union books.

☞ For CHEAPNESS, PURITY and USEFULNESS, the books of the Society have no need of any endorsement. Catalogues of its Publications, and specimen copies of its Periodicals can be had on application at 1122 CHESTNUT STREET, Philadelphia; No. 599 BROADWAY, N. Y.; or at its SUPPLY DEPOT, No. 3 Custom House Place, Chicago, Ill.

liar advantages for circulating works throughout the entire Union, are much sought after by the latter as agents for the distribution of their publications. Hence wholesale bookselling is usually combined with publishing. Hence, too, while preferring, as they manifestly do, for their own issues, works possessing some substantial merit aside from mere entertainment, their stocks are generally very miscellaneous, embracing every variety of books, from the most ephemeral to the most substantial. The fact, however, that wholesale book-selling is combined with publishing, renders it difficult, even for those who are so disposed, to furnish accurately the statistics of their own publications. But it may be safely stated that the capital employed in publishing is nearly four millions of dollars, and the annual sales exceed three millions of dollars.

As a natural consequence of the extent of the publishing interest in Philadelphia, those who are engaged in the mechanical departments of the Book manufacture are largely congregated here. There are over fifty Printing offices in the city, exclusive of the newspaper press, and as many Bookbinderies. There are four Type foundries in Philadelphia, one of them the largest in the United States, and the same number of stereotype foundries and Printing Ink manufactories. There are ten first class Paper mills within the limits of the consolidated city, and nearly sixty houses engaged in the sale of Paper and Paper-makers' materials, some of whom own mills in Delaware and Chester Counties.* We believe that the Rittenhouse mill, on the Wissahickon, was the first paper mill in the colonies of which there is any record, and the first bank note paper used for the old Continental money was made at a mill in the vicinity of the city. Of *Marble paper* for bookbinders' use,

* Mr. G. A. Shryock, lately deceased, claimed to have been the first in this, or any other country, to manufacture by machinery Paper and Boards from various kinds of straw and grass. The discovery of straw as a material for Paper was patented by Colonel William B. McGraw, of Meadville, Crawford County, in 1828. In the succeeding year, Mr. Shryock purchased a cylinder machine, and adapted it to the manufacture of Paper from this material, and his success was noticed in the American and European journals of that period.

Within two years, the AMERICAN WOOD COMPANY erected at Manayunk and Royer's ford, the largest works in the world for manufacturing Paper from WOOD PULP. The buildings occupy a space one thousand feet in length by three hundred and fifty feet in width, and cost when finished, over $500,000. Logs of wood, principally poplar, are cut into chips by large steel knives, set in revolving circular iron wheels, which have the capacity of cutting from thirty to forty cords of wood every twenty-four hours. The chips are then boiled to a pulp in alkalies, and by a peculiar process of evaporation about eighty per cent. of the soda used is saved. It is stated these mills are now producing thirty thousand pounds of Printing Paper daily.

Philadelphia supplies nine-tenths of all the fine papers that are used on extra work in the chief cities of the Union.

In the manufacture of Account Books, foreigners acknowledge that Americans excel all others; and those who are conversant with the best workmanship executed in this city, will acknowledge that our makers are not surpassed by any. Every variety and description of Blank Books for merchants, banks, and public offices, are made here, from the cheapest to those distinguished as *indestructible*, which are bound in Russia, paneled, and edged with brass, varying, in full sets, from $75 to $300. The leading establishments employ the most improved machinery, and no expense is spared to expedite and cheapen the processes of manufacture.

In Book Illustration, Philadelphia can boast of at least fifteen excellent wood engraving establishments, five eminent engravers on steel in Stipple and Mezzotint, and two of the largest Plate Printing establishments in the Union.

NEWSPAPERS—MAGAZINES.

The first newspaper published in Pennsylvania was printed in 1719, by Andrew Bradford, in association with John Copson, and entitled "The American Weekly Mercury." The second newspaper was started by Samuel Keimer, in 1728, (immortalized by Franklin, in his autobiography, as a "great knave at heart"), and bore the title of "The Universal Instructor in all Arts and Sciences, and Pennsylvania Gazette." It was a folio sheet. After the return of Franklin from England, he, in 1729, united for a short time with Hugh Meredith, and continued Keimer's paper, on a whole or half sheet, as occasion required. In 1739, as we previously stated, Christopher Sower established at Germantown a German quarterly, but changed it to a weekly in 1744, under the title of "The Germantown Gazette." What however may be called the third paper in Pennsylvania, was "The Pennsylvania Journal," issued December 2d, 1742. It was continued till 1800. "The Chronicle and Universal Advertiser" was the fourth in Philadelphia, and the first with four columns on a page. It was an influential sheet, and lived till 1773. Seven Journals in the English language, and six in the German, were started in Philadelphia before the Revolution.

The first daily paper published in the United States was "The Pennsylvania Packet, or General Advertiser," established in Philadelphia as a weekly, by John Dunlap, in 1771, but converted to a daily in 1784. At this period, Mr. David C. Claypoole became associated in its management. To him Washington presented the original manuscript of his Farewell Address, which was recently sold by his executors, through Messrs. Thomas & Sons, and purchased by Mr. Lennox, of New York,

THE ONLY

METHODIST PAPER

Published in Philadelphia.

THE LARGEST METHODIST CITY IN THE WORLD.

THE

Methodist Home Journal,

A RELIGIOUS AND LITERARY NEWSPAPER,

PUBLISHED WEEKLY AT

108 SOUTH THIRD STREET.

Rev. ADAM WALLACE, Editor and Proprietor.

TERMS: $2.50 per annum in advance.

Large circulation, and constantly increasing in the cities and towns of Pennsylvania, New Jersey, Delaware, Maryland, and all over the Union.

As an advertising medium it is unsurpassed. Limited space only allowed. ***Rates moderate.***

ADDRESS

THE METHODIST HOME JOURNAL.

Box 2850, Philadelphia.

for a sum exceeding $2,000. The first daily evening newspaper established in Philadelphia was "The Philadelphia Gazette," by Samuel Relf, in 1788. In 1790, Mr. Bache published the "Aurora," afterward purchased by William Duane. 1791, Mr. E. Bronson originated "The United States Gazette," which is now continued under the title of "The North American and United States Gazette." The other daily newspapers that flourished in Philadelphia, in the early part of the present century, were "The True American," established in 1797, by Mr. Bradford; "The Freeman's Journal;" "The Register," commenced in 1804, by Mr. Jackson; "The Democratic Press," established in 1807, by John Binns; and "Poulson's American Daily Advertiser," which was the successor of "The Pennsylvania Packet."

The first paper that habitually treated of Letters and Arts in connection with commercial and political matters, was "The Daily National Gazette," originated at Philadelphia in 1820. According to Dr. Griswold, in his history of American Literature, the establishment of this paper was an era in our national mind.

Philadelphia was the second city in the Union to establish penny papers, and the "Ledger," which we believe was the pioneer, has had for many years a uniform circulation exceeding sixty thousand copies daily. The price of this newspaper has recently been advanced to two cents, and its publisher, George W. Childs, has just erected the most extensive and splendid building for printing purposes in the United States.

At the present time twelve newspapers are published daily in Philadelphia, as follows;

Name.	Published by.	Established.	Remarks.
N. American & U. S. Gazette	Morton McMichael	1791	Republican.
Philadelphia Inquirer	W. W. Harding	1829	Republican.
Public Ledger	George W. Childs	1836	Independent.
Press	John W. Forney	1857	Republican.
Daily News	J. R. Flanigen	1848	Conservative.
Philada. Democrat, (German)	E. Morwitz	1838	Democratic.
Free Press, (German)	F. W. Thomas & Son	1848	Independent.
The Age	Welsh & Robb	1861	Democratic.
Evening Bulletin	G. Peacock & Co	1847	Independent, Successor of *American Sentinel*, established 1812.
Evening Telegraph	C. E. Warburton & Co	1864	Independent.
Evening Star	School & Blakely	1866	Republican.
Evening Herald	C. F. Rheinstein & Co	1866	Democratic.

The following papers are published weekly:—Saturday Evening Post; Methodist Home Journal; The Home Weekly; The Weekly North American; Weekly Age; Philadelphia Saturday Bulletin; Philadelphia Inquirer (Tri-Weekly); Forney's Weekly Press; Forney's Sunday Press; Fitzgerald's City Item; Dollar Weekly News; Commercial List; United States Business Journal; Scientific Journal; U. S. Railroad and Mining Register; Sunday Dispatch; Sunday Transcript; Sunday Mercury; Sunday Morning Times; Episcopalian; The Presbyterian; Christian Instructor; The Universe; The Friend; The Moravian; The National Baptist; The Germantown Telegraph; Frankford Herald; Saturday Night; Legal Intelligencer; Real Estate News Letter; American Presbyterian; German Reformed Messenger; Farm and Fireside; The New World; Keystone; and the German Sunday Gazette; Republican Flag; and Philadelphia Wochenblatt.

Every merchant and every farmer throughout the Union should subscribe for at least one newspaper from a city so important as Philadelphia; and no one who reads aright can fail to obtain an amount of practical, useful information, that will amply repay the trifling expenditure.

In addition to the Journals above enumerated or referred to, there are nearly fifty Periodicals published in Philadelphia, including Medical Legal, Scientific, and Denominational organs. Of the strictly Literary Magazines, there are four: Godey's Lady's Book; Peterson's Magazine; Arthur's Home Magazine; and the Lady's Friend. Godey's Lady's Book has been issued regularly once a month since 1830, and it is a creditable fact worthy of being noticed, that an oath has never been registered in its pages. The Philadelphia Magazines have long been celebrated for their high moral tone, and their devotion to the fine arts making exquisite illustrations a leading feature.

III.

Boots and Shoes.

Philadelphia, and Lynn, Mass., are the chief seats of the wholesale manufacture of Boots and Shoes. The latter makes cheap and common work its speciality; the former, fine boots and shoes, especially ladies shoes. A careful summary of the production in both places has been made, and it has been ascertained that Philadelphia makes the most in value, and Lynn the greater number of pairs. The operations are not, as a general rule, carried on in large manufactories, though there are over twenty manufactories, whose aggregate products exceed fifty thousand dollars annually; but the greater part of the Boots and Shoes in Philadelphia are made in shops where from five to a dozen workmen are employed, and by "garret bosses," who work in their own rooms, and sell their products to jobbers and retailers, as soon as finished, for cash. Since the introduction of sewing machines, the manufacture of *Gaiter uppers* is a distinct branch, and gives employment to hundreds of females.

The entire trade of Philadelphia in Boots and Shoes, including city manufacture and eastern work, is stated approximately at fifteen millions of dollars.

III.

Brewing and Distilling.

1.—ALE AND PORTER.

"Beer," says the author of the "Picture of Philadelphia, in 1811," whom we have before quoted, "was brewed in Philadelphia for several years before the Revolutionary war; and soon after peace, the more substantial Porter was made by the late Mr. Robert Hare. Until within three or four years the consumption of that article has greatly increased, and is now the table-drink of every family in easy circumstances. The quality of it is truly excellent: to say that it is equal to any of London, the usual standard of excellence, would undervalue it, because, as it regards wholesome qualities and palatableness, it is much superior; no other ingredients entering into the composition than malt, hops, and pure water. A fair experiment has shown, that even so far back as 1790, Philadelphia Porter bore the warm climate of Calcutta, and came back uninjured. In 1807, orders were given by the merchants of Calcutta, after tasting some of it taken out as stores, for sixty hogsheads. Within a few years Pale Ale of the first quality was brewed, and justly esteemed—being light, sprightly, and free from that bitterness which distinguishes Porter."

The reputation of Philadelphia Ale has but strengthened with the lapse of years; and at the present time the Malt liquors made in Philadelphia take precedence in every market in the Union. The qualities for which they are distinguished are purity, brilliancy of color, richness of flavor, and non-liability to deterioration in warm countries—qualities, the result in part of the peculiar characteristics of the Schuylkill water—in part of the intelligence, care and experience of our brewers, conjoined to the use of apparatus possessing all the best modern improvements made in England and in this country.

There are now twelve extensive Brewers of Ale and Porter in Philadelphia, viz., Massey, Huston & Co., Robert Gray, Wm. D. Smith & Co., Wm. Gaul, James Moore & Son, Ignatz Beckler, Christian Schmidt, Philip Guckes, J. W. Procter & Co., James Smyth, Thomas J. Martin, Jacoby, Miller & Co., and numerous smaller breweries.

The oldest Brewery in the city, is probably that on the corner of Sixth and Carpenter streets, which was built about hundred years ago, by William Gray of Philadelphia. The largest Brewery is that belonging to Massey, Huston & Co., at the Northwest corner of Tenth and Filbert streets; originally erected by the farmers of Chester and Delaware counties, Pa., and purchased from them by the Brewer's Association of Philadelphia. Within a few years the buildings have been greatly enlarged, and now the main Brewery, built in the form of a hollow square a hundred and fifty feet each way, is eight stories in height with cellars and vaults underneath the whole from eighteen to twenty-five feet in depth, that afford a storage capacity of twenty-five thousand barrels. The mash tubs in this Brewery have a capacity for infusing twelve hundred bushels of malt daily, the fermenting tuns will hold seventy-five thousand gallons, and there are storage vats capable of containing ten thousand gallons each.

Attached to the Brewery are two malt houses that have a capacity for malting two hundred thousand bushels of barley per annum. The Malt house recently erected is a hundred and forty feet in length, fifty-two feet in width, eight stories in height, with five malting floors, and has cellars, and subcellars underneath, twenty-two feet in depth, capable of storing twenty thousand barrels of Ale and Porter. The extent of business done by this firm is the best evidence of the high repute in which their products are held.

Next to the firm just mentioned, the most extensive Ale Brewer in Philadelphia is ROBERT GRAY. He is the owner of two breweries, one of which, known as the "Star Brewery," was established in 1830, by Richard M. Taylor, who was a protegé of Stephen Girard, and the other, the "Eagle Brewery," formerly owned by William C. Rudman.

Of these, the "Star Brewery," at Eighth and Vine streets, is the more extensive, and with the Malt house attached, occupies a lot containing sixteen thousand square feet. The main building has a front on Vine street of forty feet, extends back one hundred and twenty feet, and is four stories in height. It contains all the best modern apparatus for brewing, and has mash tubs that have a capacity of infusing three hundred bushels of malt daily. The Malt house attached is a four story building, forty by one hundred and twenty feet. Underneath both the brewery and Malt house are vaults of an aggregate length of two hundred and forty feet, and twenty feet in depth. The "Eagle Brewery," located at 309 and 311 Green street, though not so extensive as the former, has also a capacity for producing many thousand barrels of Ale and Porter annually. Beside these establishments, Mr. Gray has storage cellars at Third and Brown streets, and Fifth and Arch streets, capable of containing ten thousand barrels. Mr. Gray's Ales are in high repute in Philadelphia, and are extensively exported to other places.

The Brewery of Messrs. J. W. PROCTOR & Co., successors to O'Neill & Co., 934 North Third street, though not so extensive as some others, is noted for its fine Ales. This firm embarked in the business within a few years with a view of producing a superior quality of Burton and other Family Ales, and they have been remarkably successful in securing for their products a considerable reputation. The vaults in their Brewery are quite extensive, and the water used in cooling Ales during the

GEO. J. BURKHARDT & CO.

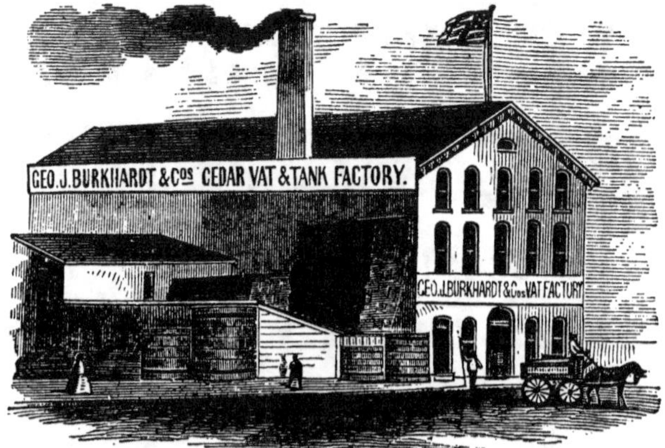

Cedar Vat & Tank Factory,

BUTTONWOOD ST., BELOW BROAD,

PHILADELPHIA.

We are prepared to make at the shortest notice,

WOODEN VESSELS

Of every description and capacity, suitable for Porter, Ale & Lager Beer Brewers, Distillers, Dyers, Chemists, Vinegar Manufacturers, etc., etc.

TANKS,

For Oil Wells and Refineries.

RESERVOIRS,

FOR RAILROADS, PUBLIC INSTITUTIONS AND PRIVATE DWELLINGS.

The above articles are made of CEDAR, which, on account of its durability, is superior to all other woods used for the same purpose.

We would call the attention of Brewers and others to our

SCOTCH MASHING MACHINE,

which is far superseding all others. Its advantages are—EASY ADAPTATION, EXPEDITION, and INCREASED AMOUNT and QUALITY of EXTRACT.

☞ Orders received for ***Iron False Bottoms, Grains Valves for Steeps and Mash Tubs, Stop-Cocks, etc.***

☞ PLANS and SPECIFICATIONS furnished for the erection of new and reconstruction of of old Breweries and Malt Houses.

summer months, is derived from an immense subterranean well twenty-nine feet in depth, and eight feet in diameter.

JAMES SMYTH has recently erected at Twenty-first and Washington Avenue a new brewery, that is said to be in many respects a model one. He has now the requisite facilities and experience to produce the best qualities of Ale and Porter.

The aggregate annual product of all the Philadelphia Ale Breweries now exceeds two millions of dollars.

2.—LAGER BEER.

The manufacture of Lager Beer was introduced into this country about twenty-four years ago, from Bavaria, where the process of brewing it was kept secret for a long period. Its reception was not a very cordial or welcome one; and about twelve years elapsed before its use became at all general. Within the last few years, however, the consumption has increased so enormously, not merely among the German population, but among the natives, that its manufacture forms an important item of productive industry. The superior quality of that made in Philadelphia has, no doubt, increased the demand, and by diminishing to some extent the use of fiery liquor, has effected partial good. Lager signifies "kept," or "on hand;" and Lager Beer is equivalent to "beer in store." It can be made from the same cereals from which other malt liquors are made; but barley is the grain generally used in this country. The processes resemble those of brewing Ale and Porter, with some points of difference, and the brewing generally forms a separate and distinct business.

There are now about thirty brewers of Lager Beer in Philadelphia, having a capital employed of $2,000,000.

The statistics of the entire Brewing business in Philadelphia, for 1866, are as follows:—

Product.

Ale, Porter, and Brown Stout, 200,000 barrels, averaging $11.00	$2,200,000
Lager Beer, 200,000, " " $11.00	2,200,000
Other Beer, say	200,000
Total,	$4,600,000

Raw Material consumed viz.:

Barley or Malt, 1,000,000, bushels, at $1.60	$1,600,000
Hops, 100,000 lbs., at 60 cents,	600,000
Total,	$2,200,000

The capital invested in Ale, Porter, and Lager Beer brewing, including *Malting*, is $4,000,000; being, it will be perceived, a larger amount in proportion to the product than probably in any other business. This arises from the necessity of occupying large plots of very valuable ground, from the extent of the buildings, and from the great number of vats and casks required. The casks alone, exclusive of vats, in use by Philadelphia brewers, cost $320,000.

GUSTAVUS BERGNER'S LAGER BEER BREWERY,

Is an excellent representative of the largest and best of this class of breweries in Philadelphia. It is located near the Reading Railroad, opposite Fairmount Park, and the buildings cover nearly a half square, fronting two hundred feet on Thompson Street, and extending back to Master Street four hundred and eight feet. They consist of an aggregation of structures, three and four stories in height, the first of which was erected in 1857, and to which additions have been made almost yearly.

The main building of this Brewery, and the whole brewing apparatus, were totally destroyed by fire on the 23d of December 1866, but the work of reconstruction was commenced immediately on the same site, and within a few months a new building was ready for operation with improved machinery, a more complete apparatus, especially in the mashing department, and a more powerful engine.

The distinguishing feature of this, as of the other Lager Beer Breweries, is the immense size of the subterranean vaults. The aggregate length of Mr. Bergner's vaults, is nine hundred and fifty feet, twenty-two feet in breadth, with an average depth of thirty feet, and divided into convenient compartments. In these vaults are deposited the casks in rows, far enough apart to permit a man to walk between them. The lower tier consisting usually of the largest casks is placed on skids or sleepers about a foot from the ground; on these two rows of casks are placed, and after these another row of smaller kegs are stowed. After a vault or compartment is filled the door is closed and straw, tan and other non conductors are placed to keep out the external heated air, the vaults are ventilated, and the temperature kept as low as possible, for should it exceed 8° Reaumur, or 50° Fahrenheit, the beer will spoil.

As stated in the Text the processes of brewing Lager resemble those employed in brewing Ale and Porter. The operations are usually commenced in December, and though the beer is fit for consumption and may be drawn within five or six weeks after brewing, the real Lager is not usually drawn until about the first of May. Like Ale it is greatly improved by age, and by being kept in a cool place. The heavy beverage in which a greater proportion of malt and hops is used, is called BUCK or BOCK BEER.

One of the peculiarities of Lager Beer, is the flavor imparted to it by the casks. The casks, previous to use, have their interior completely coated with resin; this is done by pouring a quantity of melted resin into the cask while the head is out, and igniting it. After it has been in a blaze for a few minutes, the head is put in again, which extinguishes the blaze, but the resin still remains hot and liquid; the casks are then rolled about, so as to coat every part of the interior with it; any resin remaining fluid is poured out through the bung-hole. This resin imparts some of its pitchy flavor to the beer.

Mr. Bergner has several retail establishments in Philadelphia, where his Lager is sold, the principal ones being No. 412 Library Street, and 239 Dock

3.—WHISKEY.

The consumption of Spirituous Liquors, both as a luxury and in the arts, is so vast, that their manufacture necessarily involves considerations of great commercial importance. There are said to be over three hundred persons and firms engaged at this time in the distillation of Whiskey from rye, molasses, etc., but, with one or two exceptions, their establishments are not sufficiently extensive to be called manufactories. Nearly all the houses who are extensively engaged in producing Whiskies, have distilleries located outside of the city limits, and some of them in other States. For instance, JOHN GIBSON'S SON & CO., who pay an annual revenue tax of over one million of dollars, have their distillery located on the Monongahela river, in the western part of the State. This establishment, known as the Gibsonton Mills, is undoubtedly the finest and largest distillery in the State of Pennsylvania. The buildings, which include a Malt house, are of cut stone, and worth probably $400,000. This firm employ a capital of $1,500,000 in the production of pure Monongahela, Rye, Wheat, and Bourbon Whiskies, from kiln-dried grain and barley malt, and sell their products to nearly every State in the Union.

But the branch of the general trade, in which Philadelphia is without a rival, is in her stock of fine old Rye Whiskies. Her pre-eminence in this particular is due in great measure to the efforts of the late John Gibson, who was one of the first to foresee the advantage that would result to the trade of the City, by accumulating a stock of Whiskies which, when kept years for improvement, would supplant other spirits of foreign manufacture. His success was so marked that other houses began to invest largely in the business, and at this time it is estimated that a capital of at least $10,000,000 of dollars is invested in the production and storage of Whiskies for improvement by age in Philadelphia alone, and as much more in the State. Ten years ago all of this class of Whiskies, with some few exceptions, were only known and appreciated in the South; now they are sought after and prized in every part of the country.

One of the firms that has made this branch a specialty, and probably done more than any other to quicken and benefit the general trade of the city, is that of HENRY S. HANNIS & CO. The senior partner in this firm was employed from boyhood in Mr. Gibson's store, and for seven years

street, and also several wholesale establishments obtain their supplies from his Brewery. The popularity of his Lager is attested by the extensive demand for it, while, as a citizen and a business man, he commands the respect of the community.

was nominally a partner in the house. In 1863 he left the parent house, and established a store at 220 South Front street, with a view of dealing especially in Whiskies improved by age. The times were favorable for success, as prices were rapidly advancing, and having purchased heavily, they accumulated a large stock for future requirements on favorable terms. At this time Messrs. Hannis & Co. have two stores in the city, one in New York, another in Boston, a distillery and flouring mill at Martinsburg, Va., and occupy eight entire floors in the building of the Pennsylvania Warehouse Company for storage purposes.

Besides the firms mentioned, WM. H. KIRKPATRICK & CO., HENRY WALLACE & CO., H. & H. W. CATHERWOOD, WHITE & HENTZ, J. F. TOBIAS & CO., A. J. CATHERWOOD, GIBSON & ROCKAFELLOW, and others, are largely engaged in the manufacture of pure Whiskies.

V.

Carriages.

"Comparing the state of the art of Carriage building," say the London Jurors, in their report on Carriages exhibited at the World's Fair, "of former and not very distant times, with that of the present, we consider the principles of building in many respects greatly improved, and particularly with reference to lightness, and a due regard to strength, which is evident in Carriages of British make; and especially displayed in those contributed by the United States, where there is commonly employed in the construction of wheels, and other parts requiring strength and lightness combined, a native wood (upland hickory), which is admirably adapted to the purpose. The Carriages from the Continental states do not exhibit this useful feature in an equal degree."

Comparing the state of Carriage Building in various cities, states, and countries, it will be found by those who make the comparison, that in the art of constructing light Carriages, particularly with reference to combining lightness with strength, and attaining durability in conjunction with beauty of appearance and high finish, no builders, either in this country or in Europe, have been so uniformly successful as some in Philadelphia. The quality of Philadelphia Carriages is indisputably superior. It is true that here, as elsewhere, there are carriages made, like Peter Pindar's razors, to sell; but we fearlessly claim that the general quality is above the ordinary average, and that those who desire a perfect vehicle, will be likely to attain a nearer approximation to perfection in this city than they can anywhere else. The first class builders have

studied, and know to exactness, the proper proportions of every part of a vehicle, and never use more or less material than is required.* They risk nothing to make it light; nor add any unnecessary weight to give it strength. All materials that are in the slightest defective, or that interfere in the least with the purposes of a good Carriage, are promptly rejected. The leading and important parts of a Carriage, as the wheels, axles, etc., are generally made on the same premises, and rigid supervision exercised over every part in the construction. The prominent builders have attained a high and wide-spread reputation entirely too valuable to themselves to be risked lightly through carelessness, neglect, or indifference. In addition to unremitting vigilance, combined with long experience, their efforts are greatly aided and facilitated by having at hand the very best materials, and in being able at all times to command the very best workmen. The growth of hickory, and oak, and

* WATSON'S CARRIAGE MANUFACTORY, has long been one of the most celebrated establishments of the kind, not only in Philadelphia but in the Union. It was the fortune of the late Mr. Watson, to attain a most remarkable reputation as a Carriage builder; and even now a second-hand Watson Wagon, will command at auction, a higher price than the new Carriages of many manufacturers. The care in selection of materials, fidelity of workmanship, and beauty of finish, which established his reputation, are, we are assured, maintained by his successors, whose manufactory is at 825 North Thirteenth Street. The buildings are of irregular construction, having a front on Thirteenth Street of forty feet, and extending back to Duane Street. The main buildings are of brick three and four stories in height, and divided into departments, each of which is appropriated to some special purpose in the construction of vehicles while the finished product is seen in the large salesroom on the first floor. Many of the workmen have been connected with this establishment for years, and have had a thorough training in one of the best of schools. This firm is probably the only one who manufacture their own *springs*, and in every department, the utmost care is taken to secure such materials and workmanship, as will ensure durability in their Carriages.

WILLIAM D. ROGERS, is another very celebrated Carriage builder of Philadelphia. He has constructed a number of Carriages for gentlemen in Europe, and it is related that one of his vehicles, which had been kept by its owner for a day at a hotel in Switzerland, attracted so much attention and curiosity, that the hostler charged a fee for exhibiting it and realized more money from this source in one day than his wages amounted to in six months. Mr. Rogers' manufacturing operations are carried on in a four story building, which, with the auxiliary shops and lumber yard, occupies the entire square, bounded by Sixth, Marshall, and Master Streets. His Salesroom or repository, is situated on Chestnut Street above Tenth, in a handsome edifice. forty-six by one hundred and seventy-eight feet, and three stories high. Here all ordered work is deposited for delivery, and generally as fine a collection of Carriages of all kinds, may be seen, at these rooms as can be found in America.

ash, in the vicinity of Philadelphia, is so superior for Carriage purposes, that we might say without exaggeration, no first class vehicle can be built without coming to this vicinity for the materials.

There are now about thirty establishments, great and small, within the limits of Philadelphia, that make pleasure Carriages. They have a capital invested of about $700,000—employ on an average one thousand hands, all males, and turn out an average annual product of a million of dollars

VI.

Chemicals, Paints, Glue, etc.

The manufacture of Chemicals, in the United States, may be said to date from the war of 1812. The commercial restrictions that preceded that war caused such a scarcity and dearness of Chemicals, that the preparations of the more prominent articles offered an attractive field for enterprise. Previous to that period, however, a Philadelphian had established successfully a manufactory of Sulphuric Acid. This was Mr. John Harrison, the first successful manufacturer of Oil of Vitriol in the United States, and the founder of the well-known house of Harrison Brothers & Co., the present proprietors of the Kensington and Gray's Ferry Chemical, White Lead, and Color establishment. He had spent two years in Europe in acquainting himself, as far as he could gain

Beckhaus & Allgaier, are comparatively a new firm of Carriage builders, who within a few years have attained an enviable reputation in the business. They embarked in the manufacture about fourteen years ago, with limited resources, and are now the proprietors of one of the most extensive carriage manufactories in the city. It is built of brick four stories high, has a front on Frankford Road of eighty-seven feet, and a depth in parallel wings of one hundred feet, forming a hollow square, of which a portion is utilized for seasoning lumber. The same system in the arrangement of departments and the subdivision of labor, is observed here as in other first class carriage factories. In the upper stories, the painting, varnishing, and trimming, are done; the wood work is prepared in the second story, while the smith shop, in which the iron work is forged, is in the rear of the warerooms, on the first floor. The warerooms are especially attractive, presenting as they do in a triple row a selection of Coaches, Calesches, Phaetons, etc., that for beauty of finish, and honesty of workmanship, are not excelled by any in this or any other country. It is said that as fine a coach can be purchased here for one thousand five hundred dollars, as at any carriage repository in New York for two thousand dollars. This firm is an eminently progressive one, and expends large sums for new designs, while at the same time they reproduce all the desirable styles originated by others.

access to them, with the processes used by chemists; and, after his return to America, devoted himself to the manufacture of Chemicals. How much earlier he succeeded, we have no means of ascertaining, but in 1806, he was fully established as a manufacturer of Oil of Vitriol and other Chemicals, in Green street, above Third. His leaden chamber was a small one, and capable of making about forty-five thousand pounds, or three hundred carboys of Oil of Vitriol per annum. So successful were these operations, that in 1807 he had built a leaden chamber eighteen feet high and wide, and fifty feet long, capable of making three thousand five hundred carboys per annum. The price which the acid then brought, was fifteen cents per pound.

The application of Platinum to the concentration of Sulphuric Acid, was also first attempted in Philadelphia by Dr. Erick Bollman, who had distinguished himself by a gallant and all but successful attempt, in company with Francis K. Huger, of South Carolina, to rescue General Lafayette from his guards, during his imprisonment at Olmutz. Dr. Bollman was a Dane, a man of powerful and versatile mind, a physician, a chemist, a political economist, and a general scholar. Among other pursuits, he had turned his attention to the working of crude Platinum, of which there was a considerable quantity in this country, and for which there was no demand. He had brought from France the method then lately discovered by Dr. Wollaston, for converting the crude grains into bars and sheets; and, in 1813, he had reduced it into masses, weighing upward of two pounds, and into sheets more than thirteen inches square. One of the first uses to which he applied these sheets, was the making of a Platinum Still for John Harrison, for the concentration of his Oil of Vitriol. This still weighed seven hundred ounces, containing twenty-five gallons, and continued in use fifteen years.

This early application of Platinum to the concentration of Sulphuric Acid, is highly creditable to the American manufacturer, as its use for this purpose was then a novelty in Europe.

Charles Lennig, was the first Philadelphian who largely manufactured Oil of Vitriol by putting up extensive leaden chambers, and concentrating the acids in Platinum vessels, so arranged, as to be kept constantly at work, while discharging a steady stream of concentrated acid. His works, at the present time, are among the largest in the United States, and his list of articles manufactured includes,—besides Oil of Vitriol,—Soda Ash, Alum, Copperas, Aquafortis, Nitric and Muriatic Acids.

At the present time Philadelphia contains the most extensive Chemical manufactories in the United States. Messrs. Powers & Weightman, for instance, are among the largest manufacturing Chemists in the world.

They have two establisments—one at the Falls of Schuylkill, where they make Oil of Vitriol, Aquafortis, Nitric and Muriatic Acids, Epsom Salts, Copperas, Blue Vitriol, Alum, all on a large scale. At their establishment at the corner of Ninth and Parrish streets, Philadelphia, they manufacture Sulphate of Quinine—which is their staple article—Mercurials, Morphias, and Medicinal Chemicals generally.

Their Chemicals have an enviable reputation for purity, exactness, and beauty; and the firm is well known for its liberality, fairness, and reliability. The house was founded about forty years ago by two intelligent foreigners, Abraham Kunzi and John Farr; and the reputation acquired by them has been maintained, and, if possible, increased by the present proprietors.

Harrison, Brothers & Co., established in 1806, by John Harrison, to whom we have already referred, manufacture White Lead, Red Lead, Litharge, Orange Mineral, and the Lead Chemicals generally, including White and Brown Sugar of Lead, etc.; Oil of Vitriol, Alum, Copperas, Sulphates of Alumina and Copper Pyroligneous, Acetic, Muriatic and Nitric Acids; Aquafortis, Wood Naphtha, and a full line of fine Colors in Pulp, Dry and in Oil. The productions of this house have steadily gained in reputation, and deservedly enjoy a high character for purity and genuineness.

Rosengarten & Sons, whose laboratory is at Seventeenth and Fitzwater streets, are largely engaged in the manufacture of Sulphate of Quinine, and other Vegetable Alkaloids and fine Chemicals. This house was established in 1823, by the present senior member of the firm, who was one of the first to manufacture the Vegetable Alkaloids in this country. Their articles are well known, and their laboratory is one of the most important in the United States.

L. Martin & Co., though not so old in the business as most of the others mentioned, have attained a high reputation for the manufacture of chemicals of a pure quality. Their products embrace the various preparations of *Morphia*, *Strychnia*, *Nitrate of Silver*, Acetic and other Acids, and the usual assortment of Fine Chemicals for analytical purposes. This firm have fixed a high standard for the purity of their chemicals, and allow no article to leave their laboratory of an inferior quality. Besides chemicals, Messrs. L. Martin & Co. are largely engaged in the manufacture of Lamp Black, of which their Black and Refined are especial favorites among manufacturers.

Henry Bower has a large establishment on the Gray's Ferry Road, below the U. S. Arsenal, for the manufacture of Prussiate of Potash, Glycerine, Sulphate of Ammonia, Permanganates of Potash and Soda, and

L. MARTIN & COMPANY,

140 South Wharves,

Above Walnut Street,

PHILADELPHIA,

AND

WILLIAM WRIGHT, JR.,

No. 59 Cedar Street, 2d door below the Post Office,

NEW YORK.

MORPHIA, STRYCHNIA, NIT. SILVER,

CHEMICALLY PURE ACIDS,

STRICTLY PURE

ACETIC ACID,

EQUAL TO BEST ENGLISH;

AND OTHER

CHEMICALS

UNSURPASSED FOR

PURITY AND BEAUTY.

other chemicals. Mr. Bower commenced manufacturing at this location in 1857, and from time to time has made various additions to the buildings, until they now require thirty thousand square feet of roofing. His products are distributed throughout the United States.

CARTER & SCATTERGOOD are probably the largest manufacturers in this country of Yellow and Red Prussiates of Potash. The buildings cover about one half of a square of ground owned by them between Twenty-fourth and Twenty-fifth, and South and Shippen streets.

WETHERILL & BROTHER make White Lead, Red Lead, Litharge, Orange Mineral, Nitric and Muriatic Acids, Calomel and other Mercurials, Sulphuric and Nitric Ethers, Hoffman's Anodyne, Aqua Ammonia, and other Pharmaceutical preparations.

SAVAGE & STEWART, at their "Frankford Chemical Works," manufacture Oil of Vitriol, Aquafortis, Nitric and Muriatic Acids, Aqua Ammonia, Nitrate of Iron, Muriate of Tin, Tin Crystals, Blue Vitriol, etc.

There are several establishments in the city, engaged principally in making various preparations for coloring purposes, and have been successful in attaining excellence in a manufacture where excellence is rare. The oldest color establishment is that of CHARLES J. CREASE, who makes Prussian Blues, Chrome Greens, Chrome Yellows and Reds; and besides these, he makes Nitric Acid, Aquafortis, Muriatic Acid, etc.

JOHN LUCAS & CO., also make Prussian and Ultramarine Blues, Chrome Yellows and Reds, Zinc Green, etc., both Dry and in Oil.

In these establishments, which represent a capital of two and a half millions of dollars, much the larger proportion of the best Chemicals used in the United States are made. The factories, which are in many instances immense structures, are generally located out of the city proper —at Tacony, Bridesburg, Frankford, the Falls of Schuylkill, and some in or near Camden; but the capital belongs to the city, and their products centre here as a point for redistribution. Some idea of their extent and importance may be derived from the fact, that they consume 4,400 tons of Sulphur, 1,000,000 lbs. of Saltpetre, 3,000 tons of Salt; and produce daily of Sulphuric Acid 150,000 lbs., or over 45,000,000 lbs. yearly; of Alum, 20,000 lbs. daily; of Muriatic Acid, 20,000 lbs.; of Nitric Acid, 10,000 lbs.; of Copperas, 15,000 lbs. daily; of Nitrate of Silver, 150,000 ounces annually; besides the numerous preparations before enumerated, and used in the manufacturing arts and in medicine. The consumption of *Quinine* fluctuates of course with the state of health in the West; but, it is said, that in one year 250,000 ounces were made in Philadelphia.

In addition to these and probably other manufacturers of Chemicals, there are several manufacturing chemists engaged in the preparation of Medical and Pharmaceutical Preparations. Messrs. HANCE, GRIFFITH & Co., have recently erected a very extensive establishment at the corner of Marshall and Callowhill streets, for the manufacture of solid and fluid extracts, conducting the evaporation in *vacuo* by a series of appliances among the most complete of the kind in the Union.

Their preparations are highly esteemed for their purity, strength, and reliability, and are probably not excelled by any domestic or foreign. From the preparing and grinding of the crude products, and throughout all the manipulations required, till the medicines are finished and packed for the market, the whole is performed at their laboratory, in which forty-five persons are constantly employed. Much of the machinery and apparatus which have gained a high reputation for this model establishment was designed by the senior partner, and is especially adapted to the purposes for which it is employed.

CHARLES ELLIS, SON & Co., produce at their manufactory, Seventh and Morris streets, a variety of standard Chemical and Pharmaceutical preparations. Their list includes Spread and Adhesive Plasters, Roll Plasters, Ellis's Citrate of Magnesia, and Mercurial Ointments, and the new remedies such as the Hypophosphates of Lime, Soda, Iron, etc.

Several of the Wholesale Druggists carry on, in connection with their general business, the manufacture of certain standard articles that have an extensive sale.

BULLOCK & CRENSHAW, at Sixth and Arch streets, manufacture largely the pure Chemicals for Analysis, and for the use of Schools, Colleges, Lectures, etc. They also claim to have originated in Philadelphia the manufacture of the Standard Pills of the Pharmacopœia, and other approved Recipes, and Coating them with Sugar—a branch of their business which has grown so largely, that they have erected the most improved steam machinery for their manufacture, and are daily in receipt of orders from physicians and druggists, not only from every part of the United States, but from the Canadas and Mexico. This firm also manufacture and import largely the various implements required by the practical pharmaceutist, such as Friction Presses, Copper Stills, Porcelain Evaporating Dishes, etc., etc.,

WM. R. WARNER & Co., 154 North Third street, are also very extensive manufacturers of Sugar Coated Pills and Granules, of which they sell millions annually. This firm also claim to have been pioneers in this business in Philadelphia; but without attempting to solve what

Sugar Coated Pills & Granules,

MANUFACTURED BY

BULLOCK & CRENSHAW,

N. E. CORNER OF ARCH & SIXTH STREETS,

PHILADELPHIA.

☞These Pills and Granules are accurately compounded and sugar coated in our own Laboratory, and the ingredients are, without exception, of the best quality, and therefore can be confidently relied upon.

PILLS.	*Price per bot. of 100 each*	*500 each.*
Aloes, U. S. P.	$0 40	$1 75
" Comp: U. S. P. (Pil: Gent: Comp:	50	2 25
" et Assafœt: U. S. P.	40	1 75
" et Ferri,	40	1 75
" et Mastich: (See Pil: Stomachicæ,)	50	2 25
" et Myrrhæ, U. S. P.	50	2 25
Alterative,	50	2 25
Ammon: Bromid: 1 gr.	75	3 50
Anderson's Scots,	40	1 75
Anti-bilious (Vegetable)	75	3 50
Antimonii Comp: U. S. P. (See Pil: Calomel Comp:)	40	1 75
Aperient,	90	4 25
Assafœtida, U. S. P.	40	1 75
" 2 gr.	40	1 75
" Comp:	40	1 75
" et Rhei,	75	3 50
Bismuth: Subnit: 3 gr.	75	3 50
" Subcarb: 3 gr.	75	3 50
Bismuth and Nux Vom:	1 50	7 25
Calomel, ½ gr.	40	1 75
" 1 gr.	40	1 75
" 2 gr.	40	1 75
" 3 gr.	40	1 75
" 5 gr.	50	2 25
" Comp: (Plummer's,) 3 grs.	40	1 75
" et Opii,	85	4 00
" et Rhei,	75	3 50
Cathart: Comp: U. S. P.	75	3 50
Cathart: Comp: (Vegetable)	75	3 50
Chapman's Dinner Pills,	60	2 75
Cerii Oxalat: 1 gr.	1 00	4 75
Chinoidin, 2 gr.	50	2 25
Chinoidin: Comp:	1 00	4 75
Cinchon: Sulph: 1¼ gr.	75	3 50
Cook's, 3 grs.	50	2 25
Colocynthidis Comp: 3 gr. (Ext: Coloc: Comp: U. S. P.)	80	3 75
Colocynth: et Hydrarg: et Ipecac:	75	3 50
Copaibæ: U. S. P. 3 gr.	50	2 25
" et Ext: Cubebæ,	80	3 75
Copaib: Comp:	80	3 75
Diuretic:	50	2 25
Dupuytren,	50	2 25

PILLS.	*Price per bot. of 100 each.*	*500 each.*
Emmenagogue (Mütter.)	$40	$1 75
Fel: Bovinum,	50	2 25
Ferri (Quevenne's) 1 gr.	50	2 25
Ferri (Quevenne's) 2 gr.	75	3 50
Ferri Carb· (Vallet's) U. S. P. 3 gr.	40	1 75
Ferri Citrat: 2 gr.	50	2 25
Ferri Comp: U. S. P.	40	1 75
Ferri Ferrocyanid: 3 gr	50	2 25
Ferri Iodid: 1 gr.	65	3 00
Ferri Lactat: 1 gr.	50	2 25
Ferri Pyrophos: 1 gr.	40	1 75
Ferri Sulph: Exsiccat: 2 grs.	40	1 75
Ferri Valer: 1 gr.	1 00	4 75
Ferri et Quass: et Nuc: Vom:	75	3 50
Ferri et Quin: Cit: 1 gr.	75	3 50
" " " 2 gr.	1 40	6 75
Ferri et Strychniæ,	75	3 50
Ferri et Strychniæ Cit:	75	3 50
Galbani Compt U. S. P.	50	2 25
Gambogiæ Comp:	40	1 75
Gent. Comp.	50	2 25
Gonorrhœæ,	60	2 75
Hepatica,	90	4 25
Hooper, (Female Pills) 2¼ gr.	40	1 75
Hydrargyri, U. S. P. 3 gr.	40	1 75
" 5 gr.	50	2 25
" Comp:	80	3 75
" Iod. et Opii, (Ricord's)	75	3 50
Iodoform et Ferri,	3 50	17 25
Ipecac: et Opii, 3½ gr. (Pulv: Doveri, U. S. P.)	50	2 25
Leptand Comp:	1 00	4 75
Lupulin, 3 grs.	40	1 75
Magnesiæ et Rhei (1 gr. each)	40	1 75
Morphiæ Comp:	1 50	7 25
Opii, U. S. P. 1 gr.	80	3 75
Opii et Camphoræ,	90	4 25
Opii et Camphoræ et Tannin,	90	4 25
Opii et Plumbi Acet:	80	3 75
Podophyllin et Hydrarg:	50	2 25
Potass: Bromid: 1 gr.	75	3 50
Potass: Bromid: 5 grs.	1 25	6 00
Potass: Iodid: 2 grs.	85	4 00
Quiniæ Sulph: ½ gr.	90	4 25
" " 1 gr.	1 40	6 75
" " 2 grs.	2 75	13 50

HANCE, GRIFFITH & CO.,

MANUFACTURING
Pharmaceutists and Chemists.

N. W. CORNER MARSHALL and CALLOWHILL STREETS,
PHILADELPHIA.

MANUFACTURERS OF

Medicinal, Solid and Fluid Extracts,
Sugar Coated Pills and Granules, Spread and Roll Plasters, Blue Mass and Blue Ointment, Fruit Juices, Fruit Essences and Mineral Water Syrups,

AND ALL KINDS OF

MEDICINAL PREPARATIONS.

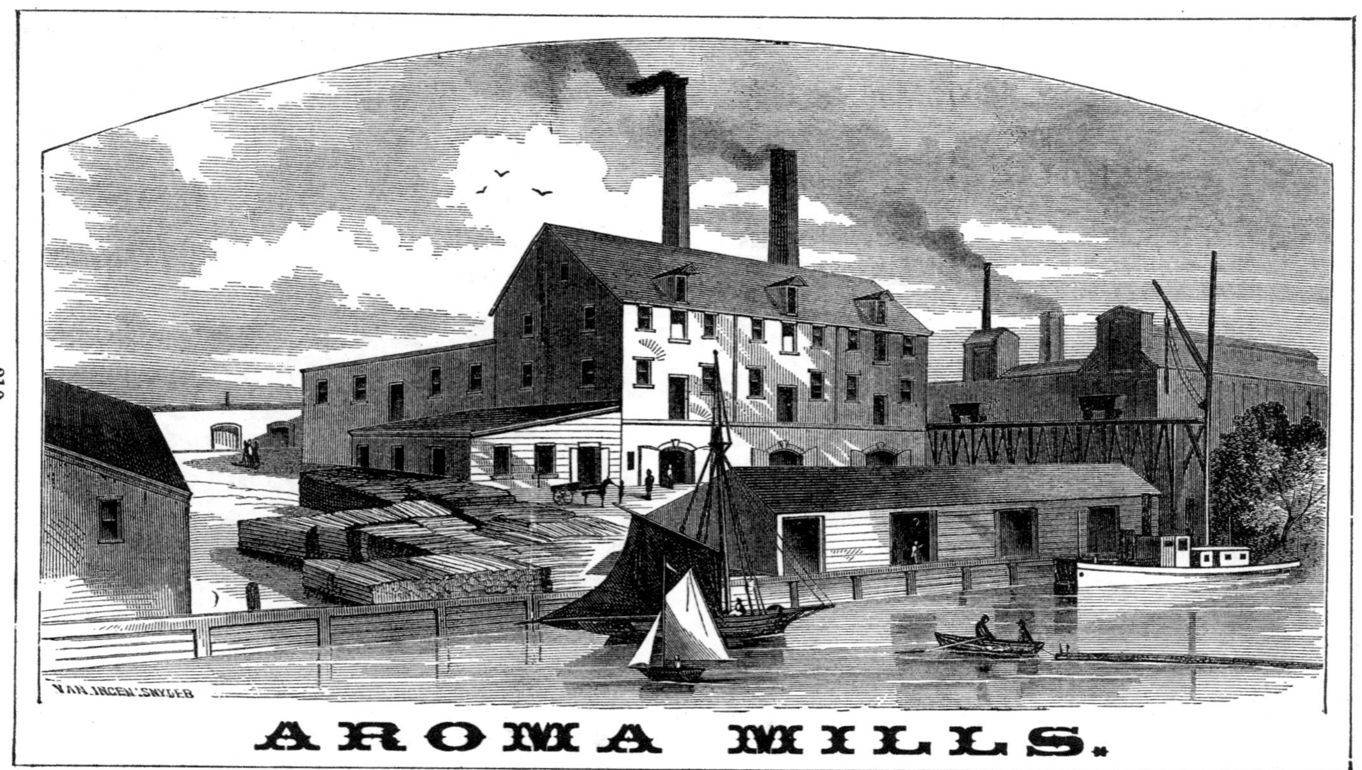

AROMA MILLS.

seems to be a vexed though unimportant question, it is quite certain that their products are endorsed as of the purest quality, and are extensively sold to the trade throughout the United States, Canada, Mexico, and South America.

This firm are also Importers and Jobbers of Drugs, Chemicals, and the various Pharmaceutical preparations that are usually found in the stock of Wholesale Druggists.

GEORGE K. RICHARDS, 67 North Second street, has a very extensive establishment for Pressing Herbs and Manufacturing Flavoring Extracts. The third and fourth stories of his warehouse are filled to the ceilings with those Herbs and Roots which have been found to possess valuable medicinal qualities, and these are cut and pressed into packages by appropriate machinery on the second floor. Usually a thousand pounds are pressed at one time. The first floor is the salesroom, filled with packages of all sorts, from pounds to ounces, conveniently arranged; while in the basement is a laboratory for manufacturing Flavoring Extracts of all kinds. Mr. Richards has almost a monopoly of the supply of Botanic Medicines, being, with one exception, the only manufacturer in Philadelphia.

The preparation of *Dye Stuffs* is made a specialty, or at least a prominent branch of their general business, by two manufacturers, viz.: BROWNING & BROTHERS, and HENRY SHARPLESS.

BROWNING & BROTHERS are the proprietors of the well-known "Aroma Mills"—a stamp which, on Extracts of Dye-Woods, is everywhere recognized as an assurance of excellence. This firm are also manufacturers of Paints, in the preparation of which they state they use only the pure Linseed Oil, and are careful to have them faithfully and finely ground.

HENRY SHARPLESS makes the usual Extracts of Logwood, Fustic and Querchia, and DANIEL MCINTYRE, at his chemical works in Chester, manufactures Dye-wood Extracts, Silicate and Arsenate of Soda, and is especially noted for Iron Liquor. Mr. McIntyre has an office in Philadelphia at 127 Church street.

WHITE LEAD—PAINTS.

The production of Paints, particularly of the Salts of Lead, which enter so largely into their manufacture, has added greatly to the chemical and manufacturing reputation of Philadelphia. Of White Lead there are three manufactories, viz.; those of WETHERILL & BROTHERS, HARRISON BROTHERS & Co., and JNO. T. LEWIS & BROS. The works of Messrs. Wetherill & Bros were established during or before the Revolution, by the grandfather of the present proprietors, who, it is said, introduced the manufacture into the United States. They are situated on the west side of the Schuylkill, employ a steam engine of eighty horse-power, and consume one thousand tons of pig lead a year. The article manufactured by this firm has always maintained a high reputation, is sent to every part of the United States, and exported to the West Indies.

Harrison, Bros. & Co. have been manufacturers of White Lead since 1806, and after Mr Wetherill, are the pioneers in its production in this country. On the completion of their Gray's Ferry works, now rapidly approaching perfection, they will have the facilities for producing about four thousand tons per annum, which we believe exceeds the abilities of any establishment in this country or Europe. The production of the Gray's Ferry factory will be in addition to that made at Kensington. In connection with the Lead works is their Color department, where they manufacture extensively Vermilions, Blues, Yellows, Greens, Dry Blacks, etc., etc.

. Private telegraph wires of this firm, connect their two factories and office in Philadelphia with their office in New York. This firm has lately shown the most remarkable energy, which does lasting credit to the manufacturing enterprise of our city. The establishment of private telegraph wires from Philadelphia to New York is an enterprise, we believe, without a parallel in this or any other country.

John T. Lewis & Brothers are the successors of M. & S. N. Lewis, who purchased a White Lead factory from Joseph Richards in 1819, and commenced the business on Pine between Schuylkill Seventh and Eighth streets. In 1849, they purchased a lot in Richmond, having a front of six hundred and twenty feet on Duke street, and three hundred and sixty feet on Huntington street, on which there was a White Lead factory already in operation. Enlarging the works, they removed their manufacturing operations to that place, where the present firm, who succeeded them in 1856, continue the business. Messrs. John T. Lewis & Brothers have now an invested capital of $350,000, employ ninety hands, and produce annually of White and Red Lead, Litharge, etc., about 4,500,000 pounds; other Paints, 1,200,000 pounds; of Linseed Oil, about 60,000 gallons, and of Vinegar, 300,000 gallons.

The capital invested in the manufacture of White Lead is nearly $1,000,000, and the annual product $1,500,000. It is to be regretted, that nearly all the raw material used is imported—English and Spanish Lead being principally employed—but it is gratifying to know that the American manufacturers, particularly those of Philadelphia, have effectually succeeded in stopping the importation of the finished product.

No painter will use the foreign if he can obtain the Philadelphia White Lead.

Another branch of the Paint manufacture consists in grinding White Lead and Colored Paints, and the Chromes and other colors in oil, in connection with the manufacture of Putty. The principal firms engaged in this business are, GEORGE D. WETHERILL & CO., JOHN LUCAS & CO., ROBERT SHOEMAKER & CO., BROWNING & BROTHERS, FRENCH, RICHARDS, & CO., and C. SCHRACK & CO. Some of these Paint Mills are most complete establishments, and have every appliance for carrying on the processes successfully and advantageously. The annual product is at least a million of dollars.*

One of the firms mentioned, Messrs. JOHN LUCAS & CO., in addition to grinding Paints, etc., at the Eagle Mills, in the city, are also the proprietors of the New Jersey Zinc and Color Works, at Gibsboro', N. J., established for the manufacture of an Oxide of Zinc, the introduction of which as a white Paint, was pronounced by the London World's Fair Jurors as one of the most remarkable events in the recent history of the Chemical arts. The "Gibsboro'" White Oxide of Zinc is pronounced fully equal to the world renowned Vielle Montagne Co.'s manufacture, and their pulp Steel and Chinese Blue and Primrose Chrome Yellow, have superseded the French and English, and are now used by the leading Paper-hanging Manufacturers in the United States. Messrs. Lucas & Co. have tanks of some twelve thousand gallons capacity, in which they make a Soluble Indigo Blue, now used so extensively for domestic purposes, and reducing the consumption of the imported Indigo. Their unfading Imperial French Green has almost driven the Imported Paris Green out of the market, having been found to be much superior in body and durability. They have erected a fire-proof Varnish-house on the English plan, and propose to introduce the manufacture of all the English Coach and Cabinet Varnishes.

* ROBERT SHOEMAKER & CO. have a factory for grinding in Oil White Lead and Zinc Paints, as well as Colored Paints, in all their variety, that produces about one hundred and twenty tons per annum. Besides this, they are extensively engaged in the manufacture of Putty and a Patent Drier for Paints, free from Shellac and Umber, and said to be superior to any Japan.

The business was established by Robert Shoemaker in 1837, and is carried on by him in association with his two sons, William M. and Richard N. Shoemaker. They have also a Wholesale Drug Store at the corner of Fourth and Race streets, which has a front on Fourth street of seventy feet, and on Race street of sixty-eight feet.

GLUE, CURLED HAIR, ETC.

These manufactures are essentially, though not nominally, Chemical. They subserve a peculiarly useful purpose, by converting substances that would otherwise be almost worthless, into products of commercial value. The refuse and offal from tanneries, morocco factories, and slaughter-houses, used in Glue and Curled-hair manufactories, are not generally available for other purposes; and, without consumption in this way, would be troublesome to remove, or prove nuisances to the community.

In Philadelphia there are three establishments engaged in the manufacture, viz.: BAEDER & ADAMSON, A. C. MILLER, and KESSLER & DELANEY.

BAEDER & ADAMSON, have the most extensive works of the kind in the Union, having recently erected a new factory, on the banks of the Delaware River, which covers ten acres of land. They manufacture here not only the ordinary products of Glue factories, but Manilla Paper for their own Sand Paper. This firm have a capital invested of $900,000; employ 400 men, to whom they pay $250,000 a year in wages; consume 6,000 tons of Coal, and produce of Glue, Curled Hair, Sand, and Emery Paper, Cow-hide Whips, Sizing for Paper-makers, and other articles, to the amount of $1,500,000 annually.

Besides their establishments in Philadelphia, Messrs. Baeder & Adamson have factories in Newark, N. J., and in Woburn, Mass.

Philadelphia has peculiar advantages for these manufactures. The climate is favorable, and the Tanneries of Pennsylvania, of which there are an immense number, furnish an abundant supply of raw material; while from South America, the importation is direct—several hundred bales of Hide Cuttings, being imported annually. The articles produced are distributed throughout the country, from the East to the West; and are exported to the West Indies, South America, and the Canadas.

VII.

Clothing—Ready-made.

Within the last quarter of a century a most important and complete revolution has been effected in the tailoring business, by the introduction of Ready-made Clothing. Some thirty years ago the only Clothing kept for sale was that which is known as "Slop Clothing," for seamen. But the inconvenience attending delays and misfits on the part of tailors—the advantages of procuring a wardrobe at a moment's notice—the ability of merchants to manufacture and supply Clothing equally as good, but much cheaper, at wholesale, than to order, led to the establishment of this as a distinct branch of business. In 1835, the wholesale manufacture of clothing in the United States was first entered into, to any considerable extent, principally in the city of New York; but many of those who then engaged in it were prostrated by the commercial disasters of '37. In 1840, the trade was re-established and increased; and since then has continued to enlarge and increase, until its present extent exceeds ordinary belief. We need, however, only point to the number of stores devoted to the business, to illustrate the popularity of the system.

One great benefit to the community, resulting from the success of the Clothing manufacture, is the immense field of employment it opens for the poor, especially for females. The poor of our large cities are thus supplied with a never-failing source of occupation. Some of the other cities have a large portion of their stock manufactured in the rural districts; but Philadelphia Clothiers deem it better policy to employ the population of their own city, and, so far as possible, to have the work done in their own establishments, being certain of having it better and more neatly done than in the country. The prices paid to employees, it is true, are not a very munificent remuneration for labor; but, by respectable Clothiers, no advantage is taken of the necessities of the helpless. Exceptional cases there undoubtedly are, in which the poor are oppressed; but we are convinced the business principles of our respectable Clothiers, accord with the principles of humanity, and that the females they employ are paid reasonably fair prices.

Coats, and finer kinds of work, except vests, are made during the dull seasons of the year by tailors, who at other times are employed in fashionable shops at higher rates. This ensures good work at cheap rates. The wages earned by these vary from eight to twelve dollars a week; but as most of them have families, the earnings of their wives and children always amount to something in addition. The cutting is a trade in itself, and requires talents of a peculiar kind. In the good Clothing warehouses, the men employed in this department are all of long experience and undoubted ability.

The goods which form the bulk of the manufacture in Philadelphia, are those styles, sizes, and qualities peculiarly adapted to the wants of distant sections—the West and Southwest. To conduct such a business successfully necessarily requires a large capital, for the manufacturing must be commenced some four months before the selling season ; and as the term of credit usually given is six or eight months, the Clothier cannot realize from his investments in a less average time than a year. The extent of the dry goods manufacture in the vicinity of Philadelphia—particularly of that class of goods which forms the raw materials of the cheaper kinds of Clothing—gives the Clothiers great advantages in procuring materials on the most favorable terms, direct from the manufactory, without charges for transportation. In several descriptions of Ready-made Clothing, therefore, the prices in this city are considerably below those in any other market.

The methods of conducting the business, as pursued by the great wholesale and retail establishments, will be found in the descriptions of the manufactories of WANAMAKER & BROWN, PERRY & Co., and BENNETT & Co., in the Appendix, while ARNOLD, NUSBAUM & NIRDLINGER, G. W. REED & Co., and GANS, WILGUS & Co., are excellent representatives of the exclusively wholesale houses.

The manufacture of *Shirts* and *Shirt Collars*, is now a distinct, organized and extensive branch of industry. In Philadelphia it furnishes at least three thousand persons with constant employment—counting solely the wholesale establishments, and those retailers who do partly a wholesale business. The Shirts made include every variety, from the cheapest—and it is claimed by disinterested persons, that the low-priced article is cheaper than that made in New England—to the best or those worth sixty dollars a dozen. M. ROSENBACH & Co., 19 North Third street, may be referred to as a representative of some seventy other wholesale manufacturers of Shirts in Philadelphia.

MRS. M. A. BINDER,

LADIES' DRESS AND CLOAK

TRIMMINGS.

LADIES' DRESS AND CLOAK

TRIMMINGS.

No. 1031 CHESTNUT STREET,

PHILADELPHIA,

IMPORTER OF

LADIES' DRESS & CLOAK TRIMMINGS.

Just received direct from Paris, by Steamship, every variety and style of

FRINGES, TASSELS, CORDS, GIMPS, BRAIDS, BUTTONS, ETC.,

In entirely new designs to suit the latest mode.

FANCY & PLAIN FANS, RIBBONS, VELVETS, GUIPURE & CLUNY LACES, CRAPE TRIMMINGS AND FRENCH CORSETS.

ALSO, PARISIAN DRESS AND CLOAK MAKING IN EVERY VARIETY.

WEDDING AND TRAVELLING OUTFITS

Made up to order in the shortest possible space of time, in the most elegant manner, and at such rates as cannot fail to please. With more convenient arrangement and extended facilities, she is enabled to give increased satisfaction in every department of her business.

SUITS OF MOURNING AT SHORT NOTICE.

At this establishment may be found the year round superior

PLAIN AND TRIMMED PAPER PATTERNS of LADIES' AND CHILDREN'S APPAREL

Of the latest and most desirable styles.

☞ *Patterns sent by Mail or Express to all parts of the Union.* ☜

VIII.

Confectionery.

The manufacture of Confectionery, in its modern development, as practiced in England and the United States, bears the distinctive artistic characteristics of French ingenuity and invention. In no other country does the preparation of sugar, as a luxury, absorb so much mental attention, and afford a livelihood to so many persons. It is a long established custom for French gentlemen to present the ladies of their acquaintance, on New Year's Day, with a box of sweetmeats; and so faithfully and generally does the custom continue to be observed, that in Paris two thousand persons find regular employment in making Confectioner's fancy boxes, the most of which are distributed on that single day. The ingenuity and invention of the French manufacturer, says some one, are inexhaustible; "Every season he produces some novelty, and for years this competition has continued between himself and his rivals, and yet there is no abatement of his ardor or his success; now his production consists of a new box; now of some intricate interlacing of fruits; now of some wonderful crystallizations, and now of some new mode of concealing the motto: but in most cases, his art is exerted tastefully to introduce a looking glass." But the competition that has existed between himself and his rivals, though it may not have abated his ardor, has induced him to resort to some very reprehensible practices. To give a more exquisite flavor to his essences, or to secure vividness and durability of color to his confections, he has not hesitated to use the most noxious and poisonous substances. An eminent English physician testifies that he detected, by post-mortem examination, the essential oil of bitter almonds in the stomach of one who had suddenly died after partaking of some French sweetmeats. To such an extent had the use of deleterious mineral substances been carried in the manufacture of Confectionery, particularly for exportation, that the French Government interfered, prescribing what colors the Confectioners might use.

In Philadelphia, the manufacture of Confectionery is carried on largely, but it is distributed among so many small concerns that it is difficult to obtain accurate statistics of the entire production. There are about two hundred Confectioners in Philadelphia, the greater part of whom manufacture to some extent and probably two thousand dollars a year would be a fair estimate of the average product of each. Within the last five years, however, there have been important advances made in this manufacture by the erection of large establishments and the introduction of steam power, and at this time there are at least four or more Confectioners who carry on the business on a sufficiently large scale to enable them to be called Wholesale Manufacturers.

Of these the oldest established is the house of GEORGE MILLER & SONS, 610 Market street. This firm have been connected with the manufacture since 1833—a period of over thirty years—and have attained an enviable reputation for producing strictly pure articles of Confectionery. There is probably no firm in the country more careful and conscientious in excluding noxious ingredients, or more successful in producing brilliant and pleasantly flavored confections that are also wholesome and free from every thing injurious. They are justly celebrated for their cream fruits, glacéd fruits, roasted almonds, and gum preparations, for which they have an established trade with all parts of the United States.

Probably the next oldest house in this manufacture, is that of J. J. RICHARDSON & CO., 126 Market street. This firm have been established about twenty-eight years, and by care in selecting good materials and fidelity in the execution of orders have built up a business that requires a four story building twenty-two by one hundred feet. The building contains all the apparatus for manufacturing plain and fancy Confections, and in it are employed about thirty hands. The store extends the entire length of the first floor, and is one of the largest of its kind in the city. Besides the various standard descriptions of Candies which they manufacture, Messrs. Richardson & Co., are also dealers in foreign Fruits and Nuts.

E. G. WHITMAN & CO., 318 Chestnut street, are also largely engaged in the wholesale manufacture of Confectionery. They employ a considerable force of hands, and produce every variety of pure Confections, including fine beverage Syrups and Preserved Fruits. This firm is another whose products can be confidently recommended for their purity, agreeable flavor, and wholesomeness.

But the largest manufacturer of Confectionery in Philadelphia, is STEPHEN F. WHITMAN, 1210 Market street. With a laudable ambition for the advancement of his pursuit, he has labored indefatigably for many years to build up an establishment in Philadelphia that would compete with any of a similar description in the world. He has sought to make fine Confections that would rival those of France, and it may be safely asserted that he has succeeded. He has recently fitted up an extensive manufactory of Chocolate and Cocoa by steam power, and his entire establishment is well worthy of a more extended description among the remarkable manufactories of Philadelphia. (SEE APPENDIX.)

Within a few years, there have been manufactories established for making CONFECTIONERS' TOOLS exclusively. The pioneer in this branch was John Gardiner; who, after procuring copies of the machines and moulds in use in the European manufactories, commenced the business of manufacturing them in Philadelphia in 1851. His step-sons, Thomas and George M. Mills, were educated by him in the art, and, in 1864, succeeded to the business under the firm style of THOMAS MILLS & BROTHER. Their manufactory, at 901 Nassau street, is now the largest of the kind, in the country, devoted exclusively to the production of Confectioners' Tools. Their products are forwarded to all parts of the United States and the Canadas, and it is a fact creditable to their skill and fidelity of workmanship, that no article of their manufacture has ever been returned, though, we understand, a warranty accompanies every invoice.

Including the ornamental branches and *pieces Montês* in particular, for which Philadelphia is famous, it may be safely asserted that the aggregate product of Confectionery, made in this city amounts to $1,500,000 annually.

IX.

The Dry Goods Manufacture.

The trade in Dry Goods, considered as a branch of commerce, is the most important of any now existing in this country. It controls a greater amount of capital, employs a larger number of persons, and distributes a greater value of commodities, than any other branch of mercantile pursuit. The list of Dry Goods merchants in our large towns is far longer than will be found engaged in the sale of merchandize under any other heading; while throughout the interior the very name of "merchant" is associated with one who, whatever else he may sell, is a Dry Goods dealer. There are certainly "merchant princes" among those engaged in other mercantile pursuits; but in capacity, energy, and aggregate wealth, the dealers in Dry Goods, as a class, are emphatically THE MERCHANTS of our day and country.

The variety of articles embraced in the term Dry Goods, is seemingly exhaustless; but the materials of which they are composed, are principally Cotton, Wool, Flax, and Silk. All of these, with the exception of the last, are natural, and leading products of this country; all of them, with perhaps the same exception, are bulky in their raw and unmanufactured state. Hence one would naturally suppose, that the mills for manufacturing them would be situated in the same country as the place of their production, if not in the same district. No one would certainly suppose, even hypothetically, that a free, civilized, and ingenious people, would rely upon foreign countries for the supply of their necessities, or be persistently guilty of the gigantic folly of going four thousand miles to mill. It is indeed difficult to reconcile such a course of conduct with the traditionary notions of American independence and American sagacity; but happily, the day is gradually passing away when any exposition of the anomaly will be necessary.

The first regular Cotton Factory established in the United States, was located in Beverly, Mass., and went into operation in 1787. In 1789, it received a visit from President Washington, then on a tour through the Eastern States. At that time the British Government, defeated in a war just closed, took its revenge in the only manner possible, viz., by prohibiting, with severe penalties, any exportation of machinery, or even drawings of machinery, from that country. A handsome set of brass models of Arkwright's machine, was secretly prepared for shipment, but was seized at the Custom House. Mr. Samuel Slater, who had served a regular apprenticeship to the business in England, came out in 1789; and although he was without models or drawings of the machinery needed, he succeeded in starting at Pawtucket, R. I., three cards and seventy-two spindles, on the 20th of December, 1790. These were the first Arkwright machines operated in this country. This first Cotton Factory started in Massachusetts, with the improved machinery, was located near Pawtucket, on the other side of the river, and commenced operations about 1795.

The first Cotton Mill established, as we are informed, in the county, now the city of Philadelphia, was situated at La Grange Place, near Holmesburg. The machinery was supplied by Alfred Jenks, who had been a pupil and colaborer for many years with Samuel Slater, and who established his manufactory of cotton machinery in Holmesburg, in 1810. The oldest established Cotton Mill, now in operation, is the Keating Mill in Manayunk, lately owned by J. C. Kempton.

The first Woolen Mill started in the State was at Conshohocken, by Bethuel Moore, a name that continues to be identified with the manufacture. It would be desirable to trace, chronologically, the successive steps marking the progressive development of the manufacture of textile fabrics, in this city; but, unfortunately, there are no records within our knowledge, containing sufficient and reliable data for the purpose. In 1824, we find a list showing there were thirty-three Cotton and Woolen

factories in the city and vicinity, worked by water or steam power; and twenty of them had no less than twenty-eight thousand seven hundred and fifty spindles in operation, and the number increasing. A few years subsequently, an English writer announced that Philadelphia *was the great seat of hand-loom manufacturing and weaving.* But beyond such isolated statements as these, the growth of this important interest seems to have attracted but little historical recognition; and we can only conjecture that it was overwhelmed by the flourish of trumpets which attends the erection of a factory in New England, though it may produce less in a month than the hand-looms of Philadelphia produce in a week.

In 1857, the then Secretary of the Board of Trade, and the editor of this work, made an investigation of the progress that had been made in this branch of manufactures, and, for the first time, made known the astounding fact that Philadelphia is the commercial centre of two hundred and sixty Cotton and Woolen factories, the greater part being located within her limits, with a hand-loom production equal to seventy additional factories of average size. In an edition of this work, published in that year, the names of the proprietors, and the location of the mills were given. We then remarked that "we are astonished by the undeniable revelation, that Philadelphia *is the centre of a greater number of factories for textile fabrics than any other city in the world.* We do not desire to be understood as saying, greater number of looms, or greater value of production; but simply what we state, a greater number of distinct, separate establishments fairly entitled to be called factories. No other city in the world, within our knowledge, is the centre of two hundred and sixty Cotton and Woolen factories, and containing, besides, hand-looms in force and production equal to seventy additional factories of average size. Moreover, we claim that Philadelphia is the centre of a larger production of indispensable domestic goods, than any other city or place in the United States. In making this claim, we do not desire to be understood as saying all descriptions of goods, but of domestic goods, indispensable particularly in the South and West." If this be true, the inference is unavoidable, that Philadelphia is the cheapest market in which the merchants of the South and West can purchase such goods. These statements lead us to the consideration of two points; first, *the description of fabrics made here,* and secondly, *the extent of the production.*

The textile fabrics made in Philadelphia might be considered as of two classes—one, designated "Philadelphia goods," and the other "imported"—the former comprising a variety of heavy articles essential in domestic use, and the other, delicate, ornamental fabrics, sold in New York, and frequently in this city, as Parisian or German goods. We, however, shall adopt for convenience the usual subdivisions, viz. : Cotton goods, woolen and mixed, Hosiery, Carpetings, Silks, etc.

1.—COTTON GOODS.

The application of the wonderful natural product, which has been called by some vegetable wool, to the manufacture of articles of utility and of ornament, is one of the most interesting records of industrial achievement. In Philadelphia, this application has principally been directed to the production of articles calculated to promote the comfort of the masses—the artisan, the farmer, and the mechanic—and very great credit is due to the fabricants for having brought many unpretending articles of this description to a high degree of perfection. *Tickings* are made in large quantities, and of a far better quality than those made in New England. They are distinguished for having more stock and less starch in them. Mr. Wallis, one of the English Commissioners to the American World's Fair, thus speaks of certain goods of this class that came under his notice. "They are 36 inches wide, 1100 reed, No. 30 warp, and No. 35 filling or weft, with 140 picks to the inch. It is scarcely possible to conceive a firmer or better made article; and the traditionary notion that really good Tickings can only be manufactured from flax receives a severe shock, when such Cotton goods as these are presented for examination." The varieties of Tickings made in Philadelphia, and its vicinity, are far more numerous than elsewhere; and the prices range from twelve to twenty-four cents—those at the latter price being a most superior article.

Of *Apron* and *Furniture Checks*, Philadelphia may be said to have the monopoly in the manufacture; none being made elsewhere, as we are informed, to any extent. They are of various grades, ranging in price from twelve to thirty cents. These goods are well known, and it is therefore needless to add that they are of the first class. A superior Check for miners' shirting is made, worth from eighteen to thirty cents.

Ginghams are made of all qualities, ranging from fourteen to twenty-five cents. These goods, for strength and durability of fabric and colors, and neatness and beauty of styles, are, at the low prices at which they are produced and sold, the cheapest article, probably, for women's and children's wear in the whole range of the Dry Goods manufacture. They are much preferable to the Scotch at the same prices, and are free from the dressing which adds so much to the apparent weight of the latter.

Of Cotton goods classed as *Pantaloonery*, *Cottonades*, etc., a great variety of kinds, qualities, and styles is made. The manufacture of these is conducted on a large scale, the production of one manufacturer alone, having reached three and a half million of yards in a year. They are now made almost entirely of fast colors, as the demand for the very low priced (of fugitive colors) is yearly diminishing. They are from twenty-five to twenty-nine inches wide, and range in price from twenty-five to thirty-five cents. Philadelphia *Cottonades* are favorites with Jobbers and Clothiers throughout the country.

Heavy wide *Brown Sheetings* are made in the vicinity of the city, probably heavier than any other in this country; some two yards wide, made of yarn, No. 14, count 50 by 56, has been specially recommended as adapted for the purpose for which they are designed. They are goods which, in consequence of the cheapness of cotton, can be produced cheaper in this country than English goods of the same quality. Heavy blue Mariners' *Shirtings*, formerly designated in the West as "Hickory Shirtings," are made largely; the prices ranging from twelve to twenty cents. *Denims* are made to a large extent, specially adapted for plantation use, being heavier than any made elsewhere. Other goods, particularly adapted to the Southern trade, and known as *Negro Plaids*, Chambrays, or Crankies, are a prominent article of production with many. *Nankeens*, twenty-eight inches wide, are made from the Nankeen cotton grown in Georgia and South Carolina; price about fifteen cents for plain, and twenty for heavy twilled. Several mills also produce *Ducks*, *Osnaburgs*, and *Bagging*, some of which is of excellent quality. *Prints* are made of all grades, from the highest to the lowest, in Madder and Steam colors; and some descriptions, as black and white, and half-mourning prints, are made here exclusively. The prices range from eight to sixteen cents—those at the latter price bear favorable comparison with the well-known Merrimacks. It is no exaggeration to say that our Calico printers are unexcelled by any. *Printing Cloths* are made at two or three factories, and the production, although limited, is quite successful—it is believed that this branch of manufacture will increase.

Cotton Hosiery will be referred to subsequently; and of the minor narrow textiles, as for instance, *Stay Binding* or *Twilled Tape*—white, black, and in colors, the production could be expressed only by millions of yards.

For the production of Cotton Yarns there are several mills; but a large quantity used by the manufacturers of Cotton goods is brought from Paterson, New Jersey, and also from Augusta, Georgia, and from other parts of the South. The production of this article, in Philadelphia, should be at least equal to the wants of the manufacturers.

2.—WOOLEN AND MIXED GOODS.

Wool is described by an eminent scientific authority, in the following lucid manner: It is a peculiar modification of hair, presenting, when viewed under the microscope, fine transverse or oblique lines, from two thousand to four thousand in the extent of an inch, indicative of an imbricated or scaly surface, on which, and upon its curved or twisted form depends its remarkable felting quality and its consequent value in manufactures. The Woolen manufacture, in its narrow or restricted meaning, applies only to Cloths made of short wool, and such as possess the quality of felting together, and elasticity; the other branch is called the *Worsted manufacture*, in which long wool, and such as possess no particular tenacity of fabric, is used. The former term, however, is rarely used in the strict sense; and, in considering the leading manufactures of Philadelphia in this department, we shall apply it according to its popular significance.

The principal varieties of Woolen goods made in Philadelphia, are *Cassimeres*, *Satinets*, *Kentucky Jeans*, *Shawls*, *Flannels*, and *Linseys*, or *Woolen Plaids*.

Cassimeres are made to a considerable extent, both all Wool, and Cotton and Wool, of various grades. The finest, in imitation of the French, are nearly equal in quality of Wool and excellence of finish to any foreign goods, while they are much lower in price. The *Satinets* range from forty-five to ninety cents, and are largely produced. *Kentucky Jeans*, of unsurpassed quality, and of great variety of colors, are a leading article of production. They are twenty-seven inch goods, of various grades, from fifteen to sixty cents. The better qualities have all wool filling. *Twills* and *Tweeds*, of various patterns and colors, and having a diversity of names, are also made in large quantities: prices from thirty to sixty cents. Most of these have all Wool filling. Philadelphia-made Jeans, Twills and Tweeds, are staple goods; and, like the Checks, Ginghams, and Cottonades, have a high and deserved reputation, especially at the West, where they are in great demand.

Shawls, chiefly all wool—both long and square, plain and fancy colors, greatly diversified in patterns, are made to considerable extent. The Medium-long Shawls bring from two dollars up to eight dollars; while the Square are from seventy-five cents to three dollars and a half.

Flannels, of various colors and qualities, both all Wool and Domet, are also largely produced. An article, all Wool, termed *Welsh* Flannel, and used largely by miners, glass-blowers, and foundry-men, for shirts, is made by several, and highly esteemed.

Linseys, or *Woolen Plaids*, are made of various qualities; some one half, others one third Wool; prices from ten up to thirty-three cents. Very large quantities are sold in the West, as far as the new Territories and the Rocky Mountains; the heaviest being used there for the clothing of laborers and backwoodsmen. They are also very extensively sold in the South for clothing for domestics; while some are used for linings. The higher grades are very superior, and all are desirable goods and in constant demand. Many are woven in hand-looms. They are largely shipped to New York, Boston, and Baltimore. A superior article of 6-4, all Wool Plaids, price about one dollar, is also made.

Of Mixed Goods there is considerable variety, principally however the product of hand-looms. *Coverlets* of Cotton and Wool, red and white, and other patterns, belong to this class, and are a favorite and serviceable article.

Damask, *Birdseye*, and *Huckaback Diapers*, from 5-4 to 11-4, both brown and bleached, are largely made. They are heavy and very serviceable goods: prices from ten and a half to twenty-six cents. Some linen Table Cloths and Toweling, of superior quality, are made on Jacquard machines. It is claimed that the *Damask Table Cloths* are equal to the very best patterns of the imported, while they are superior in durability. One firm is making Marseilles of excellent quality. *Bed Spreads*, both bleached and brown, Stair Crash, and a variety of similar goods, are also made in hand-looms.

Union Checks, half Linen and half Cotton, are made of very superior quality: price from fourteen to twenty cents.

Worsted Braid, or "*Ferreting*," occupies many looms; and *Carpet Bindings*, of Cotton and Wool, are with many leading articles of production. Of men's, women's, and children's mixed blue-and-white *Hose*, and *Half Hose*, ten thousands of dozens are annually made.

3.—CARPETINGS.

The production of Ingrain and Venitian Carpetings, in Philadelphia, s so important a branch of the general manufacture, that it deserves at our hands special and separate notice. It is also distinctive in its characteristics, both as respects the description of goods made, and the mode of manufacture. The manufacturers of Carpetings in Hartford and Lowell confine their operations, we are told, to all Wool and Worsted goods, made in super and extra-fines; while the manufacturers in Philadelphia not only make the better qualities, but go down to goods which are mixed, and sell for about fifty cents per square yard. The fabrication of low priced Carpets is said to be *exclusively confined to Philadelphia.*

As respects the mode of manufacture the business is distinctive, inasmuch as it is distributed among a large number of weavers; there being but one mill that employs power-looms. The individual manufacturers number about 140, who furnish employment to at least 2500 hand looms, the largest manufacturer having one hundred looms at work on his fabrics. Each loom will turn out, monthly, three pieces of 120 yards each, or 4320 yards Carpetings yearly; consequently, the annual production for 2500 looms would be 10,800,000 yards. The prices of Ingrain Carpetings range from 50 cents to $1 50—a low average being 75 cents, which would give an annual value of $8,100,000. The persons employed are: weavers, 2500; and all others, winders, spoolers, warpers, assistants and dyers, say 1500 more—in all 4000 persons. The average price for weaving Carpets is 18 cents, and the average earnings of weavers $12 a week, or $600 a year. The whole amount paid to weavers and others, for labor, will reach $2,400,000 per annum.

The "Glen-Echo Mills," at Germantown, Philadelphia, Messrs. M'CALLUM, CREASE & SLOAN, proprietors, is about the oldest established in the country (1830). They manufacture Three-ply, Superfines, Fines, and Venitian Carpetings. They spin and dye their own Yarns, producing the goods from the raw material. A large number of hands are constantly employed. They have over one hundred looms—some of them worked by power. The goods of their make have a reputation throughout the United States. This concern has two large warehouses in Chestnut street, between Fifth and Sixth, opposite Independence Hall; one of these being a wholesale and the other exclusively retail store. In addition to the Manufacturing, they do a large Importing and Jobbing business.

Rag and *List Carpets* are also produced to the extent of 2,000,000 yards annually, yielding, at an average of 60 cents per yard, $1,200,000. The weavers employed in this branch have frequently but one loom each, and rarely over eight. The principal manufacturer has only about twenty looms.

Within a few years, several manufactories of *Hemp Carpetings* have been established.

The persons employed in the Carpet manufacture are English, Irish, Scotch, and German; but very few Americans, as we are informed, are known to be engaged either in weaving or spinning. The economy in manufacturing would be greatly promoted, it is supposed, if there were larger mills in the City for spinning and dyeing yarns.

REEVE L. KNIGHT. HARTLEY KNIGHT.

 L. KNIGHT  

REEVE L. KNIGHT & SON,

IMPORTERS AND DEALERS

IN

CARPETINGS,

807

CHESTNUT ST.,

PHILADELPHIA.

REEVE L. KNIGHT & SON,

FEATHERS,

Bedding & Mattresses,

11

SOUTH NINTH ST.,

ABOVE CHESTNUT,

PHILADELPHIA.

4.—WOOLEN HOSIERY—FANCY KNIT WORK.

The importance of this branch of industry, and the success of the Philadelphia manufacturers, entitle it to separate notice. For more than two hundred and fifty years Nottingham and Leicester were the chief seats of the Hosiery manufacture in Europe and America. The Knitting trade had its origin in Nottingham, through the invention of the Stocking-frame, by the Rev. Mr. Lee, of that place, in 1589. At the present time, it is estimated that there are at least 50,000 Stocking-frames in operation in Great Britain, employing 100,000 persons, and producing an annual value of $18,000,000. So diversified are the articles produced in color, shape, and adaptation to markets, that one Leicester manufacturer thought he could not fairly represent his production at the Great Exhibition, in 1851, except by sending 12,500 specimens and prices. Until within the last fifteen or twenty years, America looked exclusively to foreign sources for her supply of the various articles designated as Fancy Woolen goods or Woolen Knit Work. Within that period, however, the manufacture has taken such deep root in Philadelphia—particularly in Germantown and Kensington—that the Nottingham articles no longer find any considerable sale in the American markets, or even in the Canadas. The term "Germantown Woolen Goods," is now as familiar to most dealers as Nottingham Hosiery; while the quality of the American product really is far superior to that of the foreign. The Philadelphia manufacturers have such special and important advantages over the English in the price of Wool—being able, therefore, to use much finer grades in the production of articles costing the same price—that they may reasonably anticipate a period not remote when their goods of this class will find a sale, as they certainly will receive a preference, in the English market. A few large establishments, well managed, and combining all economies, it is the opinion of competent judges, could even now export these commodities to England with profit.

The manufacture, as at present conducted, is essentially a domestic one. In Germantown, in which the production is so large as to give its name to the goods produced, there are a few extensive mills employing steam power; but the distinctive feature of the business is its hand-looms and domesticity. Fully one-half of the persons engaged in the production have no practical concern with the ten-hour system, or the factory-system, or even the solar system. They work at such hours as they choose in their own homes, and their industry is mainly regulated by the state of the larder. But the inherent, natural industry of this class of operatives, who are largely Leicester and Nottingham men, will be inferred from a visit to Germantown, and practical observation of the neatness of the dwellings, and the air of comfort that pervades all its streets and avenues.

In the City proper there are three large factories engaged in producing Hosiery, Opera-hoods, Comforters, Scarfs, &c., employing each several hundred hands, and consuming annually upward of 250,000 lbs. of American Wool. The hand-frames and machines it is almost impossible to ascertain with accuracy; but they exceed seven hundred, of which about five hundred are employed on Woolen Hosiery. The average product for each frame is about $1,650 annually; and the value of all the Hosiery and fancy Woolen goods exceeds $2,000,000 annually.

The American Knitting Mills (SCHOFIELD & BRANSON, proprietors) are among the largest in the City for the manufacture of Cotton and Woolen Hosiery. The building located at the corner of Jefferson and American streets, in the late District of Kensington, is four stories in height, and contains one hundred and twenty looms, with a great variety of improved machinery adapted to the business. Messrs. Schofield & Branson employ about two hundred hands, and produce Hosiery that competes in popularity and saleability with any manufactured in Europe.

For an account of the Great Knitting Mills of MARTIN LANDENBERGER & Co., see APPENDIX.

The foundations of the American Woolen Hosiery and Fancy Goods manufacture, it is quite evident, are laid in Philadelphia. Within twenty years, by persevering and well-directed industry, Philadelphia manufacturers have succeeded in almost excluding the foreign articles from the American market; and they certainly have succeeded in enabling merchants, from all parts of the country, to obtain in Philadelphia superior goods at less than Nottingham or Leicester prices.

5.—NARROW TEXTILE FABRICS—SILKS, ETC.

In England, the various manufactures included in the term Narrow Textile Fabrics, are known by the name of Small Wares; and on the continent of Europe the manufacturers of them are designated *Passamenteurs.* In this country, the term usually employed is Trimmings, which represents ladies' dress trimmings, carriage laces, curtain trimmings, cords, tassels, braids, fringes, ribbons, military trimmings, and numerous manufactures assimilating in character. In England, France, Germany, Switzerland, the chief seats of these manufactures, the establishments confine themselves each to a single class of goods—one making fringes,

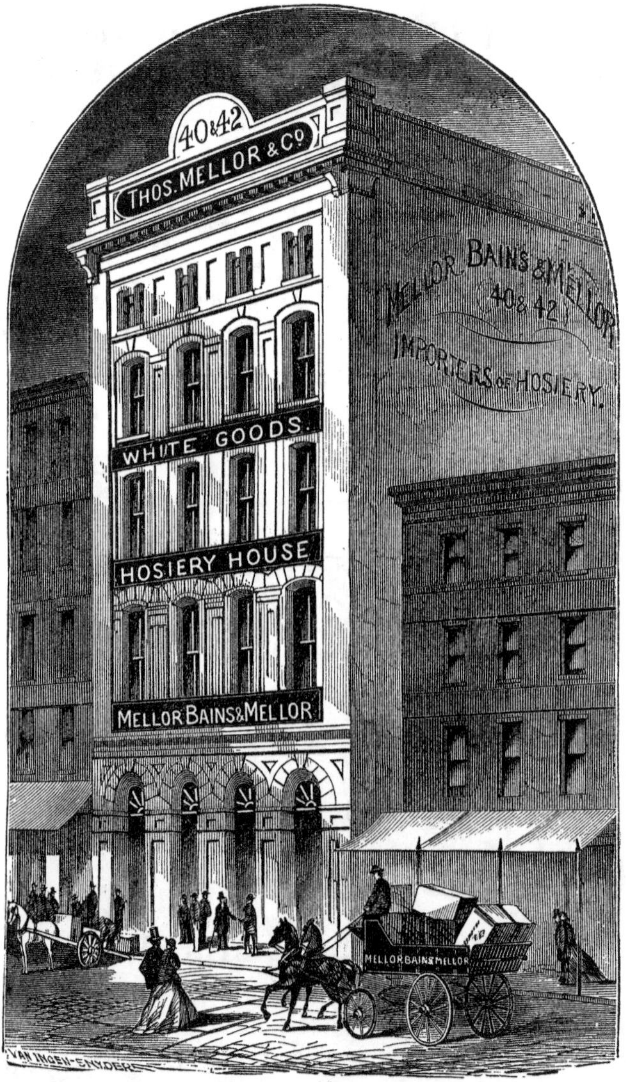

another ribbons, and so on; but here, two or more branches are often carried on by the same parties; and in the case of one firm in this city, all the above branches are united in one establishment—the largest of its kind, beyond all doubt, in the world.

Philadelphia has long been known as the principal seat of the manufacture of Military Goods and Carriage Laces; and, now, probably, one half of the whole production of the United States originates here. The branch known as "Ladies' Dress Trimmings," is comparatively of modern date in this country. Up to 1838, very little was made, being principally plain fringes, a few bindings, buttons, cords and tassels. The business, however, has become a very important item of our domestic manufactures; and since the reduction of duties on raw silk, is rapidly expanding. Patterns the most complicated are executed with facility, from designs that are original with the manufacturers. The fabrics produced here are acknowledged to be generally of better quality than the English and German; and for several years have competed successfully with nearly all articles of French manufacture.

Philadelphia is now the chief seat of the general manufacture of Trimmings in the United States. There are about thirty establishments in the city engaged in the various branches, including Carriage Laces, Regalia, and Upholstery, and one of them is the largest and most complete of the kind in the Union.

The manufactory of William H. Horstmann & Sons is the one alluded to, as undoubtedly the most extensive of its class in this country and probably in the world. The business was commenced by Wm. H. Horstmann, the father of the present proprietors, in 1815, and it is also consequently the oldest establishment of its kind in the city, if not in this country. In the infancy of its career, the manufacture was limited to a few patterns of coach laces and fringes; at the present time, it embraces a wide circle of fabrics of silk, silk and worsted, mohair, cotton, gold and silver thread, and includes some not made elsewhere in this country, besides every variety of military trimmings, including swords, drums, and metal ornaments.

In 1852, this firm exhibited a case of Silk Ribbons at the Exhibition of the Franklin Institute. We make the following extract from the report of the Judges on Silk Goods.

"By unanimous consent, the highest praise of the Committee is awarded to Wm. H. Horstmann & Sons, of this city, for their manufacture of Fancy Taffeta Bonnet Ribbons, case 1556. Indeed, your Committee must confess to having been entirely taken by surprise, on witnessing these productions of American looms, and it required convincing proof to satisfy the Committee that they were not examining the fabrics of Lyons or St. Etienne. Not only in brilliancy of coloring and weight of material, but in evenness of manufacture, they, in all respects, are equal to those which we have been so long accustomed to receive from France and Switzerland." * * * * "The merit of introducing and carrying forward to such a degree of perfection this new branch of manufacture, is due to the Messrs. Horstmann. The Committee may be deemed partial in their feelings from the fact that all its members have, for a long time, been engaged in the importation and sale of Silk Goods; but this very fact gives them additional opportunity of forming a correct judgment. They are unanimous in considering the production of the Messrs. Horstmann as one of the greatest novelty, as well as importance, in American manufacture, and are pleased to add, in corroboration of their views, that these goods have been sold in a neighboring city, through an importing house, indiscriminately with their foreign importations. Your Committee, understanding that you have a reward still higher than the usual premiums, to be bestowed in cases of extraordinary merit, are unanimous in the recommendation of its bestowal upon the Messrs. Horstmann.

"The Committee on Exhibition, in accordance with the above report, unanimously resolved to recommend to the Institute to award Wm. H. Horstmann & Sons a gold medal."

The manufactory of the Messrs. Horstmann is situated at the northeast corner of Fifth and Cherry streets, formerly the burying ground of the German Lutherans, and bought of the congregation owning the old church (built 1743), on the opposite side of Cherry street. The building forms an L, having a front of one hundred and forty feet on Fifth street, one hundred feet on Cherry street, and fifty feet wide, containing six floors. The engine house and machine shops are in a detached building in the yard. The machinery in operation in the factory is new, and much of it original.

Adjoining the manufactory on Cherry street, the firm own an additional lot, bought of the Friends, containing seventy-five feet on the street, on which there was formerly a Meeting house, that has been converted into a spacious sales-room.

Many of the most important machines, and application of machinery that are now in use in the manufacture, are indebted to the enterprise of this firm for the introduction into this country, or to their inventive genius for their origin. The Plaiting or Braiding machines were first introduced into the United States from Germany, by Mr. W. H. Horstmann, in 1824. In the year 1825, the same gentleman introduced the Jacquard machines. Gold Laces were made by power in this city several years before attempting it in the old world; and the use of power for making Fringes may be said to have been first generally adopted here. In fact it may be said that this firm was the first in any country to apply power to the general manufacture. From the report of the English Commissioners upon the industry of the United States, we extract the following paragraph, in which, after stating that Messrs. Horstmann have recently erected a very large and well-arranged factory within the city of Philadelphia, it is remarked :

"The whole establishment presents an example of system and neatness rarely to be found in manufactories in which handicrafts so varied are carried on. Female labor is, of course, largely employed in the weaving and making-up departments, and formerly in the cutting of fringes. This, however, is now performed by a machine with a circular knife, so arranged as to cut the thread on the diagonal. The double fringe, as it leaves the loom, being either run off the beam or placed upon a roller for that purpose, is divided much more exactly than it could be by hand, and at so rapid a speed as scarcely to admit of a comparison with hand labor. Any width of fringe can be thus cut, the machine being so constructed as to be easily adapted thereto."

In another part of their report these Commissioners allude to the Clinton Company, located at Clinton, Massachusetts, long known as the largest manufacturers of Coach Lace in America. The looms are of the same construction as the Brussels Carpet Power Looms. In 1857, the entire stock of goods, materials, looms, and patent rights of this Company, were purchased by the Messrs. Horstmann, and thus another important link was added to the chain, securing pre-eminence to Philadelphia as the greatest manufacturing city in the Union.

Next to Messrs. Horstmann, we believe J. C. GRAHAM is the oldest established manufacturer of Dress Trimmings and narrow Textile Fabrics in Philadelphia. He commenced business in 1850, and now occupies the four upper stories of a building, 525 Cherry street, which has a front of forty feet, and a depth of eighty feet. He employs about one hundred and forty hands, runs thirty-four Jacquard looms, several hand looms, and, in 1865, produced a value of $160,000. Mr. Graham has done much to improve and establish this manufacture in Philadelphia, and his fabrics are highly appreciated by the merchants of Philadelphia, New York, and other principal cities.

HENRY W. HENSEL has, at the southwest corner of Fourth and Commerce streets, what is probably the second largest manufactory of Dress Trimmings in Philadelphia. He occupies a six-story building, twenty-five by one hundred and twenty-five feet, employs two hundred and ten hands, runs thirty-five Jacquard, and about the same number of other looms. The machinery is of the most perfect description for the purpose for which it is designed, and is adapted not only for the manufacture of Dress Trimmings, but ribbons, neckties, window blind and shade trimmings, silk and Union beltings, picture cords and tassels, in fact, every variety of narrow Textile Fabrics.

Mr. Hensel has been identified with the manufacture in Philadelphia since 1851, and has been very successful in originating saleable patterns. Nearly all the articles manufactured by him are from designs that originated in the establishment.

Of the concerns more recently established, probably the most extensive are those of FISHER & EVANS, and J. B. CORNET.

FISHER & EVANS are the successors of Reynolds & Fisher, who commenced business in 1865. They now occupy the four upper floors of Womrath's extensive building, 415 Arch street, employ one hundred and ten hands, and produce an annual value of $150,000. They run twelve Jacquard and numerous other looms, with all the auxiliary machinery, such as Spooling and Bräiding machines, Cord Mills, etc., and produce a great variety of silk fabrics, ranging from one half inch to three or four inches in width, and also a limited quantity of woolen and cotton fabrics of the same description. Besides the usual variety of Trimmings, Messrs. Fisher & Evans manufacture Silk Chenilles, Bead Trimmings, and other hand woven work. They produce from materials in their raw state, and deal exclusively in articles of their own manufacture.

J. B. CORNET, 17 South Third street, has recently purchased the business established some twenty years ago, by Mr. Champromy. He employs about fifty persons, runs twenty Jacquard and twelve hand looms, and manufactures Ribbons, Bindings, Fringes, Gimps, Tassels, Silk Buttons, Cords, etc. His products are distributed through the usual avenues of trade, and are in extensive demand.

Fly Nets are extensively made in Philadelphia; and Regalias, etc., form nearly the exclusive business of one or two manufacturers.

The manufacture of Sewing Silks is carried on by five establishments in Philadelphia, but not as an exclusive business. It is usually conjoined with the production of what is known in commerce by the terms Singles, Tram, and Organzine.* A large proportion of the raw silk im-

* *Singles* is formed of *one* of the reeled threads slightly twisted in order to give it strength and firmness.

Tram consists of two or more threads thrown just sufficiently together to hold, by a twist of from one to one and a half turns to the inch.

Organzine, or thrown silk, is formed of two or more singles, according to the thickness required, twisted together in a contrary direction to that of the Singles of which it is composed.

WM. G. MINTZER,
Importer, Manufacturer and Furnisher of Military Society Costumes and Theatrical Goods.
131
Wm. G. MINTZER
FRINGES, TASSELS, LACES
GIMPS, FLYNETS, TRIMMINGS
REGALIA, BANNERS & FLAGS.
131 Wm. G. MINTZER 129
No. 131 NORTH THIRD STREET, PHILADELPHIA.

H. W. HENSEL,

No. 22 N. FOURTH STREET,

MANUFACTURER AND IMPORTER OF

LADIES' DRESS AND CLOAK

TRIMMINGS,

WINDOW BLIND & SHADE TRIMMINGS,

RIBBON AND TUBULAR NECK TIES,

SILK & UNION BELTINGS,

Picture Cords and Tassels,

FRINGES, GIMPS, CORDS, TASSELS, BUTTONS,

Bindings, Ornaments, Chenilles, Braids, Etc.

ported into the United States, comes from China—the Chinese silk being preferred for the pure whiteness of its color, and the strength and glossiness of its fibre. Its successful conversion into the various articles named depends largely upon the excellence of the machinery employed. In the production of Sewing Silks, our home manufacturers have been so successful, that it is supposed that the quantity now imported does not amount to five per cent. of the home production.

All varieties of Sewing Silk are made, Spool silk, Embroidery silk, Saddlers' or three-corded silks; and put up in quarter and half pound packages, or in hundred skeins, of different colors. Hundred-skeined Silk is so termed, because it is made up of from one to one and a half ounces of Silk to the hundred, measuring about ten yards in length to the skein. This article is generally sold to peddlers and jobbers. There is another description of skein made up for retailers, which measures from twelve to twenty yards in length. It is principally used by clothing houses, who find it economical to employ the larger skeins. The capital employed in the production of Sewing and other Silks, in Philadelphia, is stated at $300,000, and the annual production at $312,000. The machinery employed for spinning and twisting Silk is equal to any in the world.

The oldest established concern in this business, in Philadelphia, is that of B. Hooley & Son. The house was founded nearly thirty years ago by Messrs. B. & A. Hooley, of Macclesfield; and the present perfection attained in the manufacture of Sewing and Fringe Silks, in this city, is largely due to the enterprise of this firm. They are now making extensive improvements in their mills, with the view of improving the quality of their Silk and increasing their business, and adding to their assortment a full line of Spooled Sewing Silk and Machine Twist for Sewing Machines; they also manufacture Organzine, Tram, Floss, and all other kinds of silk for the use of manufacturers.

F. S. Hovey, 231 Chestnut street, has been engaged in the manufacture of Sewing Silks since 1843, when the business in this country was in its infancy. His long experience has given him an intimate acquaintance with every department of the manufacture, and his highly cultivated taste is manifest in the style in which his goods are prepared. His Sewing Machine Silks, and Tailors' Twists, known in the market as the "Hovacci," are acknowledged to be superior to any made in England.

6.—PRINTING, DYEING, EMBOSSING, FINISHING, ETC.

In the operation of Printing and Dyeing Textile Fabrics, the manufacturers of the United States have, without doubt, been greatly aided by the emigration of artisans from Europe. The attractions of Philadelphia, as a place of residence, have drawn hither the most skilful of these artisans—many of whom bring with them experience gained by almost unremitting attention to these departments of industry during the past half century, in England, France, and Germany. Moreover, the water and climate of Philadelphia are peculiarly favorable for success in dyeing. The influence of these natural agents has already been remarked upon; but we may refer to the fact mentioned by the English Commissioners, that in Lowell it is well known that the water of the Merrimack river, though reasonably well adapted for dyeing cottons, is not at all suited for woolens. They state, "*this question of the selection of a water site for Dyeing and Printing, is a most important one in the United States, since it is quite certain that in no country is there so great a variation in this respect.*" The principal Dye Works for Cotton and Woolen goods, in Philadelphia, are located at Frankford. The water in that locality is excellent for the purpose, and equally as well adapted for woolens as for cottons.

In the city proper there are many *Silk Dyers* and Refinishers, who have been very successful, and are deservedly celebrated. In the introductory we alluded to A. S. Mary Ainé, now succeeded by H. Tirel, who had experimented in various places, and found none so well adapted for producing desirable and brilliant results in dyeing as Philadelphia. De Laines, Merinoes, and other French goods, are consequently now largely imported in an unfinished state, and we believe at a less rate of duty, and dyed in this city in fast and exquisite colors.

The refinishing of Silks is made an almost exclusive business by a few, and so successfully performed, that old goods are made to wear the appearance of new.

7.—FACTORIES AND HAND-LOOMS.

The factory system of Philadelphia, as will probably be inferred from what has been already stated, is the result and offspring mainly of individual, unaided efforts. It owes but little, if any thing, to the advantages of associated capital; and has grown to a vigorous maturity in spite of foreign competition and unfriendly home legislation. The manufacturers having, from the beginning, directed their energies mainly to the production of useful fabrics, necessary to the comfort of the masses, have steadily worked on, aiming at substantial excellence in an unpretending sphere without attempting, until recently, to compete with others in the finer or more ornamental fabrics, or invoking the attention of the world by the erection of mammoth establishments. In the location of their factories, they have not generally been governed by any other than reasons of convenience and economy, peculiar to each proprietor; hence the factories are scattered throughout the city and its vicinity, the operatives forming no distinct class, the buildings attracting but little notice. In Frankford, and particularly in Manayunk, some show of aggregation is manifest; but in the latter place the exhibition is not very favorable for a correct observation of the beauties of the system.

The mills, though generally small, compare very favorably in machinery and amount of product with the medium establishments in New England. In Philadelphia, as well as in Lowell, several mills are often the property of one proprietor; and if we were permitted to publish statistics of individual establishments, we would enumerate one having 900 looms, 27,000 spindles, 850 operatives, and producing an annual product of 3,500,000 yards; another having 432 looms, 9,774 spindles, 38 cards, 513 operatives, and producing annually $430,000; another, having 216 looms, 8,000 spindles, 50 cards, 320 operatives, producing last year 3,272,510 yards duck, Osnaburgs, etc.; another, having 240 looms, 300 operatives, producing yearly 2,100,000 yards ginghams, pantaloonery, etc.; another, having 10,716 spindles, employing 200 operatives, and producing 750,000 lbs. cotton yarn. The Washington Manufacturing Company's Mills, at Gloucester, N. J., nearly opposite our city, and of which our esteemed townsman, David S. Brown, is President, employ 700 persons, and consume 35,000 bales cotton annually. The goods are printed by the Gloucester Manufacturing Company, who turn out about 7000 pieces of Prints weekly. It will thus be seen, there are some factories in and near Philadelphia that will compare favorably with those of any other place; but it would be highly desirable and good policy, to erect one or more calculated, from size and arrangement, to give eclat to the manufacture.

The majority of the operatives in the factories are English or Anglo-Americans. The hours for working are usually ten and a half per day; but as operations cease early on the afternoon of Saturday, the average for the week is ten hours. In New England, and many other places, labor is extended to eleven hours or more, daily. The female operatives, though perhaps less literary than their Lowell sisters, are seemingly as attractive in appearance, skilful in manipulation, and correct in deportment. The identity of interests which exists between the employer and the employed is seemingly comprehended more clearly by both, and the relations between them exhibit, on the part of the former, more paternal characteristics than is evidenced where the employers are large corporations. Some of the manufacturers are men who are distinguished for benevolent effort; and in some instances, where the factories are remote, schools and churches have been specially established by factory proprietors. The distinctive feature, however, of the Dry Goods manufacture in Philadelphia is—HAND-LOOM WEAVING.

It is a remarkable fact, that notwithstanding the rapid substitution of power for the production of textile fabrics, and the growth of large establishments from the results of accumulated capital, there is no actual decline in the number of hand-looms in operation. There are fewer looms devoted to certain classes of goods, and in certain localities, than formerly; but the aggregate of such looms now in operation is probably fully equal to that in any former period. Philadelphia is truly the great seat of Hand-loom Manufacturing and Weaving in America. There are now, within our knowledge, 4,700 hand-looms in operation in the production of Checks and other Cotton goods, Carpetings, Hosiery, etc., and it is probable that the true number approximates six thousand.

The material is furnished by manufacturers, and the weavers are paid by the yard. The weaving is done in the houses of the operatives; or in some cases a manufacturer, as he may be termed, has ten or twelve looms in a wooden building attached to his dwelling, and employs journeymen weavers—the employed in some instances boarding and lodging in the same house as their employer. Throughout parts of the city, especially that formerly known as Kensington, the sound of these looms may be heard at all hours—in garrets, cellars, and out-houses, as well as in the weavers' apartments. Among the weavers there are many very intelligent men, and some that have been employed in weaving those magnificent damasks, and other cloths, that Europe occasionally produces to gratify the pride of her rulers.

In addition to the factories located within the limits of the city, or so close to the borders as to be called in the city, there are a great number in the adjacent counties—some of them very fine and large establishments. In the counties of Chester and Delaware there are over fifty factories for the production of Cotton and Woolen goods; in Montgomery county there are twenty-one factories not included in our statement—one of which, located in Norristown, was the largest mill, it is believed, in the United States, previous to the erection of the Pacific

The Prize Medal

WAS AWARDED AT THE

GREAT PARIS EXPOSITION,

1867,

TO F. SACHSE & SON,

SOUTHEAST CORNER

EIGHTH & VINE STS.

PHILADELPHIA,

FOR

FINE SHIRTS,

OVER ALL COMPETITORS.

SPECIALTIES:

Fine French & Embroidered Bosoms, Fire, Base Ball Club, Cricket, Military, Zouave and Crimean Shirts, in all their varieties.

☞ ***Send for Illustrated Circular.*** ☜

Mill, at Lawrence; the Harrisburg Mills; the Reading Steam Manufacturing Company; the celebrated Conestoga Mills, at Lancaster; five or six mills in and near Wilmington, Delaware; three or four near Newark, Delaware; the Exton Mill, at Extonville, New Jersey; the New Jersey Mill, at Millville; and others at Bordentown, Trenton, and other places, whose head-quarters are in Philadelphia.

A careful summary of the production of these mills was recently made, and stated at seven millions of dollars; while the total production of the factories in the city and its vicinity closely approximates THIRTY MILLIONS OF DOLLARS.

This simple statement has a significance, an interest, a value to every dealer in, we may say consumer of Dry Goods throughout the Union, even to the remotest frontiers of civilization. Nearly thirty millions—probably over thirty millions—of the most useful Textile Fabrics are made annually in Philadelphia and its vicinity; and found in first hands in the warehouses of Philadelphia merchants. No comments can possibly add any thing to the force of a statement, the correctness of which all subsequent investigation will confirm, or if extended more minutely, will prove to be below the truth. We need deduce no inferences from it, for the eye of self-interest, quick in its perceptions, is generally quite as correct in its conclusions as political economy. When to the fact that thirty millions of Dry Goods are produced and controled, if not monopolized by the manufacturers and merchants of Philadelphia, we add another, viz., that the manufacturers of Old England and New England, consign every season their products to be sold in this market for what they will bring, the conclusion is inevitable, that *Philadelphia is the cheapest and best market in the Union for Dry Goods;* and fairly without a rival in those staple goods, the bulk of every stock, which, by their intrinsic value and low price, are SPECIALLY ADAPTED TO THE WANTS OF THE PEOPLE OF THE MIDDLE, SOUTHERN, AND WESTERN STATES.

X.

Fertilizers.

The manufacture of Chemical Manures in the City of Philadelphia has ttained an extent and systematic importance, within the last ten years, loubtless greater than that of any other city on the continent, and has ow become a feature of note among her varied industrial interests. The everal articles produced under the name of Raw-Bone Phosphate, Super Phosphate, etc., have an intrinsic standard, which has earned for hem a wide, popular demand, and they have taken the place, to a large extent, of the more costly Guanos, serving fully as well, in their practical application, all agricultural purposes.

The base or raw material used in the production of Super Phosphate is either the bones of animals, or the various Phosphatic Guanos, which are chiefly imported by the manufacturers themselves. Three of the four principal concerns engaged in the manufacture of this Manure in Philadelphia, each turn out over ten thousand tons per annum, and their respective brands are well known and appreciated in every State where concentrated Fertilizers are used. The firms above indicated we may name as Messrs. BAUGH & SONS, manufacturers of Baugh's Raw-Bone Phosphate, B. R. CROASDALE & Co., manufacturers of Genuine Super Phosphate ; FREDERICK KLETT, and MORO PHILLIPS.

Ten years ago BAUGH'S RAW-BONE PHOSPHATE was known to very few as a commercial Manure. It was manufactured in a neighboring town, and had for three or four years before that supplied simply a local demand. Two years subsequently the proprietors commenced the manufacture of their article in the First Ward of Philadelphia, and at that time had sale for about one thousand tons per annum. As the agricultural demand for concentrated Manures grew with each successive season, increased facilities were demanded and supplied, and in the eight years that have elapsed, the trade of this house has actually grown to the extent of over ten thousand tons per annum.

The works of B. R. CROASDALE & Co. are located at Point Breeze, on the Schuylkill River. The firm own at this point twenty-five acres of ground, on which they have erected a building having a front on the river of one hundred feet, a depth of two hundred feet, and two and a half stories in height. The machinery for grinding, etc., is propelled by an engine of fifty horse power, and about twenty persons are employed in the establishment. This firm commenced to manufacture their Super Phosphate at Gunner's Run in 1856, and removed in 1866 to their present premises, where they have facilities for producing a thousand tons per month.

Besides the establishments mentioned, whose facilities are first-class, and conducted with great system and energy, there are others who in the aggregate manufacture a very large quantity of Fertilizers of reputable name and quality. The growing preference of agriculturists for Manures of small bulk, such as the Super Phosphates produced in the Philadelphia establishments, seems each season to augment the already great tax upon the respective works in supplying the demand for this class of Fertilizers, which now amounts to not less than *two millions of dollars* per annum.

POINT BREEZE CHEMICAL WORKS.

B. R. CROASDALE & CO., PHILADELPHIA.

Why Farmers should Use Super-Phosphate.

We are aware that much deception has been practiced in the manufacture and sale of concentrated fertilizers by unprincipled dealers, who flourish for a season or two, then sink into oblivion; but our motive has been to make a business reputation that should be permanent, and we have therefore sold **only a good** and STANDARD SUPER-PHOSPHATE. From the year 1864, however, we add a guarantee of the quality of our ***Super-Phosphate,*** which should satisfy every one. By an arrangement with Prof James C. Booth, coiner and refiner of the United States Mint, in Philadelphia, and his partner, Dr. Thomas H. Garrett, every package of ***Super-Phosphate*** sold by us will be made under the supervision of these gentlemen, whose reputation, as analytical chemists, is so widely known, and the former of whom has had long experience and observation, both in the manufacture and employment of fertilizers Each package will bear the following brand:

STANDARD GUARANTEED.

BOOTH & GARRETT,

PHILADELPHIA.

The "Farmers' and Planters' Illustrated Almanac," published annually, containing many valuable recipes and engravings, which cost over five hundred dollars, with certificates from many of our prominent farmers and planters, and directions how to use the Phosphate, will be mailed gratuitously, upon application to

B. R. CROASDALE & CO., Publishers, 104 N. Delaware Avenue.

BAUGH'S

RAW BONE SUPER-PHOSPHATE,

BAUGH & SONS, Proprietors.

OFFICE, No. 20 SOUTH DELAWARE AVENUE, PHILADELPHIA.

The Great Substitute for Peruvian Guano!

CHEAP, ACTIVE AND PERMANENT.

Buy Furniture at Gould & Co.'s.

THE LARGEST, CHEAPEST AND BEST

STOCK OF FURNITURE

OF EVERY DESCRIPTION IN THE WORLD.

COR. 9TH & MARKET STS PHILADELPHIA. 37 & 39 N. SECOND ST.

SEND FOR PRINTED CATALOGUE AND PRICE LIST.

UNION DEPOTS:

CORNER NINTH & MARKET STREETS,

And 37 & 39 NORTH SECOND ST., Philadelphia.

XI.

Furniture.

Within a few years the manufacture of Furniture, or Cabinet-Making has greatly progressed in Philadelphia, both in point of taste and extent of production. In 1840, there were but few Furniture stores in the City, and they mostly small ones, keeping samples of the styles of goods, but relying mainly on orders from their customers to supply work for their employees. A Spring-seat Sofa was then a luxury—almost a novelty. The art of Veneering was just beginning to be understood. Previous to this period a crotch of Mahogany wood, (which was then mostly used for furniture,) was cut into Veneers by a narrow blade saw, drawn laterally by two men. They could not get more than four Veneers out of an inch thickness. This was a great waste of the finest class of material, and the Veneers could only be applied to flat work or very slight curves. About this time Circular Saws, some of which were seven to eight feet diameter, were introduced, and gradually improvements were made, so that at the present time, it is not uncommon to produce sixteen Veneers to the inch. Mahogany, Rosewood, Walnut, and all the finer woods, are now used in Veneering with such skill, that elliptic ogees, or oval surfaces of common wood, are covered with a thin coating of fine wood, thus reducing the consumption, comparatively, of the finer woods. In the course of time, Mahogany became scarce ; and growing in mountain fastnesses, it was procured only at a great expense. Rosewood has always been equally difficult to obtain. To supply the deficiency, the merits of American Walnut were examined, and on trial it was found equally suitable for fine Furniture. The grain of the wood, and the feathery character of the curl, (where two branches separate from the trunk,) are similar to Mahogany, except in color; the Walnut being of dark purple shade, though varying in color, according to the latitude and nature of the soil. Walnut is now used more than all other woods combined. The supply on the rich bottom lands of Indiana, and the Western States generally, is enormous, and the quality so superior that some is shipped to Europe.

All varieties of these woods—Mahogany, Rosewood, Walnut, and others, are used by the Cabinet-makers of Philadelphia. There are now nearly one hundred employers in the business, and at least ten large warehouses,* where the most fastidious tastes may be satisfied from

* The establishments of ALLEN & BROTHER, M. DEGINTHER, and GEORGE J. HENKELS, on Chestnut street, and GOULD & Co., on Market street, might be

goods already made. Philadelphia has a well merited reputation for the production of fine Furniture ; the carved work is really superb ; and the less elaborate, known as Cottage Furniture, is distinguished for excellent workmanship, high polish, tasteful painting, and moderate price. An oak Sideboard, carved by a Philadelphia sculptor, was regarded by the visitors to the New York Crystal Palace, as one of the most remarkable specimens of skill in the exhibition. The Southern demand, which is proverbially fastidious and luxurious in the choice of Furniture, is almost entirely supplied from this city. With the increasing demand for fine Furniture, there has been a corresponding improvement of taste in design ; and it may be well doubted whether France can, at this time, exhibit more magnificent displays than can be seen in the Cabinet Warehouses of Philadelphia.

It is estimated that the aggregate value of all the Furniture and Upholstery made in Philadelphia annually, approximates *three millions of dollars.*

XII.

Glass Manufactures.

Intelligent foreigners have repeatedly complimented the manufacturers of Glass in the United States—not only for excellence in the production of useful articles, to which they have hitherto given their attention principally, but also for various successful attempts that have been made in producing those rich and decorative works which belong to luxury rather than to utility. The imitations of Bohemian Glass and Opal Glass, made in several establishments throughout the Union, are considered better than a great portion of those produced in Europe. In Philadelphia, the Glass manufacture, though surpassed by many others in amount of production, is nevertheless sufficiently extensive to be called a leading pursuit. The locality, by reason of the facilities for procuring the raw materials, is one of the best in the Union. The finest qualities of sand are obtained from the adjacent State of New Jersey, and the alkali are supplied by the Chemical factories in the city.

There are, at least, fifteen manufacturers of Glass, whose headquar-

cited as representatives of the great Furniture warehouses of Philadelphia The business of manufacturing has also been greatly subdivided within a few years. For instance, Reeves & Eastburn have made the manufacture of Bedsteads a specialty, and Messrs. Uhlinger & Co., Patent School and Counting House Furniture.

A. R. SAMUEL,

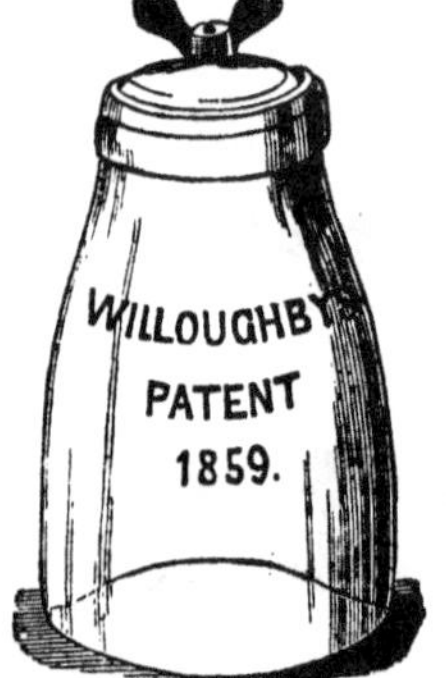

"WILLOUGHBY,"

GLASS

MANUFACTURER

AND

PROPRIETOR

OF THE

"HALLER,"

"KLINE,"

"MASON,"

FRUIT

SOUTHEAST CORNER

HOWARD & OXFORD

STREETS,

PHILADELPHIA.

"FRANKLIN."

JARS.

SOUTHEAST CORNER

HOWARD & OXFORD

STREETS,

PHILADELPHIA.

WILLIAMSTOWN GLASS WORKS.

BODINE, THOMAS & CO.

MANUFACTURERS AND DEALERS IN

Green and White Glassware,

Druggists' Vials & Bottles, Pickle & Preserve Jars, Wine, Porter and Mineral Water Bottles.

Williamstown, N. J., & 807 Market Street,

PHILADELPHIA.

J. F. BODINE,
C. E. THOMAS,

PATENT AIR-TIGHT FRUIT JARS.

W. H. BODINE.
W. R. THOMAS.

BOORSE, BURROWS & CO.

Dealers in and Manufacturers of

WINDOW GLASS,

TUMBLERS,

AND A GENERAL ASSORTMENT OF

GLASS WARE,

CASTORS,

COAL OIL CHANDELIERS, Etc.

Also, having the exclusive right of manufacturing the ***Excelsior and Heroine Fruit Jars,*** (conceded by all fruit men to be the best Self-sealing Fruit Jars in the market.)

No. 153 North Third Street,

PHILADELPHIA.

ters are in this city, though the factories of some are located in New Jersey, and outside the city limits, viz.:

H. B. & G. W. BENNERS; F. & J. BODINE; BODINE, THOMAS & CO.; BURGIN & SONS; BROWN & NICHOLSON; EVANS, SHARP & WESTCOATT; HENRY E. FOX; GILLENDER & BENNETT; HAY & CO.; THOMAS HOUGHTON; MOORE, BROTHERS; ADAM R. SAMUEL; SHEETZ & DUFFY; WHITALL, TATUM & CO., and WHITNEY & BROTHERS.

The leading business is the manufacture of Green and Crown Glass Bottles, including all kinds of Druggists' Vials, Jars, Demijohns, Carboys, etc. This kind of Glass is made of ordinary materials—generally sand with lime, and sometimes clay, and alkaline ashes of any kind; but great care and considerable experience are required, particularly in making bottles that are to contain effervescing fluids. The materials must be carefully and thoroughly fused and the thickness uniform throughout, to resist the pressure of the contained carbonic acid.

Window Glass is made in several establishments; and in addition to the various sizes and qualities, most, if not all, in this business, make double-thick and cylinder Plate Glass, suitable for coaches, pictures, and extra large windows; some of which is quite equal in quality to the English and French Cylinder Plate Glass. At some establishments, white and colored, plain and figured Enameled Glass is made.

Of the Glass manufactories located within the limits of the city, we believe the largest in Philadelphia, and, it is said, in Pennsylvania, is that of GILLENDER & BENNETT, at the corner of Howard and Oxford streets. The works cover nearly an acre of ground, and contain every thing essential for the successful prosecution of the manufacture of Flint Glass, including all descriptions of Plain, Moulded, and Cut Glass Table ware. This firm have an invested capital of $100,000, employ two hundred hands, and produce Glass to the amount of a quarter of million of dollars. Messrs. Gillender & Bennett are, it is said, the only firm in the United States that have a patent for annealing the chimneys of coal oil lamps.

Opposite to their works, is the extensive FRUIT JAR manufactory of A. R. SAMUEL, who is the only manufacturer in the country who makes this branch a specialty. He has the control of five of the most popular patents, and a furnace that is capable of turning out from seven to ten thousand gross of jars per annum. The growth of this branch of the Glass manufacture is most remarkable. In 1857, Mr. Samuel had but two competitors in the whole country, and even so late as 1859, it is estimated there were not more than fifteen hundred gross made annually. Now, there are probably fifty Jar manufacturers in the United

States, who produce an aggregate product of thirty thousand gross. Every year thousands of families are adopting the system of putting up fruits in their natural state, in preference to preserving them by the old methods.

Besides the manufacturers of Glass, above enumerated, there are several whose attention is devoted to supplying orders for special kinds of Glass, such as Tubes for Philosophical apparatus, Syringes, etc., for druggists. There is also one important and extensive manufactory of STAINED GLASS. The origin of this art is lost in the dimness of antiquity. The process employed in modern times is described as follows: After the figure to be put upon the plate is drawn upon paper, and painted as desired, it is transferred to the glass, which has been prepared to receive it. This has to be done with artistic skill, equal to that employed upon an oil painting, and requires much more care in its execution. In transferring fruits and flower pieces, all the delicate tints of the objects must be copied with the greatest nicety. The glass is then put into a kiln, and submitted to a heat almost sufficient to fuse it, which not only has the effect to add greatly to the beauty of the painting, but makes it a part of the glass itself, no power being able to remove it.

Messrs. J. & G. H. GIBSON, of 125 South Eleventh street, the firm referred to, made the Glass ceilings of the House of Representatives and Senate chamber at Washington, composed of plates having the appearance of enameled work, the coats of arms of the United States being done in rich colors, which give the effect of Mosaics set in silver. The Stained Glass made by them is considered quite equal to that of European manufacture.

XIII.

Hardware and Tools.

The term *Hardware* is one of those indefinite, comprehensive nouns of multitude, of which it may be said that it almost includes, as its name imports, every ware that is hard. Popularly it is understood to embrace all the unclassified manufactures of Iron and Steel, including all the appendages of the mechanic arts, from a file to a mill-saw; many of the details of common life, from a rat-trap to a coach-spring, articles as various in appearance, size, and uses as can well be conceived—in fact, whatever is sold by Hardware dealers. There are comparatively few large miscellaneous Hardware Manufactories in Philadelphia, like those that are common in Connecticut; but the want is supplied by a great number of excellent Hardware Stores, in whose stocks may be found every article made in the Eastern manufactories. Several of the New England companies, as for instance, the RUSSELL & ERWIN MANUFACTURING COMPANY, who are among the most extensive Hardware manufacturers in this country, have special agencies and warehouses in the city, where their products are sold at factory prices; and, besides these, there are over thirty houses engaged extensively in importing and jobbing Hardware.

The most extensive manufactory of miscellaneous Hardware, located within the city limits, is that of E. HALL OGDEN, whose warehouse is at 307 Arch street. His manufactory, located at Ninth and Jefferson streets, consists of a main building fifty by one hundred feet, with two wings, one forty by one hundred feet and the other fifty-six by one hundred, filled with all requisite machinery driven by an engine of fifty horse power. A moulding room eighty-three by one hundred feet has recently been erected, and other additions are contemplated. Mr. Ogden employs one hundred and fifty hands, and manufactures, as appears by his catalogue, one thousand two hundred and fifty distinct articles, including malleable and fine gray Iron Castings.

But the manufacture of Hardware, as carried on in Philadelphia, can only be illustrated by reference to some of its leading branches. Each of the prominent articles included in the term, such as Saws, Forks, Shovels, Files, Locks, Edge Tools, Bolts, etc., is made in separate establishments, specially devoted to the purpose.

1.—SAWS.

Every country merchant as well as every worker in wood is presumed to be familiar with the excellence of the Saws made in Philadelphia. The oldest Saw Works in the city, and, probably, the oldest now in business in the country, are those of WILLIAM ROWLAND & Co., 948 Beach street. They were established by W. & H. Rowland, in 1806, and have supplied a large proportion of the large sized Saws in use, Mill and Crosscut, varying in length from six to eighteen feet. In 1845 this firm commenced the manufacture of Steel, and now produce nearly two thousand tons annually, a large part of which they work up into Saws and Springs. Messrs. Rowland employ about one hundred hands, and use only Swedish Iron for their Spring, Tire, Toe and Blister Steel.

JOHN H. GUNNIS, 404 Cherry street, is a very old established manufacturer of Saws in Philadelphia, and his manufactory is four stories in height, the first floor being appropriated to the iron or blacksmith work, the second to the manufacture of Saw-handles, while on the third floor the hardening and tempering of Saws is executed, and the fourth is the depository for the finished article.

Mr. Gunnis produces all the standard varieties and sizes of Circular, Mill, and Cross-cut Saws, and uses only the best imported steel. His Saws are extensively used in Philadelphia, where they are best known, and are highly spoken of by all.

Near the above, at 407 Cherry street, is the Saw Manufactory of William McNiece. The building has a front of eighteen feet and a depth of fifty feet, three stories high. In this manufactory the order of operations as usually pursued in similar establishments, is reversed, the tempering being done on the first floor, the finishing on the second, while the smith work is executed on the third floor.

Mr. McNiece is the successor of Lame & McNiece, and for eleven years was foreman in the extensive works of Walter Cresson, at Conshohocken. He is a practical workman of experience and skill, and employs none but those possessing similar qualifications. His manufactures include every variety of hand, back, butcher, and wood Saws; and those for carpenter's use have a high reputation. It is said that his customers include nearly two-thirds of the retail City trade.

But the largest Saw Manufactory in Philadelphia, and one of the largest in the world, is that of Henry Disston. He occupies seven buildings with an aggregate floor surface of five hundred thousand square feet, and employs over four hundred workmen. He has recently erected furnaces for making Cast Steel and an extensive Rolling Mill, rendering his establishment complete and independent within itself. He manufactures not only every kind of Saws, but all tools for sawyer's use, including files. Among the novelties he has recently introduced are Circular Saws with *inserted teeth,* so that if one or more teeth are broken others can be readily substituted in their place.

Mr. Disston is a man of remarkable fertility of invention, and his establishment deserves a more extended description. (See Appendix.)

2.—FORKS.

There are four manufactories of Forks in the city, and one of them it is believed is the largest in the Union. Previous to 1851, the Eastern manufacturers almost monopolized the Philadelphia market; but at the present time the city Forks are so generally preferred, that special inducements alone can effect a sale of Eastern or New York made goods Forks are now exported from Philadelphia to England.

PHILADELPHIA FILE WORKS,

1514 Spring Garden Street.

JOHN C. HESS,

PROPRIETOR.

FILES AND RASPS,

MANUFACTURED FROM THE

BEST BRANDS OF IMPORTED STEEL,

AND WARRANTED

Equal to any English File in the Market.

Re-cutting done for Railroad Companies, Machine Shops, etc.

Black Diamond File Works.

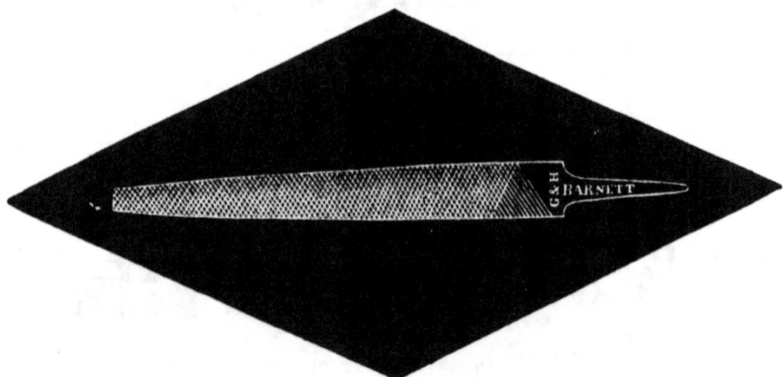

We would respectfully call the attention of Dealers to the fact that we are manufacturing *FILES* and *RASPS* from superior Cast Steel, *warranted equal to any imported.*

Orders respectfully solicited and promptly attended to.

G. & H. BARNETT,

41 & 43 RICHMOND STREET, Philadelphia.

3.—SHOVELS AND SPADES

Are made by several manufacturers, who have a capital invested exceeding a quarter of a million of dollars, and produce annually about eighty-five thousand dozen of shovels. The machinery employed in these establishments is of the most perfect description, and the manufactured product is equal in quality to the best made in the country, and far better than the shovels imported. The raw material is mainly American iron and steel, though a share of the English metal is used.

4.—FILES AND RASPS.

The manufacture of Files of various kinds in this city is becoming of some importance, and as the prejudice against American Files is fast being obliterated, it will soon become one of the leading branches of manufactures. These articles are now manufactured in Philadelphia, of a quality to vie with the best Sheffield article, and it is hoped that the mechanics and manufacturers who use this important tool, will see the importance of lending their support to those engaged in producing them, in order that the trade may be fostered, and firmly established in our midst. There are about fifteen different manufactories in Philadelphia, three of whom are largely engaged in the business, and have a large capital invested in it. Among the many difficulties with which they have had to contend, one is the large quantities in the market of Eastern Files made by machinery, and which, as a general rule, prove inefficient if not entirely worthless.*

* "The Black Diamond File Works," of Messrs. G. & H. Barnett, 41 & 43 Richmond street, are among the largest in the city. They occupy a brick building forty by one hundred feet, and employ generally thirty workmen. The Files and Rasps made by this firm are purchased principally by the dealers and machinists in Philadelphia, by whom they are highly prized; but large quantities are also shipped to the Western and Southern States.

Messrs. Barnett claim to have a new and peculiar process of tempering Files, known only to themselves, and this may be one cause of the popularity of their products.

"The Philadelphia File Works," No. 1514 Spring Garden street, of which Mr. John C. Hess is proprietor, is another of the largest establishments of the kind in Philadelphia. The buildings, which are of brick, were erected expressly for his own use, and are of the most substantial character. The *Files* produced here are made entirely from English Steel (experience having taught that the English is the only Steel from which a reliable File can be made), and Mr. Hess is now turning out Files which are fully equal to the best English brands, and at more reasonable rates. The establishment employs a large

5.—CARRIAGE BOLTS, ETC.

The production of Carriage Bolts in Philadelphia is so enormous that it would seem the establishments could supply the world. Among the pioneers in the manufacture of these bolts by machinery were Thomas and William Shields, whose manufactories are now the largest in the city. They commenced the business in West Philadelphia in 1852, producing at first about 10,000 bolts per week, but increasing largely until some five years since they dissolved the copartnership, and erected separate establishments.

The "Eagle Works," of which THOMAS SHIELDS is proprietor, are located on Pennsylvania Avenue west of Twenty-second street. The lot has a front on the Avenue of 165 feet, and a depth of 175 feet, and is in great part covered with buildings, the lower floors being used for forging, and the upper for finishing bolts. The expedition with which these articles are made is wonderful, this one establishment having a capacity of turning out 400,000 bolts of average sizes per week. Mr. Shields uses the Oliver Hammer exclusively for rounding the bar and shaping the head. The screws are cut by means of machines, each of which, operated by a girl, can finish 5,000 per day. Mr. Thomas Shields frequently employs as many as 150 hands.

The works of WILLIAM SHIELDS are located on Twenty-third street above Race street. His establishment has a front of 75 feet and a depth of 235 feet. Mr. William Shields uses both the Oliver Hammers for shaping bolts, and also a novel machine of extraordinary capacity, of which he is the proprietor of the patent. This machine is capable of producing bolts of any length or diameter at the rate of six hundred per hour, and if driven to its full capacity, will turn out eighteen in a minute, or over a thousand in an hour. His products are distributed to all parts of the United States, and are appreciated wherever known.

Both of these gentlemen are skilful and enterprising mechanics, who are entitled to much credit for having been instrumental in giving to Philadelphia the highest reputation in this manufacture. In New England, Carriage Bolts are made largely, but the quality is inferior, the manufacturers being unable apparently to prevent the crystallization of

force of skilled workmen, some of English and some of American birth. These Files are sold to all parts of the United States, and so well are their good qualities appreciated, that the demand is more than equal to the ability to supply. The manufacture of the larger sizes of Files and Rasps, such as are used by machinists and workers in iron and steel, is the prominent part of the business in this establishment.

GIRARD CARRIAGE BOLT WORKS.

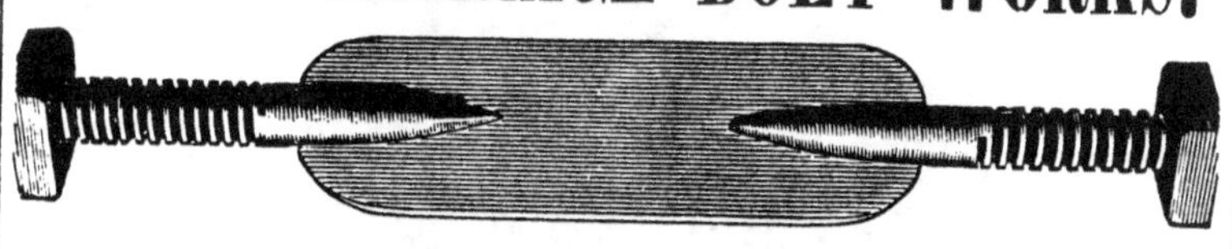

ESTABLISHED

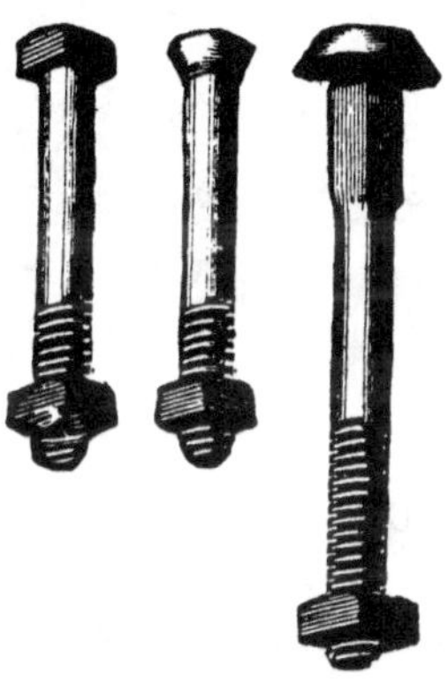

1852.

WILLIAM SHIELDS,

MANUFACTURER OF

BEST PHILADELPHIA

Carriage and Tire Bolts,

AXLE CLIPS, TURNED COLLARS, ETC., ETC.,

From best Norway Iron, and warranted.

ALSO,

Manufacturer of Agricultural and Machine Bolts of every description.

FACTORY AND WAREROOMS:

TWENTY-THIRD STREET, above Race, PHILADELPHIA.

Also, Bridge Bolts made by Patent Machinery, which is offered for sale in shop rights.

the iron, which renders the bolts brittle, and liable to snap off when used. The Philadelphia manufacturers have entirely overcome this difficulty, and the Norway Iron Carriage Bolts, made here, are the best in the world.

F. E. Townsend has recently erected a very complete establishment for the manufacture of Carriage and Tire Bolts of all kinds. Though not one of the most extensive it is one of the most complete of its kind in machinery and its appointments, in the United States. A considerable force of hands is now employed, and the products are in great demand by the first-class Coach Builders.

6.—BOLTS, NUTS AND WASHERS.

These articles have been made until within a few years by hand, but the consumption increasing so considerably in consequence of the multiplicity of Railroads, Mines, etc., required that they should be produced in a more expeditious manner, and at cheaper rates. Hoopes & Townsend, of this city were among the first to enter into this as a separate business, and after several years' experience are now able to supply the wants of Railroad Companies, Machinists, Car Builders, Bridge Builders, and others, with articles of this description, peculiarly adapted to their several purposes, as to strength, durability, finish, etc.

They claim a great advantage for their product over hand work on account of its uniformity, and also superiority over much of the Eastern work, both because the material used is of superior quality, and because it is made expressly for the purpose for which it is employed. Within the last few years they have more than doubled their capacity, and their establishment, fronting two hundred and fifty feet on Buttonwood street, extends in the form of a hollow square, back to Hamilton street. In its various departments there are numerous machines of a superior and costly description, that have a capacity for producing ten tons of Bolts, Nuts, etc., daily. Messrs. Hoopes & Townsend employ about one hundred and fifty hands.

7.—LOCKS.

It is claimed and, we believe conceded, that the ingenuity and enterprise of American mechanics have placed this department in advance of the efforts of all other nations. In the finer and more expensive class of Locks, it has been abundantly proved that the American production is superior to any other; while in Locks, Latches, etc., adapted to the wants of the builder or to commerce, the American mechanic supplies the home demand; and, to some extent, orders from abroad.

In Philadelphia there are a large number of very ingenious Locksmiths, who provide every variety of fastening essential for the security of dwellings and stores. The business is usually carried on in connection with Bell-hanging and Silver-plating; but the demand for special and patented Locks furnishes occupation to several special Lock-making establishments.*

Recently, Mr. FRANCIS JAHN, 506 Race street, who has been for several years a manufacturer of fine Swords, has commenced making French Bronzed Door Knobs, Escutcheons, and Bell Pulls. He uses as a base the best brass, which he electro-plates and bronzes in a very elegant manner. The advantage of these knobs is that they look well for years without cleaning, and cannot easily be indented or injured. They are being extensively introduced in the fine dwellings that are now being erected in this and other cities.

8.—EDGE TOOLS AND CUTLERY.

For the manufacture of *Edge* and *Hand Tools* there are two principal establishments in the city, and four others in its vicinity whose products are sold here exclusively. They employ in all about two hundred and fifty men. The Edge Tools made in the city are of good material, well finished, and in all respects fully equal to any made elsewhere.

Recently an important addition was made to the Edge Tool manufactories by the establishment of one adapted especially to the manufacture of such Tools of very extra quality. Formerly, even the shipwrights and planing machine manufacturers of Philadelphia were compelled to send to New England for their knives; but since Messrs. JENKINS & TONGUE organized their manufactory at 33 & 35 Richmond street this is no longer necessary. To develope the full capacity of steel in cutting

* A representative house of the Lock manufacturers of Philadelphia is that of J. B. SHANNON, 1009 Market street. Mr. Shannon, like most of the others, is a Bell-hanger and Silver-plater, and has also an extensive Hardware store; but his specialty is making Locks for first-class dwellings and public institutions. He has constructed Locks for a large number of Insane Asylums, where they are wanted in sets, to differ from each other, and yet to be passed by a master key by the superintendents, watchmen, and others. Similar Locks are also applied to dwellings; and one member of the family may have a key which will open all the chambers and closets, and yet the respective keys of the doors will not open any except the one to which it belongs.

Bells are so arranged in dwellings, by Mr. Shannon, that a person taken sick may communicate the need of help to any other part of the house. Invalids can thus dispense with constant attendance.

instruments, requires not only an intimate acquaintance with the metal and its manipulations, derived from practical experience, but a peculiar original faculty which very few men possess. This experience and genius one of the firm mentioned possesses in a remarkable degree, and whoever requires a tool with an extra fine cutting edge can rely upon obtaining it at their establishment. The paper-box makers bear testimony to the excellence of their shears; the dye-wood chippers, book-binders, cork-cutters (a most difficult class to suit), and those who run planing machines, ship-carpenters, shingle-shavers, etc., etc., all speak in the highest terms of their knives and other tools. Messrs. JENKINS & TONGUE also manufacture first class Hatchets.

Cutlery is made by several firms, and within a few years this branch of manufactures has been greatly increased. There are now, at least, three large manufactories of Table Cutlery in the city, and a considerable quantity is also made by the Surgical Instrument manufacturers.

Besides the articles and branches of the Hardware manufacture I have thus specially alluded to, there is an immense number of articles made in this city, and which come within the category of Hardware stock. *Wrought Nails* are made largely; and *Cut Nails* are produced at the Quaker City Nail Works, and at the Cumberland Nail Works, Bridgeton, N. J. The latter Company has about one hundred and twenty nail-machines in operation, capable of producing over one hundred and twenty-five thousand kegs of Nails per annum. The supplies of this establishment are furnished from this city, and the products disposed of in it. *Horse Nails*, chiefly for local consumption, but of superior quaility, are made by four or five persons.

Plasterers' and Bricklayers' *Trowels* are made in several establishments in this city. Messrs. ROSE, BROTHERS, at Thirty-sixth and Green streets, West Philadelphia, are the oldest and largest in the city, who make this a specialty in connection with Saddler's Round Knives. Their father commenced the business about fifty years ago, and secured a reputation so that Rose's Trowels are standard articles in all Hardware stores. The present firm occupy a building forty-eight by fifty feet, and employ thirteen hands.

The first Screw Augur made in this country, was made in West Philadelphia, and the manufacture is now carried on by the successors of the originator.

Butchers' Tools are made a specialty by AUGUST NITTINGER, Jr., 828 & 830 North Fourth street. He supplies all kinds of Butchers' Knives and Steels, and manufactures Chopping Machines, Sausage Stuffers, Lard Presses, and a great variety of similar articles.

Henry Tolman, 525 Commerce street, has made a specialty of manufacturing Pointed *Belt Hooks* and Spring *Gauge Punches*. These Belt Hooks have been extensively used in large manufactories throughout the country for the last ten years, and are rapidly superseding the use of leather lacing. When the Spring Gauge Punch is used, which puts the hooks upon a line and enables each to perform its own work, the Belt can be fastened in less than one-fourth the time usually required, thus saving time as well as the expense of lacing leather. The cost of hooks suitable for fastening a six-inch belt is only three cents.

Hammers of all kinds, from those used in watch-making to sledges and trip-hammers, are made, comparing favorably in prices with either foreign or domestic manufactures, and superior in quality and finish. *Cast Iron Butt Hinges* are made; and the Franklin Institute has pronounced the pivot-butts of one firm equal in finish to any American, and the principle better. A manufactory of *Strap* and *Reveal Hinges*, of every size and variety, has been in operation here for several years. *Brass Hinges*, *Catches*, and other Brass Building Hardware are made by William H. Dodemead, 512 Cherry street. He has also recently engaged in the manufacture of Skates. *Piano Forte Hardware*, including gimlet-pointed screws, is made very extensively by one firm, established in 1822, and said to be the oldest in the business in the United States. A great variety of *Wrought Iron Building Hardware* is made at the House of Refuge by David Brewer & Co., whose store is at 502 Commerce street.

Of small tools the variety is infinite. *Shoemakers' Tools* are made by five or more establishments, who make their tools of steel, thus securing a large sale in the Eastern markets, where cast iron is principally employed. *Steel Stamps*, *Brands*, and *Punches*, Stone-cutter's Tools and Mill Picks, Curriers' and Tanners' Knives and Tools of all kinds, Saddlers' Tools, Binders' Tools, Coopers' Tools of excellent quality, Ice Tools, and Umbrella-makers' Tools and Furniture, each employ one or more, and some of them several establishments *Skates* are made of improved construction, being fastened to the foot by springs without straps. *Sad Irons* are a leading article at a factory in West Philadelphia.

In the manufacture of the miscellaneous articles which are included in the term Hardware, it will be perceived, from what we have stated, that Philadelphia has numerous establishments, and the aggregate product is very considerable.

RUSSELL & ERWIN MANUFACTURING CO.,

RUSSELL & ERWIN MANUFACTURING CO.,

MANUFACTURERS OF

BUILDERS' HARDWARE,

RUSSELL & ERWIN MANUFACTURING CO.,

MANUFACTURERS OF

"HALE'S" PATENT MEAT CUTTERS

MANUFACTURERS OF

HOUSE-KEEPING HARDWARE,

NO. 5 NORTH FIFTH STREET,

PHILADELPHIA, PA.,

With our own make of Goods we keep a full stock of the following articles:

Scovill Manufacturing Co.'s Brass Butts; Russell, Burdsall & Ward's Carriage Bolts; Wallace & Son's Brass and Copper Wire and Belt Rivets; American Horse Nail Co.'s Nails; Putnam's Horse Nails; N. London Horse Nails; Bartholomew's, Barber's & Spofford's Braces; Augers and Auger Bitts; Safety Fuse; New England Butt Co.'s Butts; Hand, House and Gong Bells; Crossman's Chisels & Drawing Knives; American Fish Hooks and Lines; Maydole's and Warner & Noble's Hammers; Lincoln's & Stebbin's Molasses Gates; Stair Rods; Britannia Tea and Table Spoons; Coe's Wrenches; Tacks; Brads; Clout and Finishing Nails; American Screw Co's. Screws; Strap and T Hinges; Steel and Iron Squares, Slates, etc.

XIV.

Iron and its Manufactures.

It is probable that in no branch of the general manufactures of Philadelphia, is her superiority so widely known and generally conceded as in the fabrication of Metals. The abundance of Iron produced in the vicinity of the city, and its consequent cheapness, have naturally concentrated attention upon its manufactures, as well as extended its uses; while the fame of our Engineers and Machinists attracts from abroad a large and constantly increasing patronage.

In our introductory remarks we gave some statistics of the Iron production of Pennsylvania, and showed that Philadelphia is situated in the district which is entitled to be called the centre of the Iron production in the United States. Recently a very important discovery has been made of a peculiar ore, found in Schuylkill county, which promises greatly to increase the manufacture of Iron in Eastern Pennsylvania. We allude to the "Black Band Ore," so called from its similarity to the Scotch Black Band, from which the Scotch Pig Iron is smelted, and which has been found in large beds in the vicinity of Pottsville. About three thousand tons have been mined already, though the ore has thus far been smelted only with other ores in furnaces in Schuylkill and Montgomery counties.

Within the limits of Philadelphia, we believe, there are neither ore beds nor opened mines; but just beyond the city limits, in Montgomery county, ore is dug in considerable quantities, and near Phœnixville, Chester county, there is an extensive Iron mine, which is supposed to be the oldest in the United States. It was opened a few years before the Revolution, and is yet worked with much success. It is one hundred and fifty feet deep, and has been mined over sixteen acres of surface. The great Rail Mills of the Phœnix Iron Company, successors to Reeves, Buck & Co., obtain a considerable portion of the ore used by them from this mine, known as the Warick Mine.

Within the limits of Philadelphia, however, there are eleven Rolling Mills that employ over twelve hundred men, who receive annually in wages about $1,000,000. They are as follows:

The Kensington Iron Works and Rolling Mill, James Rowland & Co., proprietors.

Penn Rolling Mill, Kensington, Verree & Mitchell, proprietors.

Treaty Rolling Mill, Kensington, Marshall, Phillips & Co.

Philadelphia Rolling Mill, Kensington, Steven Robbins, proprietor.

Oxford Rolling Mill, Twenty-third Ward, W. & H. Rowland.

Fairmount Rolling Mill, Fairmount, Charles Wheeler, proprietor.

Fountain Green Rolling Mill, two miles above Fairmount, Oliver W. Barnes, proprietor.

Pencoyd Rolling Mill, below Manayunk, A. & P. Roberts, proprietors.

Gray's Rolling Mill, Manayunk, A. P. Buckley & Son, proprietors.

Cheltenham Rolling Mill, Rowland & Hunt, proprietors.

Philadelphia Spike Works, C. Winch, proprietor.

The following statement exhibits the product of the Rolling Mills in Philadelphia since 1854 inclusive, as furnished us by the American Iron and Steel Association:

Year	Tons of 2,000 lbs.	Year	Tons of 2,000 lbs.
1854	5,736	1861	15,280
1855	6,056	1862	20,904
1856	10,845	1863	27,540
1857	10,413	1864	32,006
1858	12,250	1865	26,033
1859	12,904	1866	30,000*
1860	13,454		

Probably the most extensive Rolling Mills and Iron Works in the vicinity of Philadelphia are those of the Phœnix Iron Company,* at Phœnixville, Samuel J. Reeves, Vice-President and Treasurer; of Seyfert, McManus & Co., at Reading; the Pottstown Iron Company at Pottstown, Pennsylvania, and the McCullough Iron Company.

The "Phœnix Iron Company" have several Blast Furnaces, a Rail Mill two hundred and sixty by one hundred and sixty feet, a Puddling and Reheating Mill one hundred and eighty-five by one hundred and ninety-two feet with a wing one hundred and thirty by thirty-two feet, and a Machine Shop two hundred and six by ninety feet, specially appropriated to building Iron Rafters, Flooring and Bridge work, which is one of the great specialties of this company. The offices, pattern and drawing rooms are probably the most complete of any in the country, and cost, we are informed, $26,000. The Mills contain twenty-four Heating, and twenty-two Double and twelve Single Puddling Furnaces. The office in Philadelphia is at 410 Walnut street.

The "Reading Iron Works," of which Seyfert, McManus & Co., are proprietors, include an Anthracite Furnace which is one of the largest in the country; a Foundry and Machine shop in which Cannon, weighing *forty tons*, have been cast and finished; a Steam Forge in which Marine Shafts weighing, when finished, thirty-two tons, have been forged; Tube Works for manufacturing Lap-welded Boiler Flues, and Gas and

* In 1866 these Mills consumed about 36,000 tons of Pig Iron and 54,000 tons of Coal. They have a capacity of producing 50,000 tons of finished work, annually.

STEAM FORGE.
SEYFERT, M^c MANUS & Co.
FURNACE
SHEET MILL
NAIL WORKS
ROLLING MILL
READING
IRON WORKS.
OFFICE.
SCOTT FOUNDRY
TUBE WORKS
VAN INGEN-SNYDER

Steam Fittings, with Forges and Rolling Mills for producing Iron specially adapted to this purpose; a Rolling Mill and Nail Works; a Sheet Mill, containing the celebrated Lauth Patent system of Rolls, which enables them to imitate successfully, if not to equal, Russia Sheet Iron. It would be impossible within the limits of even a page to do justice to Works of their magnitude. The office in Philadelphia, is at 28 South Seventh street.

The Works of the "Pottstown Iron Company," of which Messrs. Morris, Wheeler & Co., are agents, consist of a Blast Furnace, two Rolling Mills, and a Nail Factory, in addition to which are a Cooper Shop and a Drying House for Nail kegs, and a Smith and Machine Shop for doing such work as the Mills and Furnace constantly require.

The Works turn out from eight to ten thousand tons of Pig Iron, six thousand tons of Boiler, Tank and Boat Plates, and one hundred thousand kegs of Nails and Spikes, per annum. About two hundred and fifty men are employed. The Company own their own ore mines, and thereby controlling their product from the raw material to the finished state, are enabled to secure greater uniformity in the quality of their manufactures than would otherwise be the case.

Mr. Theodore H. Morris of Philadelphia, is President, and Messrs. Edward Bailey and William L. Bailey, of Pottstown, are Treasurer and Secretary of the Company.

We believe all the products of these Works are sold by the firm of Morris, Wheeler & Co., Sixteenth and Market streets, a house that, for nearly forty years, has been prominently and honorably identified with the Iron Manufacture.

The "McCullough Iron Company" have five Rolling Mills in Cecil County, Maryland, including one large Steam Mill employed in manufacturing Iron, which is galvanized at their Works in Philadelphia, Eleventh and Washington avenue. This Company have a large tract of land adjacent to their Works at North East, Maryland, with ore beds and coal mines, and execute all the various processes necessary to convert ore into the finest quality of black and galvanized Sheet Iron of every description.

Passing to the consideration of some of the forms into which Iron is converted, we come to a Branch of Manufactures in which Philadelphia is pre-eminent over all other Cities, viz.: that of Machinery. There are within the limits of the Consolidated City not less than one hundred Machine Shops, that have, in combination, facilities for constructing any machine that the genius of man has invented or can invent. Foremost among the shops that have tools and equipments for constructing Engines and Machines, as well as Castings of extraordinary dimensions, are the great Works of I. P. MORRIS, TOWNE & Co., and MERRICK & SONS. A brief enumeration of some of the machines that have been made at these works will establish their ability beyond question.

At the "Port Richmond Iron Works," of which I. P. MORRIS, TOWNE & Co., are proprietors, were constructed the large engines of the United States Mint; the engine for the Lake Erie Steamer, Mississippi, having a cylinder eighty-one inches diameter by twelve feet stroke of piston; two Cornish Bull Pumping Engines, for the Buffalo Water Works, each having a steam cylinder fifty inches diameter ten feet stroke; the lever beam Cornish Pumping Engine, steam cylinder sixty inches diameter ten feet stroke, at the Schuylkill Water Works; the Bull Cornish Pumping Engine, cylinder forty inches diameter eight feet stroke, at Camden, New Jersey.

The Blowing machinery for the Lackawanna Iron and Coal Company at Scranton, Pa., amongst the largest ever constructed for making anthracite Iron, the dimensions of which were given on a previous page, was built at these works, as were also the two large Blowing Engines for the Lehigh Crane Iron Company—steam cylinder of one, fifty-eight inches diameter, blast cylinder ninety-three inches diameter, and the other steam cylinder sixty-six inches in diameter, blast cylinder one hundred and eight inches diameter, both ten feet stroke of piston. The two Engines for Thomas' Iron Company, at Hockendauqua, of same size as the largest Crane Engine, were also constructed here. The beams of these engines are supported on cast iron columns thirty feet high, which rest on cast iron beds.

This firm has built a great number of smaller engines for blast furnaces, which nevertheless are ranked as first class machines, as well as others of less capacity.

The Light House of Iron for Ship Shoal, in the Gulf of Mexico, was constructed by this establishment. The height of the structure from water to focal plane was one hundred and seven and a half feet, and to top of spire one hundred and twenty-two feet; it was to be erected in water fifteen feet deep, at a distance of twelve miles from land.

Amongst the vessels belonging to the United States government, for which they furnished Engines, are the celebrated double turreted Iron-clad Monitor, *Monadnock*, which, after efficient service on the Southern coast, during the late war, made a voyage to San Francisco around Cape Horn, thus establishing the fact that vessels of her class could be relied upon for distant sea service. The Engines for the *Agamenticus*,

a sister monitor, as well as for the monitors *Lehigh* and *Sangamon*—gunboats *Itasca* and *Sciota*, double-bowed steamer *Tacony*, and the large sloops *Pushmataha* and *Antietam*, were constructed at these Works. [For a more extended account of the Port Richmond Iron Works, see APPENDIX.]

At the "Southwark Foundry," of which MERRICK & SONS are proprietors, were constructed the great Iron Pile Light-houses illuminating the Florida coast, stationed at Sand Key, Cary's Fort Reef, Coffin's Patches, Rebecca Shoal, Northwest Channel, Dry Tortugas, as also those on Brandywine Shoal (Delaware Bay) and the harbor of Chicago, besides iron lanterns for Cape Hatteras, Cape Florida, etc., and beacons for other points. The first three are among the largest in the world, being respectively one hundred and twenty feet, one hundred and twelve feet, and one hundred and thirty-seven feet high (water to focal plane), and fifty feet square, fifty feet diameter, fifty-six feet diameter at the base respectively, and weighing from two hundred and fifty to three hundred tons each. This firm made the great Gasholder frame for the Philadelphia Works (the largest in the world), being used for a gasometer one hundred and sixty feet in diameter; it weighs about one thousand tons, consisting of twelve gothic pentagonal iron towers, ninety feet high, braced apart by girders thirty-six feet long and eight feet deep, ornamented Gothic, and weighing eighteen tons each in one piece. At this time the firm are engaged in building a new Gasholder in the place of the one just referred to.

Messrs. Merrick & Sons have constructed the machinery for many of the steamers of the United States Navy, among which may be specified the *Mississippi*, paddle-wheel, two side-lever engines of five hundred horse power: *Princeton*, screw, two oscillating-piston engines of three hundred horse power; *San Jacinto*, screw, two geared-engines of four hundred and fifty horse power; *Wabash*, screw, two direct-acting engines of eight hundred horse power. Of these, the former is too generally known to need any comment, and the latter is confessedly the finest of her class in the world. Here also were made the boilers of the United States steamers *Susquehanna* and *Saranac*, eight hundred horse power each; the machinery for the surveying steamer *Corwin;* machinery and hull (iron) for the surveying steamer *Search;* and during the late war, designed and constructed the celebrated iron-clad *New Ironsides*, as also the machinery for several of the war vessels built by the Government.

For pumping purposes, the same firm constructed the great iron Elevating Wheel at Chesapeake City, Maryland, for feeding the canal. This wheel is thirty-eight feet diameter, twelve feet wide, driven by two condensing engines of great power, and elevates two millions of gallons sixteen feet high each hour. More recently, for the Midlothian Coal Mining Company in Virginia, they made a sixty-inch beam Cornish Engine, with one "draw" and three "forcing" lifts, each fourteen inches diameter, ten feet stroke, which pumps one million gallons per day from a coal pit seven hundred and seventy feet deep. They are extensive manufacturers of *Engines and Sugar Mills* for Louisiana and Cuba, and are exclusive makers of the N. Rillieux Patent Sugar Boiling Apparatus, by which white sugar is made directly from the cane juice; in Cuba by the bagasse alone, and in Louisiana by one-half the fuel ordinarily required. They are also sole manufacturers of *Nasmyth's* and *Davy's Steam Hammers*, and have built all sizes, from five hundred pounds to six tons weight of ram, and up to seven feet drop or fall. [For a more extended account of these Works, see APPENDIX.]

At the "Bush Hill Iron and Steel Works," of which Messrs. MATTHEWS & MOORE are proprietors, have been constructed the machinery for some of the largest and most complete Rolling Mills in this country. We may enumerate the Blast Furnace and celebrated Rolling Mill of the Bethlehem Iron Company, the Rolling Mill of the Abbott Iron Company at Baltimore, Maryland; the machinery for producing Steel Forgings at Lewistown, Pennsylvania; a Mill for producing Steel Rails at Harrisburg, and another of similar capacity for the Reading Railroad, in their Works at Reading, Pennsylvania. This firm are also among the largest manufacturers of Steam Boilers in this country. Their Works comprise a Foundry one hundred and thirty-five by seventy-five feet, containing three Air and two Cupola Furnaces and five large Cranes; a Machine shop one hundred and sixty-seven by sixty feet, with a Pattern shop in the second story, containing an immense stock of valuable patterns; a Boiler and Smith's shop ninety-eight by ninety feet; Steel Works one hundred and fifty by forty feet, and numerous auxiliary buildings. In this establishment about three hundred and fifty hands are employed.

The firm of MATTHEWS & MOORE dates its origin from 1846, when they succeeded Rush & Muhlenburg, who commenced business at this location, Sixteenth and Buttonwood streets, in 1816. The celebrated Oliver Evans, one of the most ingenious of Philadelphia mechanics, whose foundry was originally at the corner of Ninth and Vine streets was a partner in this firm. Messrs. William Matthews & James Moore, who comprise the present firm, commenced without the advantage of capital, but by a rare combination of mechanical skill and business talents, have built up one of the most important Machine Works in Philadelphia.

THE BRIDGEWATER IRON WORKS, FRANKFORD.

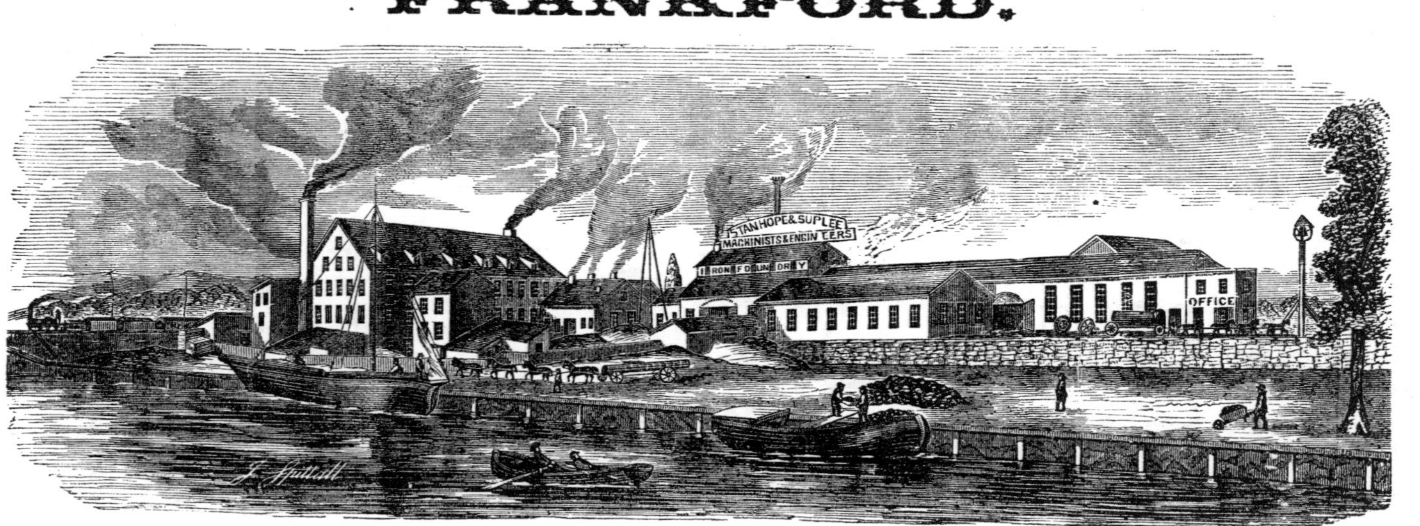

STANHOPE & SUPLEE, PROPRIETORS.

OFFICE—40 NORTH FIFTH STREET.

MORGAN, ORR & Co's. Works, at 1219 Callowhill street, is another establishment that may be referred to as representative of the excellent Machine Shops in which Philadelphia abounds. It has a front of one hundred and thirty-six feet on Callowhill street, and extends back to Noble, a depth of two hundred and sixty feet, and is three stories in height. On the Noble street front is the department for making Steam Boilers, in which this firm do a large business. The Works are equipped with all the tools usually found in first-class establishments of a similar character, and the products comprise all the varieties of work ordinarily made in Machine shops. But in addition to this Messrs. Morgan, Orr & Co., have given a large share of attention to constructing machinery for working the precious metals. They manufactured the Coining Presses of the United States Mint in Philadelphia, and those for the Branch Mint at San Francisco, California. The Peruvian Government obtained from this firm all the machinery for their National Mint, and nearly all the Coining Presses in use in South America and Mexico were made at these Works. Probably no firm in the United States has such a variety of patterns for Mint Machinery as Morgan, Orr & Co. Recently this firm have engaged extensively in the manufacture of Machines for Crushing Ores, which have been introduced into the Gold and Silver Mining Territories with marked success.

At Aramingo, near Frankford, are the extensive Machine Works of STANHOPE & SUPLEE, denominated "The Bridgewater Iron Works." They were founded in 1837, and have been in the possession of the present firm since 1857. The buildings extend over a superficial area of over 50,000 square feet, but the entire tract connected with the works is at least four acres. The main buildings, erected in two parallel ranges, comprise a Machine shop and Foundry, which with the office and Pattern room attached, have an aggregate length of 250 feet; a Blacksmith's shop 100 by 30 feet; a Boiler shop 200 by 45 feet; and numerous Pattern shops, of which the stock in a business so long established is necessarily large. These shops are provided with all the requisite tools for the construction of every description of Machinery. Since their establishment these Works have built upward of five hundred engines, varying from three to five hundred horse power; constructed the machinery for the largest Saw Mills in and about Lock Haven and Williamsport; supplied a great variety of heavy machinery for Rolling and Paper Mills, and nearly all the Machines in use in the Print Works in Philadelphia and its vicinity. Messrs. STANHOPE & SUPLEE ordinarily employ one hundred and fifty hands. Their office in the city is at 40 North Fifth street.

Having thus referred to some of the Leading Shops for the Manufacture of all varieties of Machinery, or those that may be called General Machinists, we proceed to the consideration of a much more numerous class, who have provided themselves with the necessary tools for the construction of a special and particular kind of Machinery, such as

1.—COTTON AND WOOLEN MACHINERY, ETC.

It is stated, in apparently authentic records, that the Manufacture or some parts of the Machinery necessary in the production of Textile Fabrics, was carried on in Philadelphia during the Revolutionary war. As early as 1778 it is said that Oliver Evans manufactured Wire from American Bar Iron, and invented a Machine that would work the wire into card teeth at the rate of nearly three thousand a minute, by the simple motion of turning a winch or wrench, and also a machine for punching the holes in the leather for the teeth, by which he could prick by the motion of his hand, one hundred and fifty pair of cards per day. But the first regular manufactory of Cotton Machinery, within the present limits of the city, was that established by Alfred Jenks, at Holmesburg, in 1810. Mr. Jenks had been a pupil and colaborer with the celebrated Samuel Slater, and had brought with him from New England drawings of every variety of Cotton Machinery, as far as it had then advanced in the line of improvement. He supplied the first mill started in this portion of the State of Pennsylvania, with the requisite machinery; and subsequently the Keating Mill, at Manayunk, lately owned by J. C. Kempton. In 1816 he built for Joseph Ripka, a number of Looms for weaving Cottonades. Under the universal impetus given to home manufactures during the war of 1812, Mr. Jenks greatly extended his business operations, and in 1819 or 1820 removed to his present location in Bridesburg, the increased growth of which is owing in no small degree to the personal efforts and enterprise of himself and his successors. Here, where he possessed the necessary facilities for shipping to his more distant patrons, he conveyed his old frame building from Holmesburg on rollers, which yet stands amid the more substantial and excellent structures beside it. This, however, was found too small for his increased business, and was extended by the erection of a stone building thirty feet long, now forming the north end of the present main building, which is four hundred feet in length. When the demand first arose for Woolen Machinery in Pennsylvania, Mr. Jenks answered it, and at once commenced its manufacture, and furnished the first Woolen Mill erected in the State, by Bethuel Moore, at Conshohocken, with all the machinery necessary for this manufacture

BRIDESBURG MANUFACTURING CO.

Incorporated by Special Act of the Legislature of Pennsylvania.

PAID-UP CAPITAL, $1,000,000.

President, BARTON H. JENKS. Treasurer, J. G. MITCHELL.
Secretary, SAMUEL O. SHOUSE.

Office—65 North Front Street, Philadelphia.

CARDING, WEAVING, SPINNING,

AND ALL OTHER

MACHINERY

USED IN

COTTON OR WOOLLEN FACTORIES,

WITH

MILL-SHAFTING AND GEARING.

[OVER]

In 1830, Mr. Jenks, impressed with the idea that the labor of manipulation was insufficient to supply the wants of the population, or to meet the commercial demands, invented a Power-loom for Weaving Checks, and introduced it into the Kempton Mill at Manayunk, where its success produced such excitement among hand-weavers, and others opposed to labor-saving machinery, as to cause a large number of them to go to the mill, with the avowed purpose of destroying it, from doing which they were only prevented by the presence of an armed force. This, and other improved machinery made by Mr. Jenks, soon acquired an extensive reputation, and induced the erection of larger buildings and the introduction of increased facilities. Before his decease his son, BARTON H. JENKS, became associated with him, and his genius in originating improvements in special machines, and liberal and judicious business management, advanced the concern to the first position in rank among similar manufactories either in this country or in Europe. Recently it has been incorporated by Special Act of the Legislature, under the title of "The Bridesburg Manufacturing Company," with a paid up capital of $1,000,000. A more extended account of the present Works will be found in the APPENDIX.

THOMAS WOOD, proprietor of "The Fairmount Machine Works," has a very important Manufactory of Power-looms, which has recently been considerably enlarged. During the year 1866 he employed about one hundred and thirty hands, and produced on an average forty Looms per month, or nearly five hundred during the year. His Looms are so constructed as to be adapted for use either in Cotton or Woolen factories. But the business of the establishment takes a wider range and includes the construction of Embossed *Calenders*, *Lard Oil Presses*, and all kinds of *Shafting*, *Pulleys*, *Hangers*, and *Couplings*, of which he keeps a supply on hand. His variety of patterns of pullies is especially large, and his Hangers are self-oiling and self-adjusting.

Mr. Wood has the reputation of doing thoroughly and well whatever he undertakes.

W. P. UHLINGER has been engaged since 1850, in the manufacture of Ribbon and Jacquard Looms and Rotary Knitting Machines. He recently removed to a large building on Columbia avenue, below Second street, two hundred by forty feet, three stories in height, and fitted it with a great variety of novel and ingenious machinery. He manufactures Ribbon-looms that are self-acting and, combined with the Jacquard machinery, may be propelled by power or hand. Here also are made Jacquard-looms complete, for the manufacture of any kind of figured weaving, including Table Damasks and fancy Balmoral Skirts.

To his Rotary Knitting Machines a first-class premium was awarded by the Franklin Institute, and their practical value is evident from the extensive demand for them.

A portion of the establishment is appropriated to the manufacture of Patent School and Counting House Furniture. His improved School Furniture, which was patented in 1861, has been adopted in the Schools of Philadelphia, and in many other cities. He employs in the establishment about one hundred hands, and turns out a value of $200,000 annually. Mr. Uhlinger is an able mechanic, and an ingenious inventor.

FURBUSH & GAGE, of 118 Market street, have recently erected very important Works in Camden, New Jersey, for the manufacture of Woolen Machinery exclusively. In 1863 the firm purchased ten acres of ground near the line of the Camden and Amboy Railroad, on which was a building formerly used as a Paper Mill, to which they made numerous additions adapting it to the purposes required. The Works now consist of a Foundry, Machine shop, Blacksmiths' and Carpenter Shops, and are equipped with all the best machinery for converting Iron and Wood into Carding Machines, Spinning Jacks, and Fancy Looms of various descriptions, particularly that variety known as the Crompton-loom. Their products are sent to all parts of the United States, the Canadas, Mexico, the West Indies and South America.

Messrs. Furbush & Gage have also an extensive warehouse at 118 Market street, for the sale of such supplies as are needed by cotton and woolen manufacturers.

H. W. BUTTERWORTH has a very important manufactory of DRYING MACHINES for Print works, Bleacheries, etc., at 29 and 31 Haydock street. The cylinders of these machines are made of Tinned Sheet Iron, of which Mr. Butterworth is the only manufacturer in this country. His method of Tinning Sheet Iron is original with himself, and produces a more even coating than has hitherto been obtained. The advantage of these Tinned Iron Cylinders is in costing only one half as much as copper cylinders while they are superior to them for finishing white, pink, and straw colored goods. These machines require but little power to operate them and the heat may be supplied from exhaust steam. They have been introduced into nearly all the factories in Philadelphia, and many in New England.

The establishment is a very old one, having been founded by the father of the present proprietor in 1820.

JAMES SMITH & Co. have a manufactory of Card Clothing, in which one hundred machines are operated; and besides these large establishments, there are in the city, numerous manufactories of special articles required in the manufacture of textile fabrics, such as *Shuttles*, *Reeds* and *Heddles*, etc. *Lead wire* for Looms is made at the Lead Pipe Works of Tatham & Brothers, the most extensive of their kind in this country.

2.—RAILWAY MACHINÉRY.

It is a somewhat singular fact, that the same eminent Philadelphia Engineer, to whom we referred as a pioneer in the construction of Cotton Machinery, is also credited with having built the first Locomotive Steam Engine, taking the word locomotive in its derivative signification as "self-acting."

Scott's Gazetteer, published in 1805, speaking of Oliver Evans, says:—

"He is now just finishing a machine called the *Orukter Amphibolis*, or Amphibious Digger, for the purpose of digging either by land or water, and deepening the docks of the city of Philadelphia. It consists of a steam-engine on board of a flat-bottomed boat, to work a chain of hooks to break up the ground, with buckets to raise it above water, and deposit it in another boat to be carried off. This principle he can no doubt apply to dig canals to make great dispatch. *Orukter Amphibolis* is built a mile from the water; and although very heavy, he means to move it to the water by the power of the engine. *Its first state will then be, a Land Carriage moved by steam.*"

It may also be claimed, that the first entirely successful American Locomotive was built by a Philadelphia mechanic;* while it is conceded that here many of the most important improvements in its construction and capabilities had their origin. The work-shops of this city have sent forth nearly three thousand Locomotives to perform their part in extending civilization, some of which are now thundering up mountain grades, on the long lines of the Pennsylvania Central and Baltimore and Ohio roads, while others are extending the fame of American genius in Continental Europe.

It is, however, a somewhat singular fact that, just at this time, both of the great establishments engaged in this branch are in a transitional, if not disorganized state. The *Baldwin Locomotive Works* have recently suffered an irreparable loss in the decease of their founder, Matthias W. Baldwin, while the *Norris Locomotive Works* are not at this time in operation, the proprietors having determined to withdraw from the business.

* For an interesting account of the Ironsides, the first successful Locomotive built in Philadelphia, see *Bishop's History of American Manufactures. Art. M. W. Baldwin.*

Cars are made in Philadelphia, at two establishments, which in excellence of production, if not in extent, rank among the first in the country. The Philadelphia builders have constructed Cars for more than fifty of the Railroads in the United States; and for beauty of finish, thorough workmanship, strength, and durability, their Cars have no superiors. Nearly all the Passenger and Freight Cars of the Pennsylvania Central and all for the North Pennsylvania Railroad were built by them, as well as large numbers for railroads in the Southern and Western States, and in Cuba and the British Provinces.

The locality is one of the best for this manufacture in the country; for, connected as Philadelphia is by railroads with every part of the United States, and being on tide-water, builders have every facility for convenient and cheap transportation to any part of the world; and, with Iron and Coal—the two heaviest items in their business—cheaper here than in any other shipping port in the Union, they necessarily possess unrivalled advantages for manufacturing Cars with the greatest economy.

For the manufacture of *Car Wheels*, A. WHITNEY & SONS have an immense establishment, the buildings of which cover eighty thousand square feet of ground. The moulding room is four hundred feet by sixty feet—probably the largest in this country; having two Railways extending its entire length, on which carriage Cranes are propelled, and used for removing the molten iron from the furnaces to the moulds, and the wheels from the moulds to the cooling pits. There are five large Furnaces in all—three of which communicate by tubes with an immense caldron for containing melted iron. There are thirty-six cooling pits, having a capacity for holding at a time two hundred and fifty wheels.

In 1860 Mr. PHILIP S. JUSTICE established at Seventeenth and Coates streets, very important Works for the manufacture of *Cast Steel Springs* for Railway Cars and Locomotives. The property is about one hundred feet square fronting on two streets, and is well adapted for the purposes intended. Besides Car and Locomotive Springs, Mr. Justice manufactures Power Hammers, Hydraulic Jacks and Test Guages, and employs, when in full operation, about forty hands.

Recently a Company, of which Mr. WILLIAM BUTCHER is President, erected on the Germantown Railroad near the Nicetown Station, extensive Works for the manufacture of *Steel Tires* for Locomotive Wheels and Railroad Frogs, the first it is said, ever made in this country, and similar to those made in England. The "W. Butcher Steel Works," as they are called, comprise a building for the preparation of the material which is forty-four by one hundred and forty feet, a Steel Foundry which is eighty-five by one hundred and ten feet, a Rolling Mill sixty-

two by sixty-six, a Moulding Shop eighty-five by one hundred feet, and an Annealing Shop fifty-four by eighty feet, with numerous dwellings for workmen on a tract of fiften acres of ground belonging to the Works. The Melting house has a span of one hundred and twelve feet, and is eighty-six feet in width. The huge boilers, of which there are six, run nearly the whole length of the building, and the spare heat from the melting furnaces raises all the steam necessary for the various engines. The large Engine driving the Tire Mill (which has no equal in this country), is a double Cylinder Connected Engine of enormous power, and can roll a finished Tire from the bloom in ten minutes after it is placed on the Rolls. Hydraulic Machinery is used in the pressure Rolls, and the lifting Crane, with which the huge Cylinders of Red Hot Steel are lifted and placed in the Mill, shows wonderful ingenuity.

These works, when in full operation, will employ two hundred and fifty hands, and have a capacity for producing ten millions of pounds of Steel annually. Besides Solid Steel Locomotive Tires without weld, these Works are producing Solid Steel Frogs and Crossings, which may be reversed when one side is worn, Piston Rods, Crank Pins, and Shafts and Bars suitable for Locomotive Springs—in fact, Castings of Steel to almost any pattern as solid and smooth as if Cast Iron. All the products of these works are sold by Mr. Justice at his warehouse, 14 North Fifth street.

The manufacture of the minor parts of Railway Machinery constitutes in the aggregate an important business, but it is usually carried on in combination with other machinery.

Lap-welded Boiler Flues, for Locomotives and other Engines, are made by Messrs. MORRIS, TASKER & Co., and of various sizes, from one and a quarter to eight inches, outside diameter, cut to a specific length. The reputation of this firm will be esteemed, by those who know them, as a guarantee for the excellence of every article they produce.

Railway *Turning* and *Sliding Tables* and *Pivot Bridges* are made upon a new and economical plan, and of any required length. Messrs. WILLIAM SELLERS & Co., who, by the engineering abilities they have displayed, are entitled to rank among the most eminent of European and American Engineers, make a Turn-Table of peculiar construction—the largest size being fifty-four feet in diameter. It consists of a quadrangular centre-piece or box, upon which the arms for carrying the rail are keyed in a very substantial manner. At the outer end of the arms are placed two cross-girths, carrying four truck wheels, which are intended to take the weight when the load is going on or off. The centre rests upon *Parry's Patent Anti-Friction Box;* and the power of one man is sufficient to turn the table and its load, easily, without the intervention c any gearing. They are so constructed, that water in the pit, withi eighteen inches of the top of the rail on the road, will not impa their efficiency or durability.

H. W. Hook, Broad and Hamilton streets, manufactures an *Ant Friction Journal Box* for Cars, which is remarkable for its durabilit The box is of composition—copper and tin—and is lined with *Hook Anti-Friction Metal*, which has, in addition to all the advantages of th best quality of *Babbitt Metal*, the property of resisting heat from frictio to a greater degree, than any other metal that has yet been produce Mr. Hook is also a manufacturer of Type Metal of unsurpassed hardnes

3.—MACHINISTS' TOOLS.

The excellence of the Machine Tools made in Philadelphia was re ferred to at some length in our Introductory as contributing to th manufacturing advantages of this city. There are three establishment in Philadelphia that make tools for machinists' use, which, it is nov universally acknowledged, not merely equal, but surpass in strength proportion, and workmanship, those made elsewhere in the Unite States. The firms referred to are those of William Sellers & Co. Bement & Dougherty, and Cresson & Smith. A brief account of their history and works will be of general interest.

William Sellers & Co.'s Works.

In 1848, the firm of Bancroft & Sellers commenced the manufacture of Machine Tools on Beach street, in the district of Kensington, and by superior workmanship soon made their influence felt in all branches of machine work. In 1853, they purchased a lot two hundred by four hundred feet, at the corner of Sixteenth street and Pennsylvania avenue, and erected on it a machine shop three hundred and twenty by eighty-three feet, with a foundry attached, of which the moulding floor is eighty feet square. The floor of the machine shop is in alternate sections or layers of iron and wood, resting on foundations four feet in depth, which effectually prevents vibration in the motion of the machinery. Early in 1855 Mr. Bancroft died, and since his decease the business has been carried on by two brothers, William and John Sellers, Jr., who have made important additions to the original works. In 1863, they erected a three-story fire-proof structure, one hundred and ten by fifty-five feet, forming an L with the main Machine shop, in which the light turning is

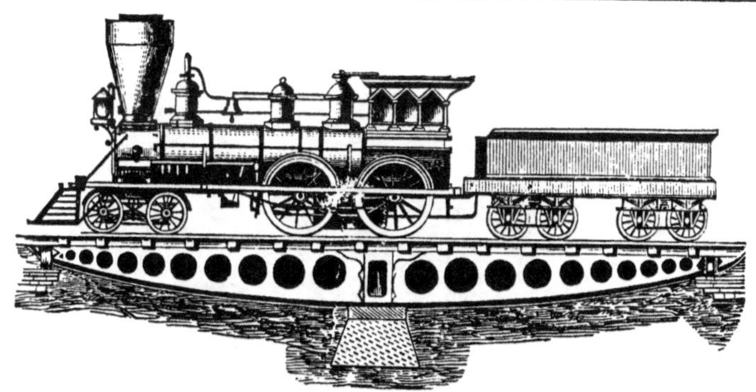

INDUSTRIAL WORKS.
21st St.
Callowhill St.

done and finished work is stored, with handsome offices, etc.; and during the present year it is proposed to erect an immense foundry, two hundred by one hundred and seventy-nine feet, extend the machine shop to four hundred feet in length, and otherwise enlarge the works, which, when finished, will be the most extensive of their kind in the Union.

The machine tools made by this firm present marked peculiarities of construction, and within a few years novelties in the arrangement of the parts have been introduced that are of a striking character. For instance, in their *Lathes*, which are of the kind known as the "Flat Top Shears," the spindles are of hardened cast steel, running in hardened cast steel bearings. The feed for turning is driven by a peculiar arrangement of friction discs, which are so arranged as to enable the workman to grade the amount of feed to the utmost nicety, between the two extremes of the slowest and the most rapid feed motion. The *Planing Machines* are driven by a spiral pinion gearing into a straight rack on the under side of the table, this spiral pinion being on a shaft which crosses the bed of the machine diagonally. This peculiar arrangement, apart from the smooth and uniform motion given to the table, enables the driving pulley on the machine to be placed with the axis parallel with the line of motion of the table, and thus allows the placing of the planers in the shop in line with the turning lathes, instead of at right angles to the lathes, as is usually the case. By this method of driving, the centre of the bed, which is usually weakened by the mass of gearing placed in it, is in their machines made unusually strong and stiff by heavy cross braces. Not only the manner of driving, but the shifting of the belts, the nature of the feed motion, and the machinery for lifting the tool on the backstroke of the machine, are all new and peculiar. An additional novelty is presented in a Planing Machine with an eight feet bed, made by this firm and exhibited at the Exposition in Paris, in the fact that the cross heads and uprights pass over the metal to be planed, instead of the metal passing under them, as in the ordinary methods of construction.

The *Bolt Cutting Machines*, made by Messrs. Sellers & Co., are also peculiarly novel and very successful. The screws are cut at one operation, as with a solid die, but when cut the revolving dies can be opened and the bolt drawn out without reversing the motion of the machine, thus saving the wear and tear of the dies, as well as the time lost in the usual backing off the solid dies. These machines of improved construction are being introduced and extensively copied in Europe, where they are taking the place of all the old style of bolt cutters.

Besides machine tools, Messrs. Sellers & Co. manufacture a variety of important specialties, such as *Shafting and Mill Gearing, Giffard's Injectors, and the Morrison Steam Hammers.* They were the first to introduce the system of manufacturing shafting in interchangeable parts, keeping a large stock of the various sizes most in demand constantly on hand. The ability to interchange the parts in any system of construction is a matter that manufacturers fully appreciate; but in shafting it not only greatly facilitates its first introduction, but it enables any subsequent alterations or repairs to be readily made, and what is of prime importance, reduces the first cost whilst it improves the article.

The "Giffard Injector," for feeding boilers, was introduced by Messrs. Sellers into this country some eight years ago, and since its introduction has been greatly improved by them, obviating the objections to which the original French instruments were liable. They have been so far improved that by the adjustment of the steam alone, the water supply regulates itself, or in other words, whatever may be the pressure of steam, in the boiler, or the quantity of steam admitted to the Injector, the water supply regulates itself to existing circumstances. An American Injector of a given size will do as much work as a French Injector of a size larger. The overflow or waste of water that formerly occurred in the French instrument upon any change of steam pressure, is entirely avoided in the new Injector.

The "Morrison Steam Hammer," an English invention, has also been greatly simplified and improved upon, by this firm, and is now one of the most effective and wonderful implements known. Any required degree of force and power being regulated and attained by a simple movement of a lever that a child could operate. About four hundred and fifty men are employed in this interesting establishment.

A Grand Prize was awarded to this firm for the Machine Tools manufactured by them and placed in the Paris Exposition.

The Industrial Works,

Of which Messrs. Bement & Dougherty are proprietors, are located in the square bounded by Callowhill street and Pennsylvania avenue and Twentieth and Twenty-first streets. Sixteen years ago a single stone shop of a somewhat substantial aspect, stood in nearly the centre of the square, owned and occupied by Mr. E. D. Marshall, who carried on the machine business, in connection with the engraving of rolls for printing calicoes and other fabrics.

This shop formed the nucleus from which has grown the present Works, nearly covering the entire square. In 1851 Mr. Marshall assc ciated with himself Messrs William B. Bement & George A. Colby, formerly from the Lowell (Massachusetts) Machine Shop, under the firm of Marshall, Bement & Colby. Under somewhat unpromising circumstances, the new firm started out in a comparatively new branch of the Machine business, the manufacture of Machinists' Tools—coming at once into direct competition with older, and by no means unsuccessful, establishments—with the single object prominently in view, of turning out the best Tools that skill and genius could produce.

In 1853 Mr. James Dougherty became a partner, the immediate result of which was the erection of a Foundry; his experience in this line, in a number of first-class establishments, having peculiarly fitted him for taking charge of this department. In 1855 Messrs. Marshall & Colby retired from the firm, and Mr. George C. Thomas entered it, though his connection with it was of short duration, for in 1857 Mr. Bement, an engineer of rare mechanical skill, and Mr. Dougherty, an experienced Iron Founder, became sole proprietors, and under their management the Works have grown until now they are scarcely second in extent and importance to any in the Union.

The main shop has a front on Callowhill street of three hundred and seventy-two feet, and with the exception of the office, is two stories in height. The aggregate floor room of the various shops, including Smithery, Brass and Iron Foundry, is about sixty-five thousand square feet, which, with the contemplated extensions, is equivalent to a one story building eighteen hundred feet in length by fifty in width; with yard room for storage of Coal, Iron and Flasks sufficient for all the requirements of the business. The location is peculiarly advantageous for obtaining Coal and Iron, which are delivered by the Philadelphia and Reading Railroad from branches entering the premises.

Among the remarkable Tools in these Shops is a Planer that will take in and plane a piece forty-five feet long, ten feet wide, and eight feet high; a Radial Drill with a swinging arm projecting ten feet; a Boring Mill to swing eight feet diameter; all heavy and massive Tools. At convenient distances, throughout the shops, are powerful Cranes attached to the columns of the building, that move with ease and safety the heaviest pieces to any required position.

The Foundry is fitted with two improved Cupolas, designed by Mr. Dougherty, capable of melting, the one twelve thousand and the other eighteen thousand pounds per hour. All the modern improvements are here combined to produce the heaviest work of the best character. A Corliss Engine of ninety horse power drives all the machinery, including blowers for Foundry and Smithery.

The character of the work done at this establishment will favorabl compare with that produced by any similar one at home or abroad; nc only in its usefulness but its general appearance. The Works are repre sented by their productions in nearly every State in the Union, as wel as in Cuba, South America, France, Spain, Austria and Russia. Thei growth and success, however, have not been due to any great or pecu liarly striking invention, though inventive genius has not been sparingly applied in the production of the model tools and appliances that have given the establishment its reputation. Among the most noticeable of Mr. Bement's improvements may be mentioned a Patent Cotter and Key Seat Drilling Machine now used extensively in Machine Shops; a Patent Pulley Turning Machine, and the Patent Adjustable Hanger and other bearings with Ball and Socket Boxes, which are not excelled by any in use. As an illustration of the ready adjustment of these Hanger Boxes it may be mentioned, that a shaft that had become accidentally bent was found to run without heating for some time, the box accommodating itself to the irregularities of the bearing with every revolution.

Among the specialties manufactured by this firm was the vertical Railway Elevator, now in use at the Continental Hotel, of which a full account is given in *Bishop's History of American Manufactures.*

Cresson & Smith

Are comparatively a new firm who, in 1859, fitted up an establishment for the manufacture of Machinists' Tools, and especially Engine Lathes. Their *Lathes* are notable for simplicity of construction, and novel in the arrangement of the gearing for screw-cutting, and the independent rack feed. The feed is positive, with three changes of thread to each change of gear on the main screw, and by the attachment of the nut to the carriage it can be thrown into gear at any point. The sides and other working parts are scraped to a perfect bearing, and every care is taken to ensure good work and durability. These Lathes are complete in every respect, with Ball and Socket Hangers, Balanced Cone Pulleys, and Steel Spindles with gun metal bearings, large and small face plates, stationary and travelling back-rests, etc. All the ball cranks and handles are of wrought iron, and the nuts, wrenches, etc., are case hardened.

Messrs. Cresson & Smith have also given particular attention to the manufacture of Shafting. Their *Hangers* in particular are notable for their neat appearance, perfect adjustability, and easy working, which is accomplished by their excellent Ball and Socket bearings, and Patent Self-Oiling Attachment.

Their *Pullies* are made in accordance with a systematized scale, and in perfect proportion, from the smallest to the largest, and with as much lightness as is consistent with strength. Their Patent Vise Coupling is being extensively adopted, from its great simplicity and efficiency.

Mr. Cresson, the senior member of this firm was formerly connected with the Industrial Works of this city, and Mr. Smith had an experience of fifteen years in the Works of the Corliss Engine Company, Providence, Rhode Island.

4.—GAS AND WATER APPARATUS.

Philadelphia, it is generally known, is the chief seat of Gas-making Machinery in the United States. Nearly all the principal Gas Works, particularly in the South and West, besides Brooklyn, Buffalo, Newport, and New Bedford, were constructed or enlarged by Philadelphia machinists; and larger Gas Castings have been executed in foundries in this city than in any other place—the Gasholder frame of the Philadelphia Works, made by Merrick & Sons, being, it is said, the largest in the world. The eminence that has been attained in this branch is, no doubt, due largely to two circumstances: first, the advantages of Philadelphia for executing heavy castings economically, because of the abundance and cheapness of Coal and Iron; and secondly, because there are establishments in this city better provided with patterns, tools, and facilities specially adapted for the manufacture of Gas Apparatus than any others in the United States. Some branches of the manufacture, which are now of great importance, had their origin here; as, for instance, the manufacture of *Wrought Iron Tubes* and *Fittings*, first undertaken in 1836 by Morris, Tasker & Morris, the predecessors of the present firm of MORRIS, TASKER & CO., who are undoubtedly the leading manufacturers in this country of these articles and Gas Fitters' Tools, while they make also Cast Iron Gas Pipes, Gas Works, Castings, Retorts, etc. [For a history of this firm, and a description of their Works, see APPENDIX.]

JOHN H. MURPHY & BROTHERS, proprietors of the "Girard Tube Works," Twenty-third and Filbert streets, are also extensive manufacturers of *Wrought Iron Welded Tubes* for Gas, Steam, Water, and Oil. This firm are the successors of Murphy & Allison, who became proprietors of the Works in 1856.

Fittings for Gas, Steam, and Water Pipes, are made a specialty at the Malleable and Grey Iron Foundries of STANLEY G. FLAGG, which are among the most extensive in the country. The "Keystone" (Cupola) Works, situated on North Front street below Girard avenue, are built in the form of a hollow square, on a lot one hundred and twenty by one hundred and thirty feet, the Moulding room occupying the greater part of the first floor of the main building, which is one hundred and twenty by fifty feet, while the Machine shop, Finishing rooms, etc., are in a four-story structure, one hundred and twenty by thirty feet, erected on the rear end of the lot. The Cupola and Annealing Furnaces are models of construction, and every department is stocked with the best machinery and tools so arranged as to secure the utmost economy of labor. Every description of light Castings from Grey and Malleable Iron is produced in these Works, but the specialty, as we have stated, is Fittings for Gas, Steam, and Water Pipes, in which, by constant care and attention, and using Iron of the best quality, the proprietor has acquired such a reputation that "Flagg's Fittings" are a standard of excellence in this branch throughout the Union.

Besides these works, Mr. Flagg has extensive Air Furnace Works at the northwest corner of Nineteenth street and Pennsylvania avenue, exclusively employed on Malleable Iron, and mainly on Castings of the larger size, such as are used by Machinists, Agricultural Implement Makers, etc., Air Furnace Iron being alone suited for such work.

McCollin & Rhoads, 1221 Market street, who have a manufactory in Montgomery county of Terra Cotta Glazed Pipe, Chimney Tops and Flues, do a large business in furnishing Steam and Gas Fittings of all kind.

Gas Metres are made in Philadelphia by five firms, and of excellence not surpassed by any in this country. In addition to every variety of wet and dry Metres, they make Photometers, Minute Clocks, Pressure Registers, Indicators and Guages, Exhausters, etc., and all kinds of similar Gas Apparatus.

5.—BUILDING AND ORNAMENTAL IRON-WORK.

The use of Iron, as a material for building purposes, must be ranked among the modern applications of this wonderful metal. The oldest Foundry in Philadelphia, devoted to the production of Building castings, was erected in 1804, and its proprietor, Mr. James Yocom, was one of the first in this country to make Iron fronts for buildings. The business now employs six Foundries, almost exclusively; and as the advantages of Iron for this purpose, combining, as it does, strength and durability, with cheapness and facility of elaborate ornamentation, become more manifest, the architectural popularity of the metal will extend.

KEYSTONE MALLEABLE & GREY IRON WORKS.
KEYSTONE MALLEABLE IRON WORKS
STANLEY G. FLAGG KEYSTONE IRON WORKS

STANLEY G. FLAGG
UNION MALLEABLE IRON WORKS
UNION MALLEABLE AND GREY IRON WORKS

M'COLLIN & RHOADS,

1221

MARKET STREET,

PHILADELPHIA,

PLUMBERS.

Every variety of ***House Plumbing*** in city and country. Water-works for farms and country seats, as light and strong galvanized ***Water-wheels,*** with pump attached. Improved self-regulating ***Wind-mills,*** Hydraulic ***Rams, Horse Powers,*** Air and Steam ***Engines*** for pumping, etc.

STEAM & GAS-FITTING.

Mills and other buildings heated by steam; Drying Coils put in; Steam Engine connections made with Boiler, and other Steam work. Buildings fitted up with Gas Pipes, and ***Portable Gas Works*** furnished. Gas Fixtures furnished, and put up with particular care.

ALSO, MANUFACTURERS AND DEALERS IN

PLUMBERS' MATERIALS.

Planished Copper Bath Tubs, Copper Bath Boilers, Pantry Sinks, Bidet Pans, Copper Showers, etc. Marbled Wash Basins, with marble tops, plain or countersunk. Water Closets, Urinals, Soap-stone Wash Trays or Sinks, and other Soap-stone work to order. Iron Sinks, Hydrants, Wash-paves, Gum Hose, Fountain Jets, and Brass Cocks of every description—rough, plain, and silver-plated.

We pay particular attention to ***Pumps*** of every kind, and have the largest assortment of the best make in Philadelphia.

With increased facilities, we are now making at our Factory, in Montgomery county, Terra Cotta glazed Pipe, Chimney Tops and Flues, and Garden Vases, and are prepared to supply them in any quantity, and on the best terms. For carrying water the smaller sizes will be found cheap and lasting, and the larger sizes are invaluable for sewers and drains.

T. H. M'COLLIN. WM. G. RHOADS.

At the present time, the firms engaged in producing Building Castings are executing work for all parts of the country, and they have a most complete and extensive stock of patterns, and every facility for the execution of orders, however difficult may be the design or configuration desired.*

Ornamental Iron Work, and especially the manufacture of *Iron Railings*, constitute to some extent a distinct business, though generally associated with Architectural Iron-work in some of its forms. The Iron Railings made in this city are of a very superior character, both as regards the construction and decorative arrangement of the parts; no expense being spared by the leading manufacturers to obtain beautiful and tasteful designs. Most of the Cemeteries and Public Squares throughout the whole country are adorned by work executed in Phila-

* The Eagle Iron Foundry,

Of which Messrs. Samuel J. Cresswell & Son are proprietors, extending from 812 to 818 Race street, occupies the site of the old United States Mint. The foundry has a front on Race street of eighty feet and a depth of one hundred and fifty feet, and is three stories in height. In the first story are the offices, the machine and blacksmith shops, and the moulding room, while the upper floors are used mainly for the storage of patterns of which the stock, by the accumulation of many years, is very large. This firm constructed the Iron fronts of the United States Court House in Baltimore, the Custom House and Post Office at Chicago, the Post Office at Milwaukie, and many in Philadelphia. They also furnished the entire Iron work for the Lebanon Valley Railroad Bridge, and for that at Beaver Meadow. But Architectural Iron and Bridge work is only a branch of their extensive business, which includes Heaters, Ranges, Plumbers' Materials, and Miscellaneous Castings of every description. Mr. Cresswell is one of the oldest Iron founders in Philadelphia, having been identified with the business since 1840, and has attained an excellent reputation. In 1852 he erected his present establishment, and in 1866 associated with him his son, Samuel J. Cresswell, Jr.

Royer & Brothers

Have recently erected, at the northwest corner of Ninth street and Montgomery avenue, an extensive and complete Foundry for the production of Architectural Iron Work. The firm consists of four brothers, all practical mechanics, who, some three years since, associated themselves with a view of making Building Castings a specialty. They now employ fifty men, and have a good supply of orders, some of considerable magnitude. All the Iron work for Dr. Jayne's Mansion on Chestnut street, all the Cast Iron for the Philadelphia Post Office, for Wanamaker & Brown's Extensive Clothing Store, for the Seventh National Bank, together with five fronts for Pittsburg and Reading, were executed by this firm.

delphia; and every city, probably every town in the Union, contains some specimen of our manufacturers' skill and taste. The English Commissioners, in their report on the Industry of the United States, referred in terms of high commendation to the Ornamental Cast Iron-work of Philadelphia, and alluded especially to the establishment of ROBERT WOOD & Co., the most extensive of its kind in the United States. The foundry and manufactory of this firm occupy almost the entire square bounded by Ridge avenue, Twelfth, Spring Garden, and Buttonwood streets, and in variety the products include everything embraced in the wide range of Ornamental and Decorative Iron Castings. The samples in their warerooms, and the illustrations in their Specimen book, numerous and curious as they are, give but a partial idea of the capabilities of this magnificent establishment.

6.—STEAM ENGINES, BOILERS, PUMPS, GAUGES, ETC.

Steam Engines are a leading article of manufacture in nearly all the machine shops of Philadelphia. There are more than twenty establishments in the city, provided with facilities for constructing any size or description of Stationary and Portable Engines; but there are none in which the Steam Engine is an exclusive article of manufacture, or none which keep a large stock of finished Engines constantly on hand. The necessity for anticipating orders has not hitherto been felt by the makers, their facilities being such as to enable them to meet the demand, as it arises, with sufficient expedition.

Probably the establishment in which the manufacture of Stationary Engines is more of a specialty than any other, is that of T. WILBRAHAM & BROTHERS, on Frankford Road and Amber streets. This firm have constructed more than five hundred Steam Engines, of various sizes, which are operating efficiently and successfully in all parts of the United States. Their Works are extensive, having a front on Frankford road of two hundred and six feet, with an average depth of sixty feet, and are arranged specially for facility in executing Engine work. The Pattern department occupies the two upper floors of a three-story building, and is filled with an assortment of modern designs. They have also in connection with the Works a Boiler department and a general Blacksmith's shop, and manufacture Shafting and Pulleys. During the late Petroleum excitement, this firm engaged largely in building Iron Oil Tanks, and constructed, among others, one that had the capacity of holding six thousand barrels of oil. In an experience of fifteen years in this city, they have established an enviable reputation as skilful mechanics and reliable manufacturers.

PEOPLE'S WORKS,

PHILADELPHIA.

HUNSWORTH & NAYLOR,

MANUFACTURERS OF

HIGH AND LOW PRESSURE

Stationary Steam Engines,

Sugar Mills, Saw Mills, Grist Mills, Marine Railways, Blowing Cylinders, Pumping Engines, Hoisting, Stamping and Mining Machinery in general. Hydraulic Presses, Cranes, Boilers, Rolls, Punching Machines, Shafting, Pulleys, Hangers, etc.

AND

MACHINERY

OF ALL DESCRIPTIONS.

Steam Boilers, Ship Tanks, Sugar Boilers, Cast or Wrought, Coolers, Pans, etc. Soap Pans and Curbs of all sizes, either Cast or Wrought. Retorts, Meters, Stills, etc. Steam Pipes, Building Fronts, Columns, etc., fitted and put up.

We are also prepared to execute all orders for HEAVY and LIGHT CASTINGS, either in green Sand or Loam, of Iron and Brass, at the shortest notice.

HUNSWORTH & NAYLOR,

Cor. Girard Avenue & Front St.

KENSINGTON, PHILADELPHIA.

JOHN HUNSWORTH. JACOB NAYLOR.

Messrs. Wilbraham & Brothers are the sole agents in Philadelphia for Judson's Patent Governor and Graduating Valves, which have been applied to and tested upon over fifteen thousand Engines.

The "People's Works," of which Messrs. HUNSWORTH & NAYLOR are proprietors, have built a large number of excellent High and Low Pressure Stationary Steam Engines. They were founded in 1836, and consist of a Foundry, Machine, Blacksmiths', Boiler and Pattern Shops, all equipped with the requisite tools to construct almost every description of Machinery. The buildings have a front on Girard avenue of two hundred feet, and cover a superficial area of 41,200 square feet of ground. Besides Steam Engines and Boilers, Messrs. Hunsworth & Naylor have constructed some large Blowing Cylinders, Pumping Engines, and a great variety of superior Saw and Grist Mill and Mining Machinery. The gentlemen comprising this firm are practical machinists and competent engineers, and have established an enviable reputation for reliability and thorough workmanship. About one hundred and fifty hands are ordinarily employed in these Works.

At the "Steamboat Iron Works," 1160 Beach street, of which JOHN HENSHALL, JR., & Co., are now proprietors, a very superior class of Stationary Engines have been made. These Works were formerly owned by Thomas B. Chapman, but in 1865 they passed into the hands of the present proprietors, who are not only practical machinists, but energetic and ingenious men. During their ownership they have constructed a variety of Marine work, and are at this time about completing two Locomotives. At these Works were built the Excelsior Brick Machines, which are now operating so successfully in the northern part of the city, and Mr. PETER J. SMITH has made an important improvement on this machine, which has been patented. This firm are now constructing a Steam Plough, of which good results are expected. They are also manufacturers of Krausch's Patent Traction Coupling for Locomotives. Messrs. Henshall, Jr., & Co. employ, at this time, about fifty hands.

Propeller Engines have for many years been a leading article of manufacture at the "Penn Works" of NEAFIE & LEVY, who have built and put in successful operation nearly three hundred of this class of Engines. They are proprietors of the patent for the "Curved Propellor," which is extensively used on the Western Lakes.

Messrs. L. B. FLANDERS & Co. have important works at Twelfth and Noble streets for the manufacture of *Portable Cylinder Borers* and *Spring Packing,* of which Mr. Flanders is patentee. His Cylinder Borers are adapted to boring out Stationary or Marine Engines of every size and form of construction, from ten to the largest horse power, without removing any part of the Machinery from its position except one or both heads and the piston. His borers have been thoroughly tested and used on several of the largest Government war vessels at the Philadelphia Navy Yard. Recently he has invented an Improved Locomotive Cylinder Borer, having a Bar and Screw of Cast Steel, with three extra Cutter heads, by means of which Locomotive cylinders and driving boxes can be bored out without removing the back-head, cross-head, or slides. This machine is fed with a constant feed and cut gears, and the clamps or cross-heads are so arranged that they may be used conveniently on Locomotive Cylinders of all sizes. Mr. Flanders' Borers are being introduced into Europe, where they are regarded as a great improvement upon the old machines.

The Patent Packing for Cylinders made at these works consists of steel springs and blocks of cast iron, or other metal, made in the form of a letter V, so that the springs between the blocks press the piston rings out equally against the inside surface of the cylinder, and also force the piston to the centre of the cylinder at the same time.

Steam Boilers are made at several of the establishments which have been already alluded to, and recently extensive works have been erected on the Gray's Ferry Road by JOSEPH HARRISON, JR., for the manufacture of a Patent Boiler, involving entirely new principles of construction that ensure, it is claimed, absolute safety from explosion. This Boiler is formed of a combination of cast iron hollow spheres, each eight inches in external diameter and three-eighths of an inch thick, connected by curved necks, and held together by wrought iron tie bolts. No punching or riveting is required in its construction, and every Boiler is tested by a hydraulic pressure of three hundred pounds to the square inch.

Patent Grate Bars for every description of Furnace, Locomotives, Marine and Stationary, are made a specialty at the Works of W. BARNET LE VAN, Twenty-fourth and Wood streets. In these Patent Bars all the objections incident to the use of thin Grate Bars are entirely overcome, and they will remain perfectly straight until burned out by long service. Thus the loss of fuel caused by the deflection which takes place in ordinary Grate Bars is prevented.*

* Mr. Le Van also manufactures Horizontal Engines that require no expensive foundations, and possess unusual capacity for the horse-power assigned to them. Wrought Iron and Steel enter largely into their composition. All of his Engines are supplied with a novel Governor with Balance Valve, which perfectly regulates the speed of any engine under the most sudden changes of

Le Van's

Patent Grate Bars.

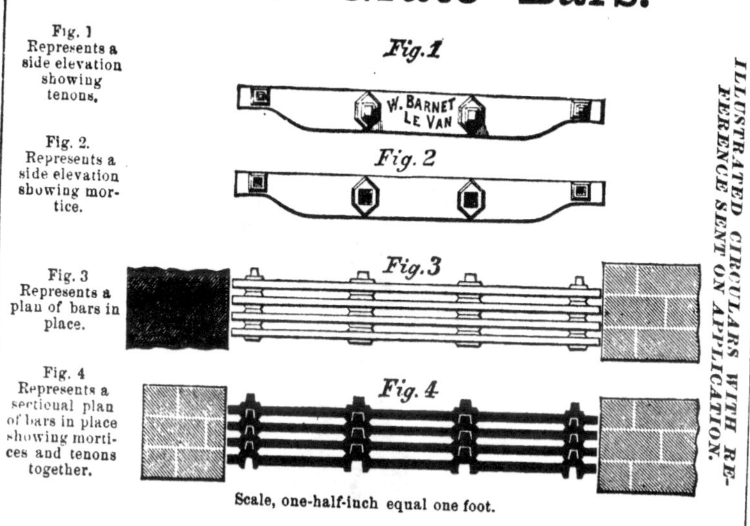

ILLUSTRATED CIRCULARS WITH REFERENCE SENT ON APPLICATION.

These bars have been thoroughly tried in every description of furnace, locomotive, marine and stationary, and are now, with perfect confidence, placed before the public as the most durable, and, in every other respect, the most efficient bar yet invented.

Thin bars are known by scientific engineers to be the best, but there has hitherto been an insurmountable difficulty in keeping them in position, from their invariable bending and giving way, requiring frequent renewal, thus causing continual expense. This difficulty the Patent Bar entirely overcomes. Thick bars only fill the furnace with iron, leaving insufficient space for the passage of air, causing the gases to escape without being consumed, which produces great loss in fuel.

The Patent Bars can be made of any thickness, from three-eighths of an inch upwards, and will not give way in the fiercest fires, but will remain perfectly straight until burned out by long service. Thus the loss of fuel, caused by the deflection which takes place in ordinary bars, is prevented. Over five hundred tons in use in this city. One set was used for over *six years*, working *day and night*, in the office of the *Philadelphia Inquirer*, another at the *Riverside Mills* for eight years.

W. Barnet Le Van.

Office & Works, N. E. Cor. 24th and Wood Streets.

There are several establishments in the city where Steam Pumps and Pumping Engines are made, but only one in which the business is a specialty. This is the "Philadelphia Hydraulic Works," of which BRINTON & HENDERSON are proprietors, situated on Evelina street east of 247 South Third street. The Works consist of a Machine shop, a Brass Foundry, Blacksmith and Pattern shops, and are arranged with special reference to the manufacture of Pumps.

One of the firm, William M. Henderson, is an ingenious inventor, and has given his attention especially to the improvement of Steam Pumps. In 1863 he patented a Direct Action Steam Pump, designed particularly to overcome the difficulties of irregularity of speed and pressure, such as occur in feeding boilers. This Pump has been extensively in use for several years in pumping not only hot and cold water, but tar, bilge water, and other matters that are difficult for any but the best pumps, and has given universal satisfaction. Recently he has patented a Steam Ram or new Boiler Feeder, which is one of the most compact and complete small machines that has ever been invented. It possesses an admirable simplicity of detail, adapting it to the use of Locomotive and Marine Boilers, Rolling Mills, Bilge Pumps, Draining Mines and Quarries, and all localities where a failure to perform would be dangerous. It is especially valuable for pumping semi-fluids or water containing foreign matter, a pocket or receiver being provided to collect any deposit, keeping it clear of the valves to be discharged by a plug at the lower end, which also answers to drain the pump in frosty weather.

Messrs. Brinton & Henderson have a capital invested of $75,000, and employ about sixty hands.

J. M. CHRISTIAN, of 418 Library street, an ingenious machinist, manufactures, among other things, improved direct acting Balance Valve Governors.

esistance. These Governors are remarkable for their simplicity of design and onstruction, and the small amount of material employed in them, which admit f their being sold at a lower price than any other good Regulator, while their ensibility is so great that they detect at once any irregularity in the motion f the crank shaft.

Mr. Le Van has recently constructed a Hydraulic Lift or Elevator for rench, Richards & Co.'s new store on Market street, which is an entire ovelty. It is operated by water, and is said to surpass all others in rapid oisting and exemption from accidents.

Mr. Le Van was educated in that great mechanical school, the "Novelty ron Works" of New York, and is an accomplished machinist and an ingenious ventor.

Lubricative Packing for Steam Engines is made largely by WILLIAM HARTLEY MILLER, at his factory 2134 Lombard street. This Packing is made in accordance with a patent issued April 4, 1865, and requires no oil. Probably an extended description of this manufactory will be inserted in APPENDIX.

2.—GAUGES AND WATER INDICATORS.

Among the ingenious men of Philadelphia there are several who have made a specialty of constructing Steam Gauges and Indicators, for the purpose of testing the true state and condition of the water in Steam Boilers. The most extensive establishment for the manufacture of these invaluable articles in Philadelphia, is that of PETER SCHOFIELD, on Frankford road and Amber streets. Mr. Schofield, in association with William Smith, has recently perfected an invention that renders practicable and efficient the Ball Indicator, which has always been regarded as embodying the true principle of a reliable instrument for testing water in Boilers. Hitherto these balls or copper spheres were imperfect in construction, being liable to sink or collapse; but in Messrs. Schofield & Smith's improvements, all the former objections are obviated. This invention consists in making the ball hollow with an outside communication by means of a cylindrical tube working in a stuffing box. As perfected, this new Indicator consists of a hollow copper sphere, seven and a half inches in diameter, which, floating on the water within the boiler, communicates by means of a hollow tube with a lever on the exterior of the boiler, where also a graduated index points out every half inch of water contained in the boiler; and should the water get below the point of safety, the indicator draws a chain connected with the steam whistle, and gives effective warning of the presence of danger. Indicators may be placed in the office as well as on the boiler, and thus the proprietor, as well as the engineer, may know at any time the true state of the water in the boilers. Mr. Schofield deserves great credit for his perseverance in improving Steam Gauges.

In connection with his Works, Mr. Schofield has a Brass Foundry for miscellaneous castings. He also manufactures Locomotive Spring Balances, and Hydrostatic Gauges that are adapted for attachment to Hydrostatic Presses, and will indicate a fluid pressure from ten to ten thousand pounds, if required.

DAVID LITHGOW, 207 Pear street is the sole manufacturer of *Grimes' Patent Water Indicator,* which has been tested for several years, and is certified to by many eminent engineers to be a reliable indicator of the

height of water in boilers under all circumstances, both on land and water, and without a single failure. It operates accurately, no matter how much the boiler may foam, or how much the vessel may be agitated by waves.

Mr. Lithgow also manufactures *Patent Mercurial Steam Pressure Gauges*, improved *Vacuum Gauges*, and *Gauge Glasses* that will stand a steam pressure of five hundred pounds and upward.

Charles Perkes, 1015 Sansom street, makes improved Slide and Vacuum Valves, Boiler Gauges, etc., in connection with a variety of Engineers' Brass Work.

Within a few years several stores have been opened in Philadelphia for the sale of all kinds of Gauges and Engineers' Supplies. The leading house of this class is that of Augustus S. Battles, 24 North Sixth, where may be found a complete assortment of Steam, Vacuum, and Water Gauges, of all the best makers, including *Ashcroft's* and *Brown's Low Water Detectors*, Steam Whistles, and Peet's Valves which give the full capacity of the pipe without the objection of constant leakage by friction.

Schmidt Brothers, 315 Vine street, are also extensive dealers in Engineers' Supplies, and manufacturers and repairers of all kinds of Steam Gauges. They are agents for Patent Rotary Steam and Ship Pumps.

7.—FIRE ENGINES.

The building of Hand Fire Engines was for more than thirty years inseparably associated with the name of one Philadelphia maker, John Agnew. Some three or four years since Mr. Agnew retired from the business, and was succeeded by Jacob B. Haupt, an accomplished machinist and engineer, the son of the well-known engineer, General Hermann Haupt. He has made various improvements in Steam Fire Engines that render those of his manufacture marvels of strength and lightness combined.

Haupt's Engines are peculiar in having but one steam cylinder and two pumps cast in one piece with the cylinder, so that the three can be bolted fast to the boiler, thereby imparting strength to the cylinder and pump castings, by which, after boring them out, a perfect line is maintained. The pumps are double-acting and brass lined, of the very best material. The suction connection is central, between the pumps, directly under the cylinder, so that both have an equal flow, or draught of water. By this contrivance, any strain of either pump is prevented. Two vacuum chambers are placed on the outside of the pumps, one on each side, to relieve the sudden jar of the suction on the engine taking its water at each end of the stroke. This general arrangement forms a very simple and beautiful design. The boilers constructed by Mr. Haupt are very simple and easily accessible for repairs. They have been known to generate sixty pounds of steam in eight minutes from fresh cold water. These engines complete with tools, suction, pipes, and appliances, weigh only from four thousand four hundred to five thousand pounds. The inventor is designing a smaller class of engines, of the same style, for rural districts, and intended to be the best for size and weight in the country.

In addition to this style he also builds two others, one of which is known as the Horizontal, and the other as the Chapman Engine. Upon both of these styles he has made great improvements. Not only is he engaged in the construction of new engines, but all the repairs of the Philadelphia Fire Department are executed at his shop.

Mr. Haupt is also sole manufacturer of the American Rock Drill, invented and patented by his father, General H. Haupt, one of which is now on exhibition at the Paris Exposition. This machine, which weighs one hundred and thirty pounds, can drill from three-fourths to one and a half inches per minute in the densest granite. It is automatic, and contains within itself the engine, rotatory, and feed arrangements.

8.—PAPER MAKERS' MACHINERY.

Philadelphia now contains one of the most complete establishments in the Union for making Paper Machinery, being provided with facilities for equipping at least twenty-five Paper Mills annually. The proprietor, Mr. Nelson Gavit, is well known as an ingenious mechanic, and has been very successful in turning out good machines, both of the ordinary cylinder and the celebrated Fourdrinier machine, which cost the Messrs. Fourdrinier $300,000 to invent, and caused their bankruptcy. The cost of one of these machines is now about $6,000, and of a cylinder machine about $2,500. One peculiarity noticeable in the machinery of this establishment is, that the shafting, of which there is about one thousand feet, turns upon *glass journals* inserted in the ordinary cast iron box, thereby avoiding a great deal of friction, and runs with much less noise, and requires oil only once in about two months. This shafting has run on glass journals for fifteen years without deterioration. About ninety hands are employed in this establishment.

ROBERT McCALVEY,

MANUFACTURER OF

HOISTING MACHINES,

No. 602 CHERRY STREET.

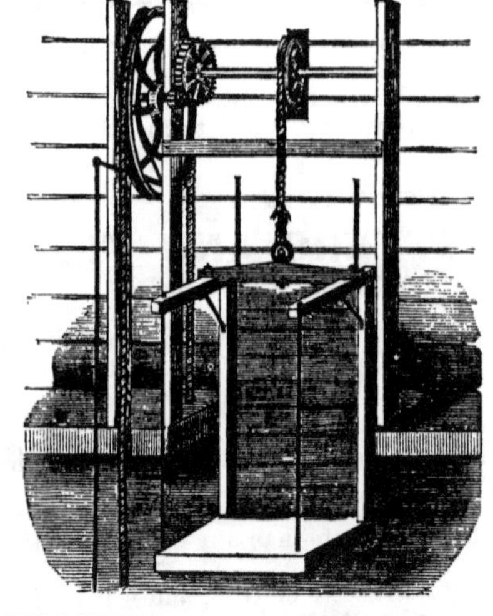

PHILADELPHIA.

DUMB WAITER HOISTING MACHINES.

Cage runs on Iron Guides on the back—the best Hoisting Machine made. In use in all the leading Dry Goods Stores in Philadelphia.

Mr. McCALVEY makes this Machine a specialty.

Cast-iron Wheels furnished for Dumb Waiters or Elevators, and put up to order.

9.—HOISTING MACHINERY.

Hoisting Machines are made at several establishments in the city, but there is one that has made this branch a specialty. We refer to ROBERT McCALVEY, 602 Cherry street, whose father established the business in 1836. Mr. McCalvey has two shops appropriated to the business, (the smith-work being done at his Works at the intersection of Columbia and Germantown avenues,) and his long experience with the consequent accumulation of patterns, necessarily gives him great advantages. McCalvey's Hoisting Machines and Dumb Waiters are in use throughout the United States, and are commended everywhere for their ease and celerity of movement, and exemption from accident.

10.—HOLLOW-WARE FOUNDRIES.

In the manufacture of Hollow-Ware, as in that of Machine Tools, the Philadelphia establishments have no equals in other cities. We believe that the oldest establishment in this branch in the city is that of SAVERY & Co., at Front and Reed streets. It was founded in 1838, and by gradual developments and additions has become one of the most extensive in the United States. The Moulding department occupies the greater part of the first floor of a three-story building, having a frontage on Front street of three hundred and thirty-two feet, with a depth of one hundred feet. The finishing is done in the second story of this building, while the upper floor is the storage-room of the light finished articles with which it is filled to the ceiling. On the opposite side of Front street this firm have a lot two hundred and eight by seventy-five feet, on which sheds have been erected for the storage of heavy castings. The stock kept on hand at this establishment is immense, and includes every variety of Hollow-ware from a diminutive saucepan to immense Cauldrons. Messrs. Savery & Co. employ about two hundred hands.

The firm that have attained pre-eminent distinction in the manufacture of *Tinned* and *Enamelled* Hollow-ware is that of STUART, PETERSON & Co., on Noble street above Thirteenth. The foundry of this firm occupies an area of sixty thousand square feet, requiring to cover it an acre and a half of Slate Roofing. The moulding floor is in the form of a square, having a superficial area of twenty-two thousand five hundred square feet. Almost three hundred workmen are furnished employment constantly, and four thousand tons of American Iron are consumed annually.

This firm, by the adaptation of ingenious machinery, have so completely surpassed all foreign competitors in the production of Tinned a Enamelled Iron ware, that articles of their manufacture have a prefere in all markets in the world. A detailed account of some of the proces and machines that are peculiar to this establishment will be found *Bishop's History of American Manufactures,* to which the curi reader is referred.

Messrs. STUART, PETERSON & Co. are also extensive manufacturers Stoves and other Heating Apparatus. [See STOVES.]

Messrs. SHARPE & THOMSON, whose Stove foundry is one of largest in the United States, have a department devoted to the ma facture of Tinned and Enamelled Hollow-ware, and produce articles t it is acknowledged, surpass the English in uniformity and durability.

11.—GALVANIZED IRON.

The principal restriction hitherto to the more extended use of Iron been its tendency to oxydation or rust, but happily mechanical ingen has overcome this difficulty. Iron is now coated with another me forming a combination impervious to atmospherical influences, known as *Galvanized Iron.*

The process of effecting this great change in this useful material, forming Galvanized Sheet Iron, is described to us by a leading fir the business, as follows: The Iron is first rolled into sheets as ordi Sheet Iron; but for the purpose of galvanizing, a selection is necess for experience has proved that Iron, though of good quality, will nc all cases combine with the zinc which is used in coating. The sh selected are rolled very smooth and well trimmed to the size requi and cleansed from all impurities by a weak acid. The effects of the are in turn removed by immersion in a tank of clear water, and then sheets are dried in an oven. The iron thus prepared is placed in co with the zinc, and, the two metals being brought to the same temp ture, combine and fuse, and form a material impervious to rust, requiring neither paint nor any preservative agent. The proper re tion of the temperature of the zinc and the iron is a point of great ni requiring in the manufacturer much previous experience.

The firm to whom this material is indebted for much of its pr popularity and even intrinsic value, and who, we understand, were first to introduce the manufacture of Galvanized Sheet Iron, are Mc McCullough & Co., now succeeded by the McCULLOUGH IRON COM of Philadelphia. Their Works include five Rolling Mills in Mary besides the Galvanizing Works at the corner of Eleventh and Was

JOHN FARREL. S. C. HERRING. W. S. CUNNINGHAM.

FARREL, HERRING & CO.,

629 CHESTNUT STREET,

(JAYNE'S HALL,)

PHILADELPHIA,

SOLE MANUFACTURERS IN THIS STATE OF

HERRING'S FIRE-PROOF SAFES.

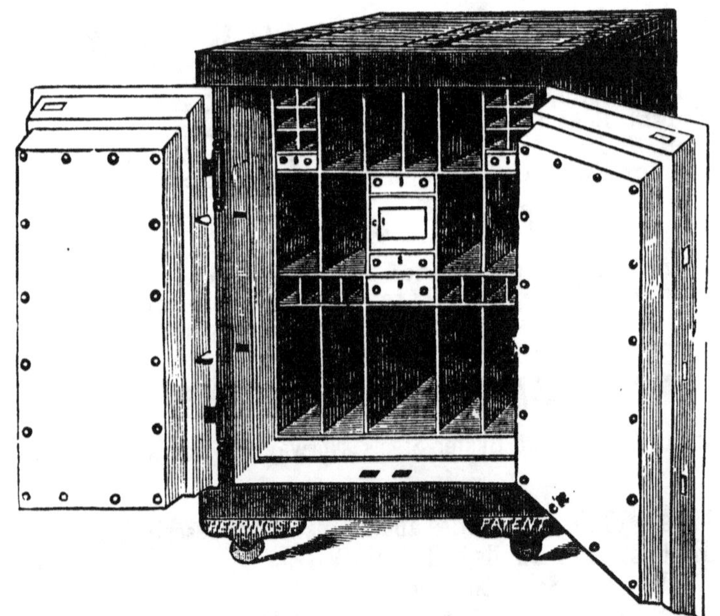

Which Received the Medal at the World's Fair, London and New York.

☞ These Safes are Warranted Free from Dampness.

Also, Manufacturers of HALL'S PATENT POWDER-PROOF LOCK, likewise awarded a Medal at World's Fair. The only BANKERS' CHESTS, lined with Patent Franklinite.

BANK VAULTS, BANK-LOCKS, STEEL CHESTS, etc.

ton avenue, in Philadelphia, and are believed to be the most extensive of the kind in the Union. For a history and description of them, see McCullough Iron Company, in Appendix.

The Galvanized Iron of this firm has been tested by the eminent chemist of the Mint, Professor Booth, who pronounced it equal to that of English manufacture; and in certain tests by sulphuric and other acids, it proved superior. Its applications are necessarily almost as numerous as Iron itself, being available wherever exposed to corrosive influences, and specially adapted for Roofing, Iron-work for Ships, Water and Gas Tubing, Window Shutters, Telegraph Wire, etc.

XV.

Jewelry, Silver and Plated Wares.

Philadelphia has long been the chief seat in America for the conversion of the precious metals into coin. The United States Mint was established in this city in 1793, and, up to the close of the year 1866, the entire coinage amounted to $541,174,959.

At the present time, Philadelphia is also one of the principal points for the manufacture of Gold and Silver Plate, and works have been produced in both these metals that would do no discredit to the master goldsmiths of Europe. Mr. Wallis, in his Report on the Industry of the United States, remarked that while this manufacture is carried on, more or less, in all the large cities, in Philadelphia it partakes of the character of a settled trade, "there being some twelve or fourteen establishments in which a considerable number of persons are employed, and the productions of which are of a varied, but, for the most part, of a useful, as well as an ornamental character." Many of the most magnificent services of Gold Plate, Silver Trumpets, Horns, etc., which have been presented to distinguished citizens and societies, in different parts of the country, were executed at workshops in Philadelphia. The productions include besides Gold and Silver Plate, every variety of fine Diamond and Pearl Jewelry, Gold Chains, Gold and Silver Pencil and Pen Cases, Spectacle Frames, etc.

The manufacture of *Gold Watch Cases* and *Dials* is also a large and growing business, consuming a large amount of the precious metal. English and Swiss Silver Watches are almost invariably imported complete and ready for sale, but Gold Watches are usually cased here. William Warner, of this city, was the first American manufacturer of Watch Cases, having established the business previous to 1812; and

Philadelphia continues to be now one of the chief seats of this manuf ture. In purity of Gold, in excellence of workmanship, and in elabora ness and beauty of ornamentation, it may safely be said, the cases mε in Philadelphia are not surpassed by any.

The manufacture of *Gold Leaf* is also a considerable business in wh over two hundred persons are employed. The malleability of Gold such that it may be beaten into leaves one two-hundred-and-eigh thousandth of an inch in thickness; in other words, a pile of two hu dred and eighty thousand leaves will measure but one inch in thickne Gold Leaf is made into books containing five hundred Leaves, the leav being three and three-eighths inches square; so that each book contai *five thousand six hundred and ninety-five* square inches of Gold Le sufficient to carpet a small bed-room, and yet the weight of Gold is l than four pennyweights.

The most extensive Gold-beating establishment in Philadelphia—a we believe, the largest in the United States, is that of Hastings & C 148 North Fifth street. This house was established in 1825 by Matth Hastings, who was succeeded by his son Robert E. Hastings, the he of the present firm. They employ seventy hands, and have facilities producing two hundred and fifty packages or twelve thousand five hu dred leaves weekly. In the process of hammering or beating, me branes of parchment, vellum, and gold beaters' skin (a peculiar su stance prepared from the outer membrane of the large intestine of t ox), are interposed between the hammer and the Gold.

2.—SILVER AND PLATED WARES.

The *Silver Ware* made in Philadelphia, it is claimed and genera acknowledged, is at least equal, in workmanship and design, to the ve best made in this country; while many connoisseurs, who have visit the most celebrated silversmith shops in the old world, ascribe to no of them precedence over those of Philadelphia. The articles are gen rally made of a fixed standard, several degrees purer than coin, an consequently, possess great intrinsic value, aside from mere workma ship.* The manufacture of Spoons and Forks by machinery is large

* Messrs. Bailey & Co., the leading Jewelers and Silversmiths of Philad phia, claim the distinction of having first introduced the use of Silver of t full British standard, say from 925-1000 to 930, the American standard bei but 900. One advantage of thus raising the standard is, that it successful secures the trade from importations of Silver from England; for purchasers a assured, by a full guarantee, of receiving Silver as pure as that stamped by t

FRANCIS JAHN,

No. 506 Race Street,

MANUFACTURER OF EVERY DESCRIPTION OF

REGULATION SWORDS,

AND OF

French Bronzed Door Knobs, Escutcheons, Bell Pulls, Etc.

☞ The attention of LOCKSMITHS' and BUILDERS throughout the country, is called to this new style of KNOBS, etc. They are made of Brass, Electro-plated and Bronzed, and superior to any ever made for fine doors. Will look well for years without cleaning.

ELECTRO-PLATING in GOLD or SILVER done in the best manner.

HINDERMYER'S

DRAUGHT COOLING

Syruping Apparatus

FOR SODA-WATER.

This new apparatus, shaped as is here represented, can be made to any size. It is manufactured of

BRITANNIA METAL,

Heavily Plated with Silver.

Being *double* throughout, it is a non-conductor, and a great economizer of ice. On the inside are

Ten Britannia Metal Syrup Cans

with an *anti-drip faucet* to each can. The inside chamber contains the ice and 30 feet of

Pure Block-Tin Pipe,

through which the Soda-Water is drawn. The apparatus is *stationary*, and so constructed that the soda-water and syrup faucets are within easy reach of the operator at all times. This apparatus is so constructed that the cans and faucets can be taken out at any time to be cleaned or repaired. I also get up a *Marble Drawing Apparatus*, with syrup cans and faucets, constructed on the same principle. The marble apparatus made to any *size* or *shape* required.

For further particulars, address the manufacturer.

JOS. HINDERMYER,

Nos. 911 & 913 Vine St.

PHILADELPHIA.

N. B. Also, all kinds of Soda-Water apparatus manufactured for using Soda, Whiting and Marble Dust; also, all styles of Counter Fixtures on hand or made to order.

CHAS. LIPPINCOTT,

916 FILBERT STREET, PHILADELPHIA,

MANUFACTURER OF

Fountain Mineral Waters & Syrups,

ALSO, ALL KINDS OF

SODA-WATER APPARATUS.

AGENT FOR

Star Spring Saratoga Water.

HARVEY & FORD,

LEDGER PLACE, SECOND ST., Below ARCH.
And 147 FULTON STEEET, NEW YORK.

BILLIARD BALLS.

BAGATELLE, POOL & TEN-PIN
BALLS,

MOUNTINGS,
UMBRELLA, WHIP & CANE

BRIAR AND OTHER WOOD PIPES.

Latest Styles Ivory and Bogwood Jewelry, and every description of Fancy Goods—Ivory, Bone, and Hard Woods.

carried on, the shops being provided with "rolls," and all other improved machinery that has as yet been introduced. A great deal of Silver-ware made in Philadelphia is retailed in New York as Parisian.

Within the last few years, since the discovery of the process of Electro-plating, the wares produced in an inferior metal, but covered over with a film of silver, have become quite popular. This is comparatively a new process, not having been brought to perfection until 1840, when the discovery of the cyanide solution enabled manufacturers to deposit any required amount of silver on base metals; but its progress has been very rapid, and there are now several firms in Philadelphia who now devote themselves exclusively to the production of Plated Wares. It is one of the advantages of Electro-plating that all ornaments however elaborate, or designs however complicated, that can be produced in Silver, are equally obtainable by this process.*

English government. This improvement in the quality of Silver also renders the manufactured articles more beautifully white, susceptible of higher polish, and less liable to oxidation and consequent discoloration.

The house of Bailey & Co. has been in existence over thirty years, and its reputation at the present time throughout the Union is unsurpassed by that of any other similar establishment.

The variety of costly goods kept in a store like that of Bailey & Co. is a marvel to those who enter this miniature world for the first time. Besides Silver and Plated Wares to which we have alluded, this firm have the most complete assortment of Diamonds and other precious stones, fine Jewelry, and Gold Watches. The manufactories of London, Paris, Vienna, Geneva, Rome, and Florence, are annually visited, and their choicest products brought to Philadelphia. Besides these, the Fancy Goods Departments are exceedingly comprehensive, and include Fans ornamented with real Lace, and varying in price from $25 to $100 each; Gilt Ornaments of various kinds, Choice Music Boxes, French Clocks, Real Bronze Figures and Ornaments, etc. Extensive as the collection now is, it is probable it will be far surpassed when the firm take possession of their new Marble store, now being erected at the corner of Twelfth and Chestnut streets, which will have a front of forty-four feet and a depth of two hundred and thirty-nine feet. This will be fitted up in a style surpassing any thing ever before attempted in this city.

* The oldest establishment in this business in Philadelphia is that of Harvey Filley & Sons, 1222 Market street. Mr. Filley has been identified with the manufacture of Plated Wares since their first introduction, and is entitled to very great credit for upholding the standard, producing none but first-class goods, on which buyers can implicitly rely. For style, finish, and durability, the Plated Wares manufactured by Harvey Filley & Sons have no superior, and their reputation for fair dealing is attested by a long list of customers.

3.—SODA WATER APPARATUS.

The increasing consumption of Soda Water has induced several persons to embark in the manufacture of the Apparatus as a specialty. We believe the oldest of these manufacturers is CHARLES LIPPINCOTT, 916 Filbert street, whose Fountain Soda Water is known throughout the city. He has been an established manufacturer for nearly thirty years. JOSEPH HINDERMYER'S Apparatus is so minutely described in another place that further comment is unnecessary.

XVI.

Lamps, Chandeliers, and Gas Fixtures.

In nearly every Exhibition of American Manufactures which has been held in the last quarter of a century—we presume it will be conceded by every one—the most attractive, artistic, and brilliant feature of the display was the Chandeliers, Candelabras, Girandoles, etc., made and deposited by Philadelphia houses. The manufacture of these, as a branch of American industry, is of quite modern origin, and is mainly indebted to Philadelphia enterprise and skill for its present development. Previous to 1830, the whole trade in Chandeliers was in the hands of foreign importers. Now, the American market is entirely supplied by home manufacturers; and if we may judge from the frank acknowledgments of intelligent foreigners, and their unequivocal testimonials to the excellence in design, workmanship, and finish, displayed in the products especially of Philadelphia workshops, we anticipate a period, not remote, when Lamps and Chandeliers from this city will compete successfully with those of Europe, in European markets.

The pioneer establishment in this manufacture, and the one which, in extent, is now confessedly without an equal in Europe or America, is that of CORNELIUS & BAKER. Founded about a half century ago, it has grown from a small workshop, employing two or three journeymen, to be an immense factory, requiring as its motive power several hundred workmen and two large steam-engines. The operations are conducted in two extensive buildings located in different parts of the city, but they

Mr. JOHN BOWMAN has recently fitted up a very fine store for the sale of Silver and Plated Wares, at 704 Arch street. We believe all of his Wares are *triple plated* in his own manufactory, in which a large force of skilful hands are employed. The assortment of Tea Sets, Urns, Pitchers, Castors, etc., presented in his new store, is superb.

are so managed in order and system as to constitute but one factory. The Cherry street factory is an immense structure, five stories high, built in the form of a hollow square, and is entirely fire-proof. The floors are of brick, the stairs and window-sash of iron, and the roof of slate and iron—not a pound of nails nor a particle of wood having been used in its construction. As many as eight hundred persons are frequently employed in this manufactory.

Nearly all the Chandeliers in the various Capitols of the United States were manufactured by CORNELIUS & BAKER; and the Apparatus, with its two thousand five hundred burners, that lights the House of Representatives and Senate Chamber at Washington. The Gas Fixtures in the Academy of Music at Philadelphia, were also made by this firm. The Chandelier, hanging in the Auditorium, is said to be the largest in the world, being sixteen feet in diameter and twenty-five feet long, with two hundred and forty burners.

Messrs. MISKEY, MERRILL & THACKARA are the successors of Archer & Warner, who commenced the manufacture of Lamps and Chandeliers in Philadelphia in 1848. They have a large and well-arranged factory on Race street near Fourth, where every branch of the business is carried on, from making the most ordinary Gas Fitting to constructing the most costly and magnificent Chandelier. Probably no firm is more liberal to artists who can aid them in producing graceful and effective designs for Lamps, Chandeliers, and Gas Fittings. Among the specialties manufactured in these works is the magnificent bronze balustrading of the stairways of the new Capitol building at Washington, which is not merely a wonderful specimen of workmanship, but is altogether more elaborately elegant than any thing of the kind which is to be seen in Europe. The designs represent alternate groups of infants, eagles, and serpents, pursuing and pursued through wreaths of foliage and flowers, and they consequently comprise almost all the curvilinear forms and intricate traceries which lay the heaviest tax upon the skill of the draughtsman and the founder.

VANKIRK & MARSHALL, 912 Arch street, are comparatively a new firm—though established in 1860—who, like the others mentioned, have been remarkably successful in this branch of manufactures. Their manufactory is in Frankford, and comprises several three-story buildings that cover an entire square. The Foundry is thirty by two hundred feet, the Rolling mill is of nearly equal extent, while the Finishing department is an immense structure, having a front of sixty feet and a depth of three hundred feet. About four hundred hands are employed in the establishment.

In the salesrooms, at 912 Arch street, may be seen specimens of the firm's productions in Chandeliers, Bronzes, and Gas Fixtures, which, for beauty of design and accuracy of workmanship, will compare favorably with any made in this country or in Europe.

1.—COAL OIL LAMPS, LANTERNS, ETC.

Probably no article has attained such vast commercial importance within the same brief period of time as Coal Oil. As an illuminator it is nearly equal to gas in brilliancy, and on the score of economy takes precedence of gas. Its popularity for illuminating purposes has been greatly stimulated by the various ingenious contrivances that have been made for burning it, and large establishments have been erected for the manufacture of Coal Oil Lamps.

Probably the largest of them in Philadelphia is that of HENRY COULTER, at Oxford and Hedge streets, in Frankford. Seventy-five persons are employed in this manufactory, and the department for drawing and stamping, which is thirty-five by two hundred and twenty feet, contains two hundred presses. Here are manufactured the well-known Paragon Coal Oil Burners, which are adapted to burning all kinds of mineral oils with scarcely a possibility of accident. Zinc Jar Tops are also largely manufactured, amounting in some seasons to ten thousand gross. Mr. Coulter was the pioneer in the introduction of American Coal Oil Burners into Europe, and continues to supply the European markets with large quantities. His salesrooms are at 56 and 58 South Second street.

WILHELM & NEUMANN, 919 Race street, are the leading manufacturers of Street Lamps and Globe Lamps for Cars, Ships, and Fire Engines. All the Street Lamps used in Philadelphia, amounting to over six thousand, were made by this firm, and many in other cities and States. They occupy a four-story building, of which the first story is used for the Stamping department and salesroom, the second for the manufacture of Hotel and Street Lamps, and Tin Lanterns, while Brass Lamps and German Silver Car Trimmings are made in the third, and the Japanning is done in the fourth story.

This firm have also a Brass Foundry at Eighth and Cherry streets, where their castings are made, and also the silver domes for Fire Engines, which are a late and attractive novelty.

27

XVII.

Lead Pipe, Bar Lead, Shot, etc.

Philadelphia has two large establishments that are engaged in the manufacture of the various articles of which Lead is the principal raw material, viz: THOMAS SPARKS, and TATHAM & BROTHERS.

The Shot Works of THOMAS SPARKS are the oldest established of the kind in the United States. The corner stone of the "Southwark Shot Tower," as it was then called, was laid July 4th, 1804, by Bishop & Sparks, and the building pushed rapidly to completion. The Patent Shot manufactured by this firm became very celebrated, and was in great demand during the war of 1812. Since 1838 the present owner, a nephew of the original founder, has been connected with the manufacture, and since 1854 he has been the sole proprietor. The manufactures include Shot, Drop and Buck Shot, Round and Conical Bullets, and Bar Lead, of which about four hundred tons are made annually, though the capacity of the works is fully equal to three thousand tons per annum.

TATHAM & BROTHERS have long been the leading firm in the United States in the manufacture of Lead Pipe. They were the first in this country to introduce the process of manufacturing Lead Tubes by Hydraulic Pressure, having purchased the patent from the English inventor in 1840. The evident superiority of this process over the old method of the draw bench soon supplanted in the market all tubes made in that way. Since the date of its introduction the manufacture has been much improved, and pure Block Tin Tubes are now made with the greatest facility.

The latest improvement which this firm have presented to the public is a lead tube encasing another tube of pure block tin. Tubes made by this process are double the strength of those made from pure lead, though the cost is not increased. Messrs. Tatham have secured the use of this new patent, and are now prepared to supply any demand for these Tubes. An imaginary difficulty presented itself upon the first introduction of the Tin Encased Tube, in regard to the possibility of soldering the tubes without melting the tin. This has been overcome by the use of a solder that melts sooner than tin, and the joints are now made quite perfect.

Messrs. Tatham have recently erected a large fire-proof building at No. 226 South Fifth street, where they now carry on their manufacturing operations. This building is an architectural curiosity of which we may give some account in the APPENDIX.

XVIII.

The Leather Manufacture.

The manufacture of various kinds of Leather, particularly Sole Leather and heavy Upper Leather, has long been a leading pursuit in Pennsylvania, and a large portion of that produced in the interior of the State finds a market in Philadelphia. At the present time there are ten tanneries within the city limits which tan about forty thousand hides annually, principally slaughter hides, besides calfskins; and the amount of Leather produced in the city alone, exclusive of Morocco, exceeds two millions of dollars. The tanneries are generally located in the northern part of the city, the pioneer one in that section having been erected by Jonathan Meredith, grandfather of the Hon. William M. Meredith, an eminent lawyer, and late Attorney-General of the State. In addition to the tanning interest, there are over thirty curriers who carry on the business of finishing Calfskins, Harness, and Upper Leather.

But the branch of the Leather manufacture in which Philadelphia may claim a decided pre-eminence is that of *Morocco*. At least *one and a half millions* (1,500,000) of Goat skins are annually converted into leather in Philadelphia; and the excellence of quality is no less remarkable than the quantity. The Goat skins are chiefly obtained from the East Indies, and three-fourths of the whole amount imported into the United States are brought to Philadelphia.

The skins are principally imported into Boston, brought to Philadelphia to be made into Morocco, and many of them again returned to Boston to be converted into shoes. Boston and New York are both largely supplied with Morocco from Philadelphia, and also the principal cities in the West, from Pittsburg to St. Louis. The climate and peculiarities of water in Philadelphia seem admirably adapted for this manufacture; and with the aid of the highest skill, attracted hither as the chief seat of the manufacture, contribute to produce results that are apparently not attainable anywhere else in this country.

XIX.

Marble, Slate, and Soapstone.

Marble is so extensively used as a building material in Philadelphia, that to supply the local demand would constitute in itself an important trade. This city is, however, also a principal depot for the supply of the adjacent States, extending even to the South and West; while the skill of her artists and sculptors in Monuments and Mantels is known and acknowledged throughout the whole country.

There are now in Philadelphia over sixty yards in which Marble is worked into various useful and ornamental forms, and nine steam mills for sawing and preparing it. As a striking exception to the rule of frequent removals, characteristic of American enterprise, we may state that the ground now occupied by Messrs. Struthers & Son, on Market street above Tenth, has been used uninterruptedly as a Marble Yard since 1798. This firm are the successors of J. Struthers and Son, who commenced business about the beginning of the present century. The Marble work of nearly all the elegant and costly public buildings, for which Philadelphia is distinguished, was prepared in this venerable and celebrated yard—as, for instance, the United States Mint, the Custom House (formerly the United States Bank), the United States Naval Asylum, the Philadelphia and Western Banks, Philadelphia Exchange, the Girard Buildings, the Farmers' and Mechanics' Bank, and many others. The present proprietor, Mr. William Struthers, the surviving partner of the old firm, has added largely to the evidences of Philadelphia skill in the departments of Monumental Sculpture and Architectural decoration. He constructed the Sarcophagi that invest with increased interest the burial places of Henry Clay and John M. Clayton, and has sent various Art objects in Marble not only to all parts of the United States, but to England, the West Indies, China, and Syria. He was the first, it is said, to introduce into Philadelphia the sandstone from the Albert and Pictou quarries, and has induced many of the leading merchants and most wealthy citizens to adopt it in the ornamentation of their warehouses and dwellings.

Mr. Struthers has recently erected a four-gang Steam Mill on Walnut street wharf, at the Schuylkill, for sawing the Marble which is finished at his Works, No. 1022 Market street.

John Baird's Marble Works

Include two Steam Mills, one of them the largest that has yet been erected in the United States. Mr. Baird has been identified with the working of Marble since 1841, when he commenced the business on Ridge avenue above Spring Garden street, and has achieved a distinction and success that place him foremost in the pursuit throughout the Union.

In 1846 he built a Steam Mill, adopting a new principle in the application of power to the Sawing of Marble known as the Crank action, or Pitman motion, the superiority of which over the old Pendulum mills was so manifest, that all the mills in Philadelphia, and, with few exceptions, all in the United States, have been rebuilt or remodeled on this plan. He also originated and adopted various implements in feeding the saws and regulating their motion, which made his mill the wonder of the trade for its efficient action. It is probable that no Marble Mill of equal capacity in the world has performed so much work within the same period of time. For many years Mr. Baird has been the principal consumer of Italian marble in the Philadelphia market, his purchases for a series of years amounting to twenty-four thousand cubic feet annually. This he has converted into various beautiful forms for the ornamentation of dwellings and cemeteries, and some of the most beautiful specimens of the Phidean art of which this country can boast are the products of his workshops. It has been his practice not only to secure the best native and foreign artists in carving and designing, and to stimulate their ambition by rewards and liberal remuneration, but to encourage the study and practice of both these arts by establishing schools for the benefit of his apprentices. The fruits of his enterprise in this respect may be seen in his Mantel warerooms, and in the Monuments and Tombs which adorn our Cemeteries. His warerooms contain upward of one hundred and thirty different patterns of Marble Mantels, made from all varieties of marble, common and rare, from the clouded Pennsylvania to the Carrara statuary. The designs in most instances are original, and the carving on the most costly renders them worthy of a place among the *chef-d'œuvres* of the art. In Monumental Art, the triumphs of the proprietor of these Works are written on the Cemeteries of Philadelphia, the Mausoleums of the South, and the resting-places of the dead throughout the Union. Whether it be a people's testimonial of gratitude to heroes who sacrificed their lives in battling with the pestilence, or a contribution from patriotic mechanics to the pile erecting in honor of Washington, or the more numerous and diversified mementos of affection to departed relatives or friends, the same wealth of resources, the same masterly execution, are visible in all.

In 1865 Mr. Baird purchased a tract of three and a half acres of land at Locust street wharf, on the Schuylkill river, and proceeded to erect on it a Steam Marble Mill, which exceeds in magnitude any thing heretofore attempted in this branch of manufactures. The building is two hundred and fifty-five feet long, seventy-five feet wide, and when completed, according to the original plan, it will contain eighteen gangs of Saws and eleven Rip Saws, which will be capable of sawing one hundred thousand cubic feet of Marble in a year. The machinery includes all the latest improvements, such as the *Tingley Patent Feed Motion*, and *Merriman's Patent Gang*, and others original with the proprietor. But the special feature of this mill is the adoption of the *English Steam Travelling Crab*, by the use of which a block of marble can be taken from the hold of a vessel at the wharf and placed directly under the Saw gang, without the intervention of manual labor. This remarkable machine has been employed for carrying heavy weights in factories of a different description, but its use in a Marble Mill is, we believe, original with Mr. Baird.

The erection of this Mammoth Mill is destined to mark the beginning of an era whose influence will be felt in all the wide ramifications of the Marble business. It is the first attempt to introduce into this trade the principle of subdivision or limitation to specialties which, in the Furniture and other branches of manufactures, has operated greatly to the advantage of consumers. It is well settled that those who have machinery and facilities for producing at one operation a hundred setts of any article, whether of Wood, Iron, or Marble, can produce them at one-fourth less cost than those whose capabilities are limited to one piece of the same size. By the system which he has introduced into his Works, the completeness of the machinery, and the consequent economy of time and material, Mr. Baird will be enabled to supply Marble workers with Grave-stones, Plinths, and other stock of standard sizes, nearly finished, at a little, if any, advance upon the price they have heretofore paid for the same material in the rough. Marble direct from Leghorn can be brought to his wharf, and placed under the saw gangs, at a saving of all transit duties and transhipment charges that have heretofore amounted to more than the original cost in the block. Here, then, will be the great depot of Finished Marble in America, to which the dealers from all sections of the country will resort for supplies which they can obtain with advantage to themselves and benefit to the city.

The erection of this Mill will have another important influence in directing the attention of capitalists and artizans to the advantages that the banks of the Schuylkill present for manufacturing operations. Mr. Baird is universally acknowledged to be one of the most intelligent as well as enterprising and successful of the business men of Philadelphia.

His practical sagacity is proverbial and attested by the accumulation c a large fortune, the result of his own unaided foresight and industry. H has made a careful study of the respective advantages offered by differei localities for manufacturing operations, and has decided that the bank of the Schuylkill, below Chestnut street bridge, are superior to them al It is probable that the west bank of that river, from Chestnut street t Gray's ferry, is the best site in America for the location of Iron Work and large Steam Manufactories of all kinds. There the proprietors o such works can obtain at all times a supply of the best workmen withou the complications that result from the relations of landlord and tenan where dwellings must be provided for the operatives. It is always bes that manufacturers should have no other relations with their work men than simply that of employer and employee. There, also, manu facturers are not only independent of the extortion of Transportatio Companies, but have the advantage of numerous competing lines. Th river itself is always accessible for vessels not drawing more than twent feet of water, while along its banks is the Junction Railway, a neutra organization, that connects with the four great lines of railway leadin to the North, West, and South. How important this consideration i has been painfully realized in many instances by manufacturers in Ne England, who are dependent upon one railway for the transportation o their raw materials and finished products; and this circumstance, com bined with other advantages, confirms the correctness of Mr. Baird' judgment in declaring this to be the best site in America for larg manufactories.

Besides those mentioned, Messrs. JACOBY & PRINCE, EDWIN GREBLE L. THOMPSON & Co., J. E. & B. SCHELL, ELI HESS, and JOHN V. VANDERBILT, have Steam Mills for sawing and finishing Marble.

Formerly, a large portion of the Marble used for building purposes i Philadelphia was obtained from quarries in the vicinity of the city, bu these, in most instances, have attained an extraordinary depth, amounting in some places, as in the Hitner quarry, to two hundred and forty feet; and the working of them is consequently attended with much expense. Now, nearly all the American Marble used in Philadelphia is obtained from quarries in Massachusetts and Vermont, and supplied tc the trade by the principal jobbers—SAMUEL F. PRINCE, J. K. & M. FREEDLEY, and FREEDLEY, MACDONALD & Co., all of whom are alsc proprietors of quarries.

WM. STRUTHERS. JNO. STRUTHERS.

STRUTHERS & SON,

Marble and Sand Stone

1022

MARKET STREET,

AND

Steam Mill, Walnut St. Wharf, Schuylkill.

J. A. BAILY, SCULPTOR. STUDIO AT THE ESTABLISHMENT.

MONUMENTS,

MANTELS, STATUARY,

AND

BUILDING WORK IN GENERAL.

Designs for Monumental Work furnished.

2.—SLATE, SOAPSTONE, ETC.

In our introductory remarks we adverted to the Slate Quarries in Lehigh county, and stated that the very best qualities of Slate are obtained in Pennsylvania. About 1846 the first Quarries were opened in the vicinity of Slatington Station, now on the Lehigh Valley Railroad, and an idea of the rapid growth of the consumption of Slate may be obtained when we state that while in 1847 the shipments from the point above named amounted to only thirty-eight squares (a square covers one hundred feet), in 1865 they amounted to sixty-four thousand squares, and in 1866 it is estimated they were not less than one hundred and thirty-five thousand squares.

Among the heaviest operators in Lehigh county is the LOCKE SLATE COMPANY, whose office is at No. 1010 Market street, Philadelphia. In addition to the Quarrying of Roofing Slate, this Company has two factories for the manufacture of their Patent Oval and their new style frame School Slates, with a capacity to turn out eight thousand cases per annum, each case averaging twelve dozen Slates. They give employment to one hundred and fifty and sometimes two hundred persons; and the handsome and thrifty little village of Slatedale owes its existence to their enterprise.

Of *Soapstone* the manufactures are quite limited, though the material obtained at quarries now within the corporate limits of Philadelphia, about two miles above Manayunk, is of the best quality. One of these quarries is the oldest in the country, having been opened before the Revolution; but until it came into the possession of its present enterprising owner, SAMUEL F. PRINCE, its value was scarcely appreciated. Its products now amount to about twelve thousand tons annually, and are disposed of principally to Iron manufacturers along the Schuylkill, and in New Jersey, and to mills that grind the material into powder.

This stone, hitherto little used for economic purposes, is adapted for many—particularly for fire-stone, kitchen sinks, wash-tubs, bath-tubs, and especially for baths, and sizing rollers used in cotton mills. For the last purpose it possesses the advantage of not being affected by the acids ordinarily used in sizing, and of not warping, contracting, or expanding, by changes of temperature and moisture.

Both Slate and Steatite, or Soapstone, have not attained their maximum of appreciation, and offer excellent opportunities for the enterprising to establish new manufactures.

XX.

Oils.

In the manufacture of Linseed Oil there are three mills employed, generally possessing very improved machinery, and the largest of them makes three thousand gallons of Oil and twenty-five to thirty tons of Oil Cake per day. The material used for making the Oil is Flaxseed, or Linseed, imported from Calcutta—that being the only point from which the best article can be obtained. In addition to the regular establishments making this Oil, the manufacturers of Zinc Paints, Colors, White Lead, etc., frequently make sufficient Linseed Oil to supply their own necessities.

There are eight concerns in the manufacture of *Lard, Tallow,* and *Red Oils*, of which the annual product amounts to nearly as much as that of Linseed Oil.

An Oil is also made from Resin, or Gum of the Pine tree, and known as Resin Oil. The Gum contains several ingredients which, when submitted to the process of distillation, are separated, and all are useful, but are of entirely different natures, being Oil, Acid, Naphtha, Pitch, and Tar. The bleaching and pressing of Sperm and Whale Oils is another considerable item connected with the Oil business of Philadelphia. One of the firms—of which there are four—is successor to the parties who claim to have first introduced the process of the Chemical bleaching of Oils in this country.

The manufacture of Railroad and Cart Greases, made of Resin and other Oils, is carried on to a considerable extent. These Greases, which are now extensively used for the oiling of machinery, vehicles, etc., have grown much into favor of late.

Within a few years the Refining of Petroleum has become a considerable business in Philadelphia, and there are now several large establishments engaged in it, being principally located on the Schuylkill river. In 1866, nearly two millions of gallons were refined in Philadelphia, and there were shipped from this port to various ports in Europe 568,119 barrels of Refined, and 124,423 barrels of Crude Petroleum.

GEORGE W. PLUMLY,

MANUFACTURER

OF EVERY DESCRIPTION OF

PAPER BOXES,

SOUTHEAST CORNER

FOURTH & BRANCH STS.,

PHILADELPHIA.

A LARGE STOCK OF

DRUGGISTS' PILL AND POWDER BOXES

KEPT ON HAND.

HOWLETT & ONDERDONK,

PATENT MACHINE

PAPER BAG AND

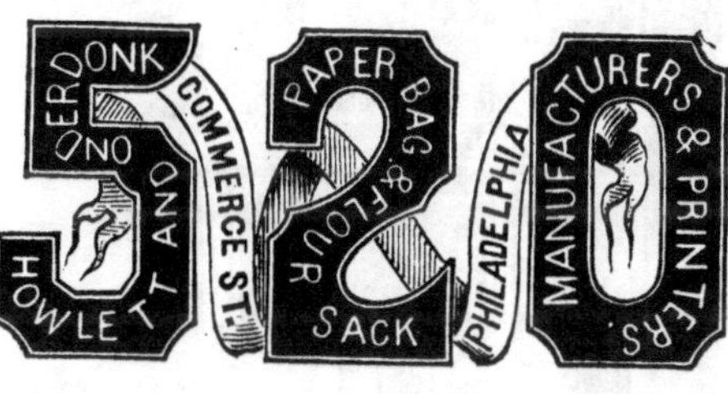

FLOUR SACK

Manufacturers and Printers,

520 COMMERCE ST.,

PHILADELPHIA.

THE MOST LIBERAL DISCOUNT TO THE TRADE.

XXI.

Paper, Paper Collars, Bags, Envelopes, Etc.

The extent of the Publishing business in Philadelphia necessarily makes this city one of the great Paper Marts of the Union. Along the Brandywine, in Delaware and Chester counties, the Paper Mills are very numerous, and some of them very long established, as, for instance, the Ivy Mills, which produced the Paper on which the old Continental Money was printed, the first Bank Note Paper made in America. Of the Mills located within the city limits the most celebrated are the Wissahickon Mills, owned by Messrs. CHARLES MAGARGE & Co.; and the most extensive are the Wood Paper Mills, before referred to, and represented by the firm of JESSUP & MOORE. Having, however, alluded to the statistics of this trade in another place, we can pass to a consideration of a few of the forms into which paper is manufactured.

2.—PAPER COLLARS.

It is estimated that no less than fifty tons of Paper are now consumed weekly in the manufacture of Paper Collars. As the country is largely indebted to a Philadelphian for the introduction of this new article of commerce, it may not be inappropriate to give in brief the history of this invention. On July 25th, 1854, Mr. Walter Hunt of New York received a patent for Collars made of two pieces of paper cemented together and indented by serrated rollers, claiming this to be a new article of manufacture. Previous to the issue of the letters patent Mr. Hunt he disposed of one-half his interest in the same to John W. Ridgway, of Boston, for four thousand dollars; and subsequently sold the other undivided half-interest for three thousand dollars to E. H. Valentine & Co., who commenced the manufacture in the third story of a building, No. 408 Broadway, New York. The community, however, did not regard these new collars with favor, and the patent was frequently sold and resold with diminished value, none of the purchasers seeming pleased with their bargain. In 1858, Mr. William E. Lockwood, a young dry goods merchant of Philadelphia, purchased the interest formerly owned by Valentine, removed the machinery to Philadelphia, and commenced the manufacture in the Keystone Mills, near Fairmount. His experience during the first six months was by no means encouraging, and he was compelled to suspend operations, as the stock accumulated for want of a market. When one considers that millions of these collars are now made and sold annually, it seems incredible that less than ten years ago required persistent effort to overcome the prejudices of the communi against them. This Mr. Lockwood succeeded in doing, by a judicio system of advertising, and setting forth their advantages, in which was aided by the strong argument, that these patent collars can be wo much longer than an ordinary laundry-dressed collar without losing th gloss and brilliancy, can be sold at a price less than the cost of washi and starching linen collars, and what was convincing to those prejudic in favor of cleanliness that they were in reality wearing the identic material used in a linen collar, only more clean and pure in its prese form than ever before. His accumulated stock was then soon dispos of, his factory reopened, and in less than two years his facilities f manufacturing were found inadequate to supply the demand, and removed to his present location, 255 and 259 South Third street, whe he has now probably the largest and most complete establishment of th description in the United States. (See APPENDIX.)

There are now three Paper Collar manufactories in Philadelphia, th of W. E. & E. D. LOCKWOOD, and the KEYSTONE COLLAR COMPA (VAN DEUSEN, BOEHMER & Co.), 627 Chestnut street, being the mc important. These factories turn out millions of these useful artic annually.

3.—PAPER BOXES AND BAGS.

The demand for Paper and Pasteboard Boxes by Confectioners, Je ellers, Pill Makers, Perfumers, Stationers, and Merchants generally, so great that their production forms an important branch of manufa tures. In Philadelphia it is believed the business is carried on to greater extent than in any other city. There are over thirty differe establishments in Philadelphia in which Paper Boxes are made, that e ploy in the aggregate over four hundred hands. The manufactory GEORGE W. PLUMLY, Fourth and Branch streets, occupies a six-sto building, and is the most extensive of its kind in the United States. T cheapness with which boxes can be made is remarkable; some of ve neat appearance can be made at about five cents per dozen; and yet ea is composed of several separate pieces, and each has to be many tim handled, covered with colored or fancy paper, labelled, and packe Although most of the manipulations must be done by hand, yet withi the last few years a great variety of machinery has been invented f the purpose, which gives increased facilities in the various operation One machine invented by Mr. Plumly for cutting round pieces for th

ends of boxes operates with such rapidity, that five thousand may be cut with it in an hour.

Another form in which large quantities of the coarser kinds of Paper are consumed is in the manufacture of *Paper Bags* and *Flour Sacks.* One firm (Messrs. HOWLETT & ONDERDONK, 520 Commerce street) have a capacity of turning out a half million of these Bags daily. Orders are often filled at this establishment for millions of Paper Bags at a time. This firm have also a machine, patented by Mr. Howlett, which enables them to produce Satchel Bottom Sacks, of large sizes, at less cost than any in the market. With one exception, the establishment of Howlett & Onderdonk is said to be the largest of its kind in the world.

4.—ENVELOPES.

It is stated in the British Post Office Statistics that of three hundred and forty-seven millions of letters which were posted in one year, more than three hundred millions were enclosed in Envelopes. In this country the habit of using coverings for epistolary communications is even more general than in England, and, consequently, the proportionate consumption of Envelopes must be larger. In Philadelphia the manufacture of Envelopes has not hitherto been prosecuted with any remarkable degree of vigor; but during the last year the Messrs. LOCKWOOD, to whom we referred as manufacturers of Paper Collars, have embarked in it with characteristic energy, and it is probable Philadelphia will soon have the most gigantic Envelope manufactory in the United States, if not in the world. They have secured a proprietary interest in certain letters patent covering both the form of the Envelope and also the machinery for manufacturing them, at a saving of thirty per cent. over any previous method, and have fitted up an establishment for manufacturing them at the rate of a half million daily. In their new machine, which has a capacity for production quadruple that of any Envelope Machine previously invented, the paper passes in at one end, and the Envelope, *gummed and completed,* comes out at the other, one hundred and seventy-five and two hundred a minute. Besides the facilities for rapid production, this firm have another great advantage in a patented form of cutting, by which they are enabled to obtain nineteen Envelopes out of a sheet that would cut only fourteen of the old pattern, and probably, in close competition, they would consider this saving of paper a sufficient profit. It is also claimed for these Patent Envelopes that they are neater in appearance, more evenly gummed, less liable to be opened than any others, and to Printers and Lithographers they offer advantages in the saving of type, inasmuch as they are of uniform thickness throughout; while in the old form of Envelopes there are five thicknesses of paper in some places and only two in others. In connection with their manufactory, Messrs. Lockwood have ten Steam Power Presses and one hundred and fifty fonts of new type, and are prepared to print one hundred and forty thousand Envelopes per day.

XXII.

Saddles, Harness, Whips, Etc.

The manufacture of Saddlery in this country is distinguished from that in any other part of the world by the immense variety of styles and qualities which are produced. We are informed by a leading manufacturer, that of Saddles there are probably not less than five hundred various styles and qualities, with a proportionate quantity of Bridles, Bridle Mountings, Martingales, Girths, Circingles, Stirrup Leathers, Saddle Bags, Medical Bags, etc. Of Harness, for Coach, Gig, Dearborn, Sulky, Stage, and Omnibus, there are, perhaps, three hundred styles and qualities; while in coarse Harness, for Carts, Drays, Wagons, and Ploughs, there is also great diversity.

It is a fact well known to persons who are familiar with the history of Industry during the past few years, that the Saddle and Harness-makers of Philadelphia have invariably carried off the "palm" at local Exhibitions and Fairs; and the fact that the Prize Medal was awarded to a Philadelphia firm at the World's Fair, in London, cannot be unknown to any observant person who has traversed Seventh street, north of Chestnut. The special causes conducive to superiority in the Harness manufacture are manifold: all the raw material consumed, especially the Leather and the Hardware, are made here of the very best quality; the workmen have permanent employment, and the manufacturers have an established reputation for faithful work, which they are determined to maintain. The solvency and character of the trade in Philadelphia, enable them to buy at the very lowest rates; and the system of manufacturing involves much less ostentation, and, consequently, less expense than in many other cities where the sales-house and factory are distinct and separate establishments, even if owned by the same parties. In this city, the goods are generally manufactured and offered for sale under the same roof. The ingenuity of the manufacturers, too, has been repeatedly and successfully called into exercise, and the very best of the new styles of Saddles made in the North were first originated and introduced

SADDLERY
STOCK LARGE, VARIETY UNSURPASSED IN THE UNION
PROFITS SMALL, PRICES LOW COMPARISON INVITED.
631 631
KNEASS & CO.
631
MARKET ST.
SUCCESSORS TO M. MAGEE & CO.
Wholesale & Retail Manufacturers of
SADDLERY & HARNESS
HORSE FURNITURE, &c.
WHERE THE LARGE
STANDS IN THE DOOR.
Factory 622 Commerce St.
HARNESS
& HORSE FURNITURE
LONGACRE

by one of our large houses; while improvements upon the old English styles render those made in Philadelphia in several respects superior to the foreign. In the new styles, of the Spanish and Mexican order generally, the utmost care is taken to guard against injury to the horse, and also to produce (which they have, beyond all other places) the most comfortable and pleasant Saddle, for both horse and rider. Hog-skin continues to be the principal Leather consumed in the best Saddles, on account of its softness and capacity for exposure to the sun and rain; though Buckskin is also frequently used for the seat, and for the horns of Ladies' Saddles in particular.

The firms of W. S. Hansell & Sons, Kneass & Co., E. P. Moyer & Brothers, Samuel R. Phillips, and Lacey, Meeker & Co., have assorted stocks of Saddles and Harness, of their own manufacture, as complete and extensive as can be found in the United States. The two first named are old and well-known firms, who have also branch houses in New Orleans. Messrs. Moyer & Brothers have been established on Market street for fifty-six years. Mr. S. R. Phillips is the gentleman referred to as the recipient of the Gold Medal at the London Exhibition; while Lacey, Meeker & Co., 1216 Chestnut street, are a branch of one of the oldest Saddlery manufacturing firms in the United States.

Besides these firms, Philadelphia has at least one very extensive house for the sale of Saddlery Hardware and Carriage Materials. We allude to Scott & Day, successors to Wm. P. Wilstach & Co., 38 North Third street. The store of this firm is two hundred feet deep, and five stories in height, and is crowded from the basement to the attic with goods adapted to the wants of Saddle, Harness, and Carriage Manufacturers. Here may be found every thing included in the term Saddlery Hardware, Stirrups of all kinds, Saddle Trees of every variety, Enamelled Leather for Carriage Makers', Springs and Axles of the best makes, Felloes and Spokes of all sizes, Norway Iron Carriage Bolts; in fact, several thousand different articles adapted to the wants of the trade in the Southern, Western, and Middle States. The members of this new firm were formerly partners in the well-known house of Wilstach & Co., and have had twenty years' experience in the business.

2.—WHIPS.

The manufacture of Whips, though entirely distinct from that of Saddles and Harness, has relations with that trade so intimate, that it may properly be considered in the same connection.

The Whip Manufactory of Charles P. Caldwell, in Mantua, West

Philadelphia, is the oldest established and one of the most extensive in the State. He manufactures all kinds of Whips, from those which sell at $2 a dozen to those fancy Presentation Whips worth $100 apiece; and has received numerous testimonials for excellence, including one Gold and two Silver Medals from the American Institute in New York. Fully one-half of his products are sent to New York and Boston. Canes are also made in this manufactory and the materials and mountings are often exceedingly rare and costly.

XXIII.

Soap and Candles.

"The quantity of Soap consumed by a nation," says the celebrated Liebig, in his Familiar Letters on Chemistry, "would be no inaccurate measure whereby to estimate its wealth and civilization. Political economists, indeed, will not give it this rank; but, whether we regard it as a joke or earnest, it is not the less true that, of two countries equal in population, we may declare with positive certainty that the wealthiest and most highly civilized is that which consumes the greatest weight of Soap." It is not, however, merely by the quantity consumed of this important article, that the distinguished chemist would establish its claims to represent the civilization of a people. The vast train of chemical, manufacturing, and commercial operations called into existence for its economical production, and the cheaper, more extended, and altogether new arts and processes incidentally growing out of these, would, even with political economists, entitle it to this rank.

The materials used in making Soap are alkalies, and fatty substances, or oils, both of animal and vegetable origin. Of the former, Potash, Soda, and a small proportion of Lime, are employed. The artificial production and cheap supply of Soda, the alkali chiefly used, from common salt, introduced about the beginning of the present century, has since that time completely revolutionized the business in Europe and in this country, and probably within the last twenty years quadrupled its amount.

In this city, there are about thirty-five establishments engaged in the manufacture of Soap; and few branches of our manufactures have grown more rapidly with the prosperity of the city. We have been assured that there is more Soap now made here in one month, than there was twenty years ago in a whole year. At that time we were greatly dependent upon New York, New England, and Western Soap-makers;

THE LARGEST

Soap Manufactory

IN THE

UNITED STATES.

McKEONE, VAN HAAGEN & CO.'S

WAREHOUSE—32 SOUTH FRONT ST.,

PHILADELPHIA.

FAMILY, CASTILE AND RICHLY PERFUMED

SOAPS,

IN GREAT VARIETY.

WARRANTED EQUAL TO THE BEST IMPORTED IN

QUALITY AND PERFUME.

and Colgate's Soap of New York crowded every store; but now our own manufacturers supply our market, to the exclusion of nearly all competitors, and have besides large supplies for exportation. A few of them manufacture almost entirely for exportation to the West Indies, South America, etc. They make all the varieties in common use, and some make Soap of superior quality. Palm Oil is extensively employed for making Soap and Stearine Candles; and Olive Soap of remarkable power, soluble in strong brine, and therefore well adapted for marine use, is an important article made here.

The oldest Soap manufactory in the city, and probably the oldest in the country, is that of G. & T. H. DALLETT, Tenth and Callowhill streets. The Dallett family have been engaged in the business for three generations, a period of sixty years; and the present firm have the advantage of their predecessors' connections, as well as their own long personal experience. This firm manufacture Fancy as well as Common Soaps, and Mould and Dipped Candles, and have an enviable reputation for reliability, and fair dealing, which have been crowned with a well-merited business success.

The most extensive Soap manufactory in the city, and probably in the United States, is that of Messrs. McKEONE, VAN HAAGEN & Co., on Callowhill street, opposite Fairmount. The lot on which it stands fronts on Callowhill street one hundred and seventy-five feet with a depth of two hundred and sixty feet, which is in great part covered with buildings of two or three stories in height. In one of these are made the Family, and in another the Toilet Soaps, in large quantities. Nearly one hundred hands are constantly employed in this manufactory.

The warehouse and office of the firm are at 32 South Front street, whence vast quantities of Soap are distributed to supply the demands not only of the city trade, but of the Eastern, Western, and Southern States, in all of which their productions are highly appreciated.

Messrs. JONES, EAVENSON & SON, are also extensive manufacturers of Soap, at the Quaker City Manufactory, 315 North Twentieth street. They produce Olive, Oleine, American Castile, and Fancy Soaps, and, to some extent, Tallow Candles. Their business is conducted in a three-story brick building, having a depth of ninety feet; and two of their soap pans have a capacity of producing at a charge forty thousand pounds of Soap each.

In this, as in most other branches of manufactures, there is a subdivision of reputation, and some firms are more successful than others in producing particular kinds of Soaps. For instance, LINDLEY M. ELKINGTON, of 116 Margaretta street, has attained a marked distinction for the manufacture of *Palm Soaps.* The business was established by his father, George M. Elkington, more than forty years ago, and by strict attention to purity, and maintaining the original standard of quality, Elkington's Palm Soaps have attained a high degree of popularity, and are in extensive demand. Besides Palm Soaps, Mr. Elkington manufactures various Family Soaps that are said to be of more than the usual weight.

Soaps *for fulling purposes* are a specialty with Mr. ALEXANDER McCONNELL, 1322 North Fifth street. Mr. McConnell commenced the manufacture of Soaps in Philadelphia in 1848, and, by straightforward and fair dealing, has established a flourishing business, supplying largely the city trade. Besides Fullers' and other Soaps, Mr. McConnell makes Tallow Candles, and imports Sal Soda and other chemicals.

A pure *White Linen Soap* is a specialty with Mr. WILLIAM SIMPSON, of 2930 Market street. This is said to be the only Patented Soap made in Philadelphia, and is very economical, from the fact that it will not waste away. This Soap is as hard in two days as Resin Soap is in six months, and is adapted alike for private families, laundries, and manufactories of fine woolen cloths. Every bar of this Soap that is genuine is stamped with the impress of a Harp, which Mr. Simpson has adopted as his trade mark.

The manufacture of CANDLES is so very generally associated with that of Soap, that the branches may be considered inseparable. The advances that have recently been made in chemical science have wonderfully influenced the manufacture of both articles; and by the separation of constituents, purification, distillation, pressure, and other arts and appliances known to the initiated, it is possible to attain very remarkable results from very unpromising materials. The most impure fats, as well as Palm and other oils, may be made to yield, by the skilful Candle-maker, a product from their solid portions but little inferior to those made from wax, which is too expensive for ordinary use. *Dip Candles* are nearly obsolete; but the manufacture of Moulded Tallow Candles is still an important part of the business of nearly all the Soap-makers.

XXIV.

Stoves and Heating Apparatus.

Five large Foundries in this city are devoted exclusively to the manufacture of stoves; while two others make stoves, together with miscellaneous castings. The capital invested in the manufacture is nearly $1,000,000, and the product about fifteen thousand tons annually. The designs, in many instances, are remarkable for their elegance, and the establishments are not surpassed in facilities or in extent by any others. The cheapness of the raw material, and mildness of the winters, enabling the manufacturers to continue operations without cessation throughout the year, are marked advantages, and the fineness of the castings induces professed manufacturers in other places to obtain their supplies from this city. The varieties made here embrace almost every description, from the old Franklin Stove, and the *Ten-plate* Wood Stove, down to the most modern styles and patterns, including Gas Cooking Stoves. In originating patterns and beautiful styles, the Philadelphia manufacturers and Stove pattern-makers have been remarkably successful; and Stoves from this city have been shipped to Oregon, California, Australia, and Europe; while in our own markets no others can compete with them.

Probably the largest establishment engaged in the casting of Stoves in the city is that of SHARPE & THOMSON, successors to North, Chase & North, who were prominent Stove Founders in the city for twenty years. This firm have in connection with their works six acres of land, located between Second street and Moyamensing avenue and Mifflin and McKean streets. The buildings are of brick, and include two of immense size. The Moulding Room, which, it is supposed, is the largest in the United States, has a front of one hundred and one feet on Second street, and a depth of three hundred and five feet on Mifflin street, containing over thirty thousand square feet. In this are two cupola furnaces, capable of melting eighty tons of iron per day. The Finishing Department and Wareroom occupy another extensive building two hundred and sixty-eight feet long, and sixty-eight feet wide. Besides these immense structures, there is a building seventy-one by eighteen feet, in which the offices connected with the foundry are located, a Pattern Store Room, one hundred by eighteen feet, and numerous auxiliary buildings necessary to the successful prosecution of the business on an extensive scale. There is an Artesian well on the premises, which supplies water for the engine and boilers. Almost three hundred men are employed, and twenty-five hundred tons of No. 1 American iron are converted into Stoves annually.

Besides Stoves, Messrs. SHARPE & THOMSON manufacture Tinned and Enamelled Hollow-ware of a fine quality, as has elsewhere been stated, and, in addition to their manufactories, the firm have a large five story store, at 209 North Second street, which is one hundred and fifty feet in length, and filled with samples of their various productions, presenting an assortment of unsurpassed attraction. The salesrooms are in charge of Mr. CHARLES SHARPE, while the Manufacturing Department is under the supervision of Mr. EDGAR L. THOMSON, an experienced founder.

The "Excelsior Stove Works," on Girard avenue and Marshall street, of which Messrs. ISAAC A. SHEPPARD & Co. are proprietors, are fitted up with a full assortment of modern patterns and appliances for the manufacture of Stoves, Heaters, Ranges, and Hollow-ware. They employ from seventy to eighty men, and melt from twelve to fifteen hundred tons of iron per year. The gentlemen in charge of these works are men of large experience in iron founding, and as a result of their experience, they have found they can produce castings of greater strength and tenacity from a proper mixture of brands of American Iron, than from Scotch Iron, and consequently they use only American Iron of the best brand.

Besides Stoves, etc., Messrs. Sheppard & Co. make castings for Plumbers, in fact, all kinds of light and fine castings.

Messrs STUART, PETERSON & Co., whose works have been previously referred to, produce about five hundred Stoves a week, or twenty-seven thousand in a year. They make all kinds and styles of Stoves that are in demand, and employ six persons whose business is exclusively to originate and prepare new patterns and designs of Stoves and Heating Apparatus.

The firms of ABBOTT & NOBLE, COX, WHITEMAN & COX, and LEIBRANDT & McDOWELL have extensive Stove Foundries, and in addition there are probably fifty stove makers who get their castings from founders, and finish them in their own shops.

Besides the Stove Founders, there are several firms who make a specialty of manufacturing Stove Shovels, Pipe, Elbows, and Coal Hods. The principal houses in this branch are those of GEORGE GRIFFITHS, WILLIAM EASTERBROOK, and McCOY & SNELL, all of whom are practical manufacturers, and keep extensive stocks of the finished articles in store.

2.—HEATING APPARATUS.

In a climate where artificial heat for a considerable portion of each year is requisite for domestic comfort and business convenience, the sub-

ject of warming buildings in the best and most economical manner, has necessarily engaged the attention of many of the thinking minds of the country, and a great variety of ingenious contrivances for the purpose has been presented. Philadelphia has her full complement of these devices, and her manufactories of Heating Furnaces are numerous, successful, and important. We shall not attempt to do more than allude to the peculiar points presented by a few of the prominent manufacturers.

One of the pioneers in introducing improved methods of heating in this city, is CHARLES WILLIAMS, whose establishment at 1132 and 1134 Market street, is among the most extensive in the city. He employs about forty hands in manufacturing the *Culver* and *Golden Eagle* Hot Air Furnaces, which have been thoroughly tested and approved as efficient for the purpose for which they were designed. In these furnaces a moist and perfectly natural and healthy atmosphere is obtained before the air reaches the rooms to be warmed by means of a Vapor Pan, and other ingenious contrivances.

Mr. Williams also manufactures Ranges of all descriptions, and *Collins' Ventilator*, which is used in the principal public buildings of Philadelphia, New York, and Boston. They are also extensively employed in ventilating railroad shops and engine houses, more than a thousand of them having been ordered by the Pennsylvania Central Railroad.

Another old and celebrated establishment in this branch is that of D. MERSHON'S SONS, 1209 Market street. The founder of this firm was engaged since 1840 in constructing, improving and setting up Furnaces for warming and ventilating Private and Public Buildings, and, as the result of his long experience, devised what is known as the *Mershon Soft Air Furnace*, which is adapted alike to the use of anthracite or bituminous coal, coke or wood. This firm have recently invented and patented a novel and very convenient arrangement for regulating the fire from any story in the building, without going to the cellar. With this labor-saving invention, the furnace need be visited but once during the day, after which any of the inmates, even a child, can control the fire from any of the upper rooms with the greatest ease. A scale is engraved on one of the side plates of the regulator, and an indicator slides up and down on this scale by means of which the heat can be regulated to any number of degrees required.

Messrs. Mershon's Sons have a store one hundred and sixty feet in depth, and employ in their manufactory from thirty to forty workmen.

Messrs. J. REYNOLDS & SONS, northwest corner Thirteenth and Filbert streets, are manufacturers of what is known as "Reynolds' Air-Tight Gas Consuming Heater," with *Patented Dust Screen*, and *Grate Bar Rests.*

These Heaters are made of the best *Wrought-iron Plates, ¼ inch thick* next to the fire, well riveted together, so as to make it perfectly tight, preventing the escape of either gas or dust, and avoiding all sand and putty joints. By being made of wrought-iron, and riveted together in this manner, the expansion and contraction of the several plates or parts are equal, and consequently the Furnace is always air-tight. The great distinguishing feature between this and all other heaters is the *Round Air-Tight Door, Turned and Ground,* with beveled joint or edge on the door and frame, and brass slide and hinge. The heater is regulated entirely by this ground Air-Tight Door, no dampers being used, the upper or feed door being at all times kept closed, except when supplying coal, and the fire controlled by the ground brass slide in the lower round or air-tight door—opening it to increase the heat, and closing to reduce it. By this means no current of cold air is allowed to pass in the upper or feed door, over the fire, thereby cooling it and reducing the temperature of the radiating surface.

The advantages of the attachments of the Patent Dust Screen and Grate Bar Rests are fully set forth in a pamphlet published by this firm, which also contains a considerable amount of valuable information on the subject of Heating.

Mr. Jesse Reynolds is an ingenious man who has had an experience of nearly twenty years in the construction of Heating and Ventilating apparatus. All the work, from the crude inception until the Heater is set up, is performed by the employees under the firm's direction, and satisfaction is in every instance guaranteed.

Messrs. Andrews, Harrison & Co., 1327 and 1329 Market street, manufacture both Warm Air and Steam Heating apparatus, in which they use the Harrison boiler as the steam generator, and cast-iron sections with serpentine flanges as their radiators. But the specialty for which this firm is distinguished is the manufacture of *Low-Down* and *Raised Grates,* suitable for either wood or coal fires. Mr. James Andrews was, we believe, the inventor of what is known as the Low-Down Grate. When desired, their Grates can be constructed with an air-chamber in the rear of the niche, with heating capacity sufficient to warm one or two Bed-chambers over the Grate, during the Spring and Fall Seasons, and even during the severest winter producing sufficient caloric to overcome the chilliness or dampness so often experienced in retiring or rising.

For warming and ventilating large buildings, such as stores, churches, school-houses, asylums, etc., as well as dwellings, the apparatus known as Gold's Patent, and manufactured by James P. Wood & Co., No. 41 South Fourth street, has probably no superior in this country. This

HOUSE WARMING.

ANDREWS, HARRISON & CO.,

1327 & 1329 MARKET STREET,

Fronting on East Penn Square,

ARE MANUFACTURERS OF

LOW-DOWN & RAISED

FIRE GRATES,

Which are unequalled for beauty of design and finish. These Grates are not only highly ornamental for Drawing Rooms, Parlors, Chambers and Offices, but possess superior advantages as thorough ventilators. When desired they can be so constructed as to warm one or two bed-chambers over them, during the spring and fall seasons, and in the severest winter, producing sufficient heat to overcome the chilliness or dampness so often experienced on retiring or rising. The large number of these Grates now in use in the United States and Canada, attests the high appreciation in which they are held by the community.

Messrs. ANDREWS, HARRISON & Co , are also manufacturers of a *Steam Heating Apparatus*, using the Harrison Boiler as the steam generator, and cast-iron sections with serpentine flanges as their radiators. The boiler is safe from explosion—efficient, durable, and economical in the use of fuel, and its application to the purposes of House Warming is a success that cannot be too highly appreciated. This firm also manufactures *Warm Air Furnaces*, *Ranges*, *Bath Boilers*, etc.

Messrs. ANDREWS, HARRRISON & Co. employ in their establishment men of mechanical skill and experience, and by their care and attention in the collection of a new and unique assortment of *Patterns*, offer superior inducements to purchasers.

☞ *Harrison's High and Low Pressure Steam Boilers* constantly on hand.

CHARLES WILLIAMS,

Heating and Ventilating Warehouse,

1132 & 1144 MARKET STREET.

CULVER'S IMPROVED

WARM AIR FURNACES,

FOR MASONRY.

Gold Eagle Heater,

One of the best Portable Heaters ever offered. Six sizes—requiring no brick work—will heat the largest size dwelling house.

Ranges of all sizes, Low Grates, Fire-place Heaters, Registers & Ventilators for all description of ventilating.

WHOLESALE & RETAIL.

apparatus combines the two methods of heating, hot water and steam, and thereby secures the uniformity and endurance of the hot water surface with the expansive power of steam, able to multiply itself to any extreme of temperature. A brief description of this celebrated patent will be of interest to many.

The Steam Generator or Boiler is composed of cast-iron similar sections, each a complete Boiler in itself; by using few or many, the capacity may be increased or diminished to suit the size of the building to be warmed. The fire-chamber and flues are formed in the boiler entirely surrounded by water, the water and steam chambers of all being united. The boiler enclosed in an air-chamber with Radiators, forms part of the heating surface of the building, and affords easy access for cleaning and repairing. The entire apparatus is located in the basement, or cellar, where fresh pure air from the outside is warmed, and thence conveyed to the different apartments; the construction of this apparatus, and its action on the atmosphere, produce a circulation which renders the ventilation complete, and the warming thorough. The Radiators are also composed of cast-iron similar sections, studded with nearly one thousand points or projections, and bolted together in groups of from ten to fifty, as may be required and supplied with steam; the condensed steam returning by a separate pipe to the boiler, thus making the circulation. A Pedestal Radiator of beautiful design, intended for halls and locations where flues cannot be formed to convey the heat, and where a tangible direct heat may be required in the apartment, is also formed of similar sections, placed on an open base, and capped with an open cornice, covered with fret work or slabs of marble as may be desired, its internal arrangement being the same as the Radiator just described, with alternate steam and air passages, its entire height and the outside neatly corrugated, forming the handsomest as well as the best Steam Radiator ever made.

XXV.

Sugar Refining.

The introduction of Machinery and Steam has effected an entire revolution in the processes of purifying and refining Sugar, and this combined with the substitution of aluminous finings in place of bullock's blood, which supplied a fertile source of deterioration, has wonderfully increased the quantity of production, and raised the standard of quality. The Raw Sugar, from the West Indies, is imported in cases and hogsheads; from the English Islands in hogsheads; from South America chiefly in bags, as also from Manilla and the Mauritius. These latter bags are double, and made from the leaves of reeds, plaited or woven into suitable material.

The first operation of the Refiner, after removing the Sugar from the hogshead, boxes, etc., is dissolving the Sugar in a pan by means of steam passing through a perforated pipe in the bottom of the pan. The color is then extracted from the solution by means of chemical and mechanical means, when it is passed to what is known as the vacuum pans, heated by steam, for the purpose of being boiled. By this means the liquor is so concentrated that the Sugar is only held in solution by the high temperature, so that on cooling a rapid crystallization takes place, which produces that uniform fine grain, such as is required in Loaf Sugar. The syrup, after boiling sufficiently, is poured into the moulds, which are of the funnel or sugar-loaf form, for the purpose of assisting the separation of the mother-liquor. The syrup or liquor which runs from the mould is again boiled, from which the lower grades of Sugar is produced. The syrup coming from this second process is sold for molasses. The production of molasses is about one-fifth from each hogshead.

The art of Refining, it is believed, has attained a higher standard in this country than in any part of Europe, and the excellence of this manufacture is not approached by any imported article. Within a recent period, our own city has advanced greatly in both the quantity and quality produced. A few years since, but a single Refiner had a name here, and a well deserved one; now, several others are approaching the high standard with rapid strides. We would instance particularly, FICKEN & WILLIAMS, ROGERS & MITCHELL, HARRISON, HAVEMEYER & Co., who with DAVIS, McKEAN & Co., successors to J. S. Lovering & Co., NEWHALL, BORIE & Co., TAYLOR, GILLESPIE & Co., and E. C. KNIGHT, constitute the principal firms engaged in the business. Messrs. FICKEN & WILLIAMS have greatly enlarged their works and perfected their processes, and the new houses of HARRISON, HAVEMEYER & Co., and E. C. KNIGHT have introduced all the latest improvements.

The buildings used by these firms are very extensive, and the combined steam power amounts to over twenty-five hundred horse power. The number of men employed in the different works is about eight hundred, and the amount of Raw Sugar imported and used, will reach a thousand hogsheads per week, from which nearly five thousand barrels of Loaf and the different grades of Clarified Sugar are produced, the greater portion of which is for the Philadelphia market. Each barrel of Sugar weighs about two hundred and forty pounds.

The value of Refined sugar made in this city in one year, taking for the basis of calculation the data given above, with the prices which

FICKEN & WILLIAMS,
STEAM SUGAR REFINERS,
DOUBLE LOAF,
A CRUSHED,
CRUSHED,
Patent Cut Loaf,
CHIP CRUSHED,
Pulverized,
LOZENGE DO.
GRANULATED.
A COFFEE,
COFFEE,
B Coffee,
COFFEE,
C COFFEE,
X YELLOW,
YELLOWS.
Crown, Willow and Fifth Streets,
PHILADELPHIA.
J. Haehnlen Philad.a

ruled in 1866, and a working period of ten months would be quite $20,000,000, and the business of the year, including Molasses, amounted to $21,000,000. The Refineries here are of sufficient capacity to produce $30,000,000 annually, if constantly in operation.

XXVI.

The Tobacco Manufacture.

The consumption of Tobacco by both sexes is so vast that it may fairly be classed among the necessaries of life. A single city claims to do a business in Tobacco of $100,000,000 annually; twenty-five thousand of her inhabitants are employed in receiving, selling, sampling, and manufacturing the article; over sixteen hundred distinct firms are engaged in it exclusively. Four of her fine cut manufacturers employ fifteen hundred operatives, and three of her cigar factories six hundred more. Excepting cotton alone, it is the most important export; and in freights, duties, and revenue, it does a larger business than any other staple. A small town in the upper part of Missouri contains two tobacco factories, which turned out in one season (so it is reported) seven millions of pounds of Chewing Tobacco. A large part of this, as of the products of St. Louis, Louisville, Cincinnati, Chicago, Detroit, etc., found its way to the Eastern markets, and helped to swell the trade here. The Tobacco crop for this year is estimated at four hundred million pounds, and, as the raw material is greatly enhanced in value by the process of manufacture, it is safe to say that the entire value of the article, in its raw and manufactured state, will not fall much below two hundred millions of dollars annually.

It is estimated that there are about twenty-five hundred places in this city where Tobacco in some shape is sold. The receipts of manufactured and leaf Tobacco in this market, from the South and West, may be set down at about fifty million pounds per year. The imports are comparatively light, owing to the excessively high rates of duty, and the exports are also light, most of the dealers engaged in the exporting of Tobacco being found at other points. During the year 1865, there was manufactured in this city a little over a million of dollars' worth of Tobacco, about a half million of dollars' worth of Cigars, and over a quarter million of dollars' worth of Smoking Tobacco, as appears from the books of the Internal Revenue Department. It is quite evident, however, that this statement gives hardly an approximation to the exact state of the trade, especially of the Cigar manufacture. A few years since a leading tobacconist made a personal investigation of the extent of the business, and asserted that there were about one thousand manufacturers of Cigars in Philadelphia; thirty of whom employ from ten to sixty-five hands, and the others from one to five and six; and that the whole number of employes, journeymen, and girls, was fully four thousand. If each hand made fifteen hundred Cigars per week, a minimum amount, the weekly production would be six millions, or three hundred and twelve million Cigars per year. It is quite probable that since the great increase in cost fewer Cigars are made in Philadelphia than formerly, but the aggregate product amounts to not less than $2,500,000. Besides, several of the largest manufacturers have their factories outside of the city limits. For instance, Mr. STEPHEN T. GREENLEY, of 333 Chestnut street, a leading Cigar manufacturer, has his factory at Sellarsville, Pennsylvania, where he employs one hundred hands. Messrs. L. BAMBERGER & Co., who are among the largest wholesale dealers in Domestic Cigars, are also supplied principally from the interior of the State.

The art of making Cigars here has been brought to great perfection. Cigars made by JAMES DALEY have brought as high as $120 per thousand. The Cigars are, for the most part, made with the greatest care, packed and branded in close imitation of the finest Havanas, and so flavored as to puzzle good judges to tell them from the imported—not for the purpose of deception, but from a desire to turn out the very best article that can be made. The Tobacco grown in some portions of our own State is coming more and more into favor, as increased care is taken in its culture. Many prefer it to the Connecticut Seed Leaf, though prejudice reduces its market value considerably below the latter.

Of *Snuff*, there is also a large quantity made in the city. The house of W. E. GARRETT & SONS has been established for a century, and their Snuffs are widely and favorably known. Mr. A. RALPH, formerly with this firm, has recently established works for manufacturing a superior quality of Scotch Snuff.

The style of Chewing Tobacco most popular here is the black, sweet "Navy." It is a little singular that the people of different cities should have their own local peculiarities in the matter of Chewing Tobacco. Thus, in New York, "Fine Cut" is the popular style; in Baltimore, "Fine Bright," Hard Tobacco is preferred; in Boston, something sweet but not black is desired; and in the West the taste is divided between "Fine Cut," "Bright," and moderately sweet goods. Here in Philadelphia, nothing can be too sweet or too black to suit the great mass of chewers.

Philadelphia is favored in having a large number of excellent houses

J. RINALDO SANK. WM. M. ABBEY. JOSEPH BROOKE.

J. RINALDO SANK & CO.,

TOBACCO

AND

General Commission Merchants.

UNITED STATES BONDED WAREHOUSE,

Nos. 31 N. Water St. and 30 N. Delaware Avenue,

PHILADELPHIA.

ALSO,

AGENTS FOR THE SALE OF

COTTON, RICE, NAVAL STORES,

SOUTHERN TIMBER AND LUMBER.

☞ *Consignments solicited of any of the above articles of SOUTHERN PRODUCE. Liberal advances made and prompt sales rendered.*

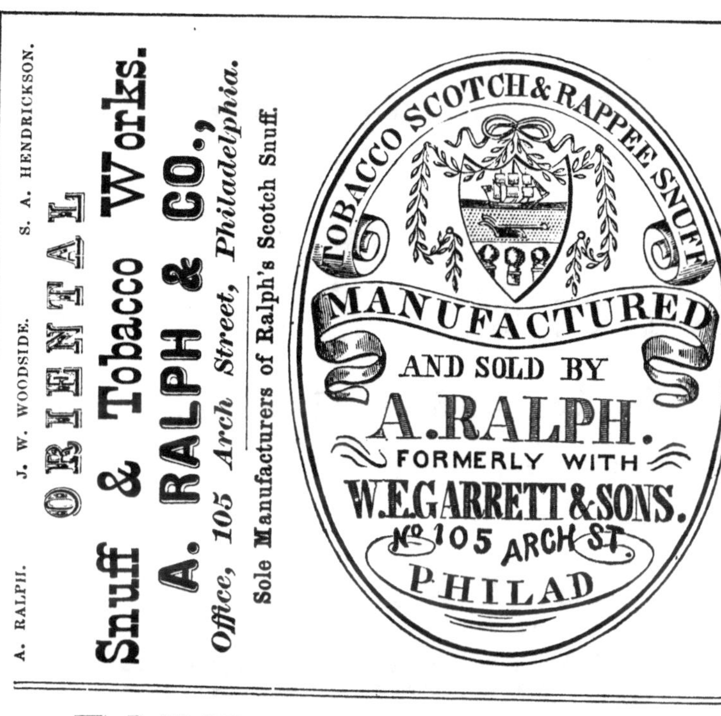

JAMES DALEY,

N. E. Corner THIRD AND RACE STREETS.

FINE SEGAR

MANUFACTURER.

West End Branch. 2570 Callowhill.

Factory, 1921 Callowhill Street.

in the general trade noted for their uniform courtesy, liberality in making advances, and their promptitude in returning accounts of sales. Among the commission houses the firms of BUCKNOR, McCAMMON & Co., J. RINALDO SANK & Co., S. FUGUET & SONS, GEYER & HISS, DOHAN & TAITT, BOYD, FOUGERAY & Co., L. BAMBERGER & Co., McDOWELL & DUNCAN, and JOHN DOUGLASS, are well worthy the attention of those who are seeking a good market; and among the wholesale dealers in Manufactured Tobacco, Messrs. B. A. VAN SHAICK, HICKMAN, HOLL & Co., J. KINSEY TAYLOR, and LOUIS B. GRIM, are fair representatives of the best houses in Philadelphia.

XXVII.

Umbrellas and Parasols.

The manufacture of Umbrellas and Parasols has long been a prominent pursuit in Philadelphia, and the establishments include some that are without a rival in this country. It is probable there are more than a hundred places in Philadelphia, where Umbrellas and Parasols are made to some extent, but the very extensive establishments are limited to four or five. The causes that have contributed to the supremacy of Philadelphia in this manufacture, are principally those which have led to a like result in other branches; but there are also special and particular reasons for the superiority. The *sticks* and *metal mountings* made in Frankford, a populous suburb of the city, are unsurpassed for excellence and efficiency. The *stretchers*, made from the best Pennsylvania Iron—the wire, drawn at Easton, and formed, forked, and japanned at the House of Refuge, under the superintendence of a firm in this city—are tougher, and less disposed to rust or oxidize, than any in the world. The mechanical genius of the manufacturers has also been active, and a number of very important improvements, which facilitate the manufacture, have originated here. Of this description we might instance the "Sorting Machine," invented by the elder Mr. Sleeper, for adjusting the strength of the ribs, or *setts*, worked by balance-weights, and by determining the strength of each rib, ensures the perfect and regular shape of their goods.

WRIGHT BROTHERS & Co.'s Umbrella manufactory, it is believed, is the most extensive of its kind in the world. It was founded in the year 1820 by four brothers, who came from England to the United States in 1816, and embarked in the business of preparing Whalebone for Umbrellas, to which they added, in the year named, the manufacture of the finished article. An account of this important manufactory will be found in the APPENDIX.

Erasmus J. Pierce, who was succeeded by WILLIAM A. DROWN & Co., was one of the pioneers in this manufacture. Previous to the war of 1812, he was engaged in the business in Baltimore; but his residence in this city dates, we believe, from 1815. At that time, forty Umbrellas per day was a large product—fully as much as the demand would warrant. Mr. Pierce retired from active participation in the business about 1836, and at the time of his retirement was accounted among the very largest manufacturers. His successors, Messrs. Wm. A. Drown & Co., have a large manufactory in Frankford in connection with their warehouse, 246 Market street, and are noted for the fine styles of Umbrellas and Parasols which they produce. Their goods are well known throughout the entire country.

BENEDICT MILLER, of 39 North Sixth street, is one of the oldest practical Umbrella manufacturers now engaged in the trade. He has been uninterruptedly connected with the pursuit since 1842, now a quarter of a century, and has established an excellent reputation for good workmanship. His Umbrellas and Parasols are especially sought after by the city dealers, who appreciate substantial, combined with elegant, workmanship.

The circumstances that have contributed to the development of the Umbrella manufacture in Philadelphia, we have in part already alluded to. The establishments in Frankford for the manufacture of Metal Mountings, Tips, etc., are deservedly noted, and supply not only the manufacturers of this city but of New York. The Ivory and Bone Turners, and Carvers, perform their part well in ornamenting the handles. In the manufactory of HARVEY & FORD, undoubtedly the most extensive in the United States, one hundred and fifty operatives are often employed, and their carved Ivory-work successfully rivals the finest of England and France. There is also a very extensive establishment devoted exclusively to preparing *Whalebone* and *Rattan* for all the purposes to which these articles are applied.

XXVIII.

Wagons, Carts, Drays, Etc.

Within comparatively a few years the demand for Wagons of a peculiar construction has elevated the business of Wagon-making into the rank of manufactures. The wheelwright and the blacksmith are no longer able to supply the combined wants of the United States Govern-

JOHN L. FULTON,

COACH & WAGON BUILDER,

Twentieth and Pine Sts., Philadelphia.

REFERENCES.

J. BINGHAM, Sup't Adams Express, Phila.
F. LOVEJOY, Gen'l Sup't Harden Express, Phil.
CLARK & CO., Special Freight Agt., Pittsburg.
GEO. BINGHAM, Agt. Adams Express Co., "

☞ *Repairing Promptly and Neatly Executed.* ☜

UNION

STEAM MILL,

SPOKE AND HANDLE

MANUFACTORY.

JOHN G. DAVIS & SON,

FRONT STREET,

ABOVE LAUREL,

PHILADELPHIA.

WILSON, CHILDS & CO. PHILADELPHIA PLANTATION & ROAD WAGON WORKS.

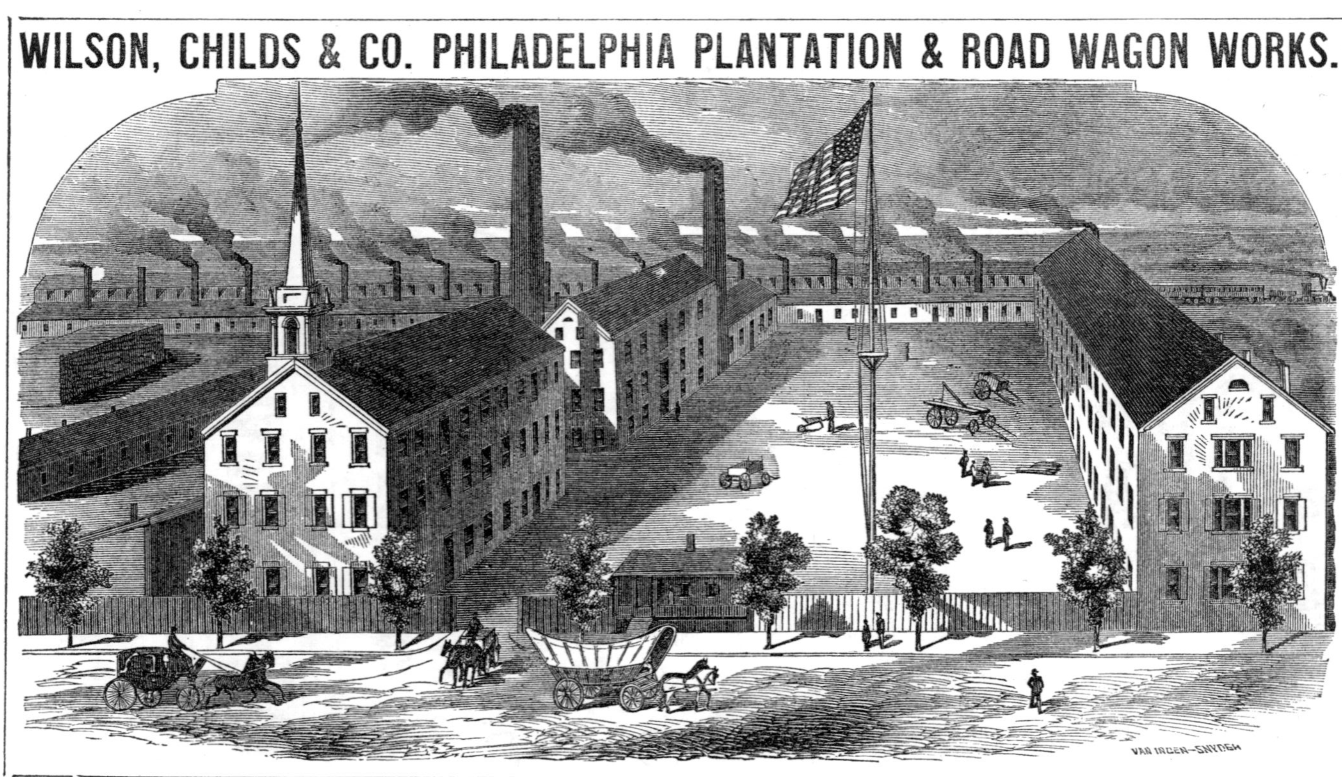

ment, Express Companies and Emigrants; and establishments are required that can purchase lumber and iron in large quantities, and which are provided with all the requisite machinery and appliances for turning out heavy vehicles with expedition and rapidity. The excellence of the timber furnished from the forests of Pennsylvania, New Jersey, and Delaware; the reputation of the builders, established even prior to the time when Conestoga Wagons transported all heavy goods from the Eastern States to the West; the present facilities of the Wagon-making establishments, and the immense stock of well-seasoned lumber always kept on hand, make Philadelphia, at the present time, the best and principal seat of the Wagon manufacture.

The Philadelphia establishments send their products to almost every part of the South, including Texas; and even Mexico obtains a second edition of our old Conestogas, which are there drawn by mules. The United States Government has for some time obtained its Army Wagons principally from Philadelphia, inasmuch as those made here have been found to be the most serviceable on the Western frontiers, where only those of the best material and construction can withstand the abrasion of severe use. For a description of one of the largest Wagon-making establishments in Philadelphia, see in Appendix, WILSON, CHILDS & CO.

Philadelphia has also the largest manufactory of *Spokes and Felloes* in the United States. The proprietors, Messrs. JOHN G. DAVIS & SON, have, in fact, two manufactories; one at 1028 North Front street, and the other at 1528 Frankford road. The former is seventy by one hundred feet, on a lot two hundred feet square; the other is sixty by one hundred and fifty feet—both three stories in height. These factories have machines capable of producing a million of spokes annually, besides plough-handles, carriage-poles, and felloes. The stock of these articles in a finished state, kept on hand at all times, is immense. Messrs. Davis & Son employ about one hundred hands. The timber for the spokes is hickory and oak—principally hickory—and the felloes, poles, etc., are made of hickory, ash, and oak, all of which are well-seasoned before being used.

XXIX.

Wood-Working—Building Materials, Etc.

The working of Wood in Philadelphia, and throughout the United States, is especially remarkable for the application of labor-saving machinery, by which the most important results are attained from apparently very simple means. All the implements and machinery designed for the purpose, and in use in this country, from an ordinary axe to a planing machine, are far in advance of those in use in Europe. A house in Liverpool is now importing the best American Wood-working machines, and making great efforts to introduce them generally into England.

The abundant supply of Lumber in Pennsylvania, and the sources of that supply were stated in the Introductory. It is probably our duty to describe the facilities that are in use in the various and numerous Wood-working establishments; but we are reminded that the leading branches of productive industry have already consumed more than their allotted space; and we are satisfied that nothing like justice could be done, within narrow limits, to a subject so comprehensive. It must suffice, therefore, for us to say that the machinery in the various Planing Mills, Sash Factories, Turning and Scroll-sawing Establishments, etc., is truly remarkable for its efficiency; and that those establishments occupied in preparing the various parts of Wood-work required in Buildings, can supply builders at a much cheaper rate than the latter can produce them in their own workshops, without the aid of such machinery.

MISCELLANEOUS MANUFACTURES.

I.

Artificial Teeth and Dental Materials.

Philadelphia was the original seat of the manufacture of Porcelain Teeth in the United States, and over one half if not two thirds of all the Artificial Teeth now used in this country are made in Philadelphia. One manufacturer, Mr. S. S. White, produces annually nearly one million and a half of Artificial Teeth. His products are exported to Europe, South America, and the West Indies, and in all markets they take precedence over the European manufacture.

Mr. White is the successor of Jones, White & McCurdy, who for many years were the leading manufacturers of Porcelain Teeth in this country. Many of the important improvements relating to the shape, articulating and enamel surfaces, originated with them. Mr. White received the Prize medal at the Paris Exposition, for Porcelain Teeth, over all European manufacturers.

The oldest of the established firms now engaged in this manufacture is that of T. G. Armstrong & Son, 520 Arch street. The senior of this firm commenced the business in 1836, and many of those who are prominent manufacturers, gained the rudiments of their knowledge in his establishment. His long experience has necessarily given him a thorough familiarity with the best methods of making Porcelain Teeth, and his products are in favor in both continents.

McCurdy & Hoover, as successors of Samuel W. Neall, who established the business in 1840, are the next oldest manufacturers of Porcelain Teeth. The present firm are succeeding well in maintaining the reputation acquired by their predecessors.

Johnson & Lund, 27 North Seventh Street, are with one exception the most extensive manufacturers of Porcelain Teeth in the United States. During the last year they employed forty hands, and produced eight hundred and thirty-two thousand Artificial Teeth. In 1865 they received a Prize medal, at the World's Fair, Stetlin, Prussia, and their products are now sold largely in Europe. This firm have a branch office in Chicago.

The firm of H. D. Justi & Co. is composed of H. D. Justi and Charles

L. Orum, who were formerly connected with the firm of Orum, Armstrong & Justi. The specialty of this firm is known as "Justi's Star Section," though they manufacture all other kinds of Artificial Teeth. Their stock of patterns is very large, and adapted to the wants of the dental profession in all parts of the United States.

The PHILADELPHIA DENTAL MANUFACTURING Co., incorporated by Act of the Legislature, is managed by gentlemen formerly connected with the firms of Rubencame & Stockton, and W. A. Duff & Co. By the union of these well known firms, the Company has secured not only a large stock of moulds and patterns, but an important accession of skill and practical experience. The following are the officers: President, W. A. Duff; Secretary, T. H. Stockton, Jr.; Treasurer, John R. Rubencame; Superintendent, Dr. J. J. Griffith.

Besides manufacturing Artificial Teeth, all of these firms keep an assortment of Instruments, Gold Foil, Operating Chairs, and other materials required by dentists in the practice of their profession.

There is also in Philadelphia the oldest established and most extensive manufactory of Gold Foil for dentists' use, in the United States. We refer to that of CHARLES ABBEY & SONS, 228 Pear street. Mr. Abbey, the founder of the firm, has been uninterruptedly engaged in manufacturing Gold Leaf and Gold Foil since 1816. His connection with the manufacture of Gold Foil, to which he has exclusively given his attention for the last quarter of a century, may be said to cover the entire period of its use for purposes of dentistry. When he commenced, dentistry was scarcely known as a distinct profession. This firm, consisting of himself and two sons, prepare all the Gold Leaf they use, not purchasing from refiners, as is customary; and that they make a most superior article is proved by the fact that "Charles Abbey & Sons' Foil" is quoted in the trade's circulars at a higher price than that of any other manufacturer. They supply all the prominent dentists in the United States, and receive orders from England, France, Germany, Sweden, and from other parts of Europe.

Philadelphia has also several excellent and extensive manufactories of Dental Instruments. HORATIO G. KERN, 25 North Sixth street, is one of the oldest established in this branch, and is now probably the largest manufacturer. For thirty years he has been enlarging his facilities and acquiring experience, and at this time stands at the head of his profes sion in this country. While manufacturing a variety of Surgical Instru ments, he has achieved reputation especially in those for dentists' use. He has a special instrument for every tooth in the human mouth, an furnishes Dental Cases at prices varying from $50 to $500. His Dent

Forceps, which are a specialty, are regularly exported to the West Indies and frequently to Europe.

Messrs. SNOWDEN & BROTHER, whom we will notice in connection with Surgical Instruments. are also extensive manufacturers of Dental Instruments.

II.

Artificial Limbs and Surgical Instruments.

Within the last twenty years, so many and great improvements have been made in the construction of Artificial Limbs, that some eminent surgeons have declared that they would prefer these substitutes to some natural legs they endeavor to save. During the last few years, the demand for Limbs has been so great that their manufacture has become an important item in the aggregate industry of the city. Philadelphia, it is generally known, has the largest manufactory of Artificial Limbs in the United States; but its proprietor, Mr. Palmer, has been so enterprising in making his inventions and facilities known, that it is not necessary for us to repeat a more than "twice-told tale," and we can therefore proceed to notice some of the excellent manufacturers whose modesty is their chief fault.

Of this class is Mr. RICHARD CLEMENT, 829 Chestnut, who, for eighteen years, was foreman in one of the largest establishments, where he had ample opportunity to acquire a thorough knowledge of what are merits and defects in the construction of Artificial Limbs. During his novitiate, he made the models that gained for the United States the Grand Prize Medal at the World's Fair, in London, in 1851. For the last three years he has been conducting business for his own account, and is now manufacturing "Clement's Improved Artificial Leg," which is giving comfort to thousands of the unfortunate, and has received the approbation and commendation of the most eminent surgeons.

MASON W. MATLACK, 145 North Seventh street, has had an experience of eighteen years in the manufacture of Artificial Limbs, and has attained a position entitling him to the highest rank among this class of manufacturers. Mr. Matlack has kept pace with all the late improvements in this branch, and has achieved a marked success. In addition, he is also a manufacturer of apparatus and machines for deformities and ununited fractures, in which he has also been very successful.

Mr. Matlack conducts business as the successor of John F. Ord, who was long and favorably known as a maker of Artificial Limbs.

H. A. GILDEA, 312 South Fourth street, has achieved a distinction, especially in the manufacture of Artificial Arms, though he also makes Limbs that are highly recommended for their ease and durability. Every Board of Surgeons appointed by the Surgeon-General of the United States have testified to the superiority of Gildea's Arm. Recently, he has perfected a model of a *hand*, which is a marvel of ingenious mechanism, combining strength and lightness, with beauty of form.

Mr. D. W. KOLBE, 15 North Ninth street, who for many years has been an established manufacturer of Surgical Instruments, is now also an extensive maker of Artificial Limbs. Mr. Kolbe has a contract with the State of Georgia, and also with the United States Government, for the manufacture of his Limbs, which have been approved by the surgical authorities.

There are also in Philadelphia several firms who make a specialty of constructing *Trusses*, *Splints*, *Bandages*, and various appliances for the remedy of physical injuries and deformities. One of the largest establishments in this specialty is that of RITTER & PENFIELD, 112 South Eighth street, who employ a large number of hands in manufacturing Trusses, Abdominal Supporters, Suspensories, etc. Within a few years, very great improvements have been made in the construction of these articles tending to promote the comfort and convenience of those who are required to wear them.

Elastic Stockings, Belts, Knee Caps, and Suspensory Articles for the treatment of varicose enlargements of the veins, dropsical swellings and rheumatism, which until lately were imported, are now made in this city, of a quality superior to the foreign. They are made of Vulcanized Indian Rubber Thread, which first receives a covering of cotton or silk, and is then woven into a porous and elastic fabric, either with silk or cotton of the required size and shape; and when applied to the parts for the purpose named, or as a retaining apparatus over spirits, etc., they exert, without lacing, a gentle and equable pressure, and form a neat and convenient application pervious to air and to the perspiration, and are very durable.

2.—SURGICAL INSTRUMENTS.

Philadelphia has become noted for its manufacture of Surgical and Dental Instruments, partly by reason of the number of its eminent Colleges of Medicine, which have made it the chief seat of medical learning in this country, and partly by the superior skill of the Instrument maker; most of the improvements made in these Instruments, in this country, having originated in this city. Nearly the whole of the West and

CLEMENT'S
Patent Improved Artificial Legs.

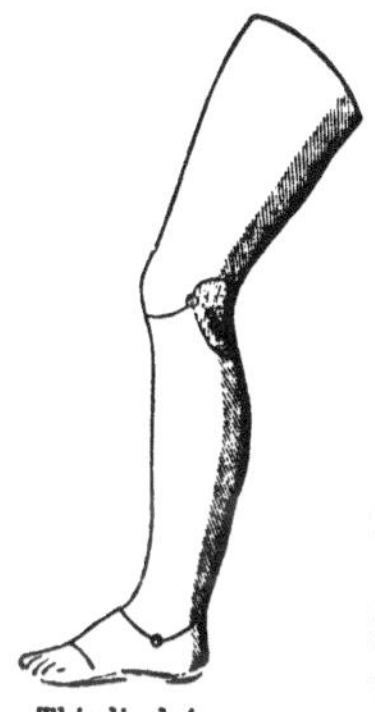

This invention stands approved by *every* Surgeon who has examined it, many of whom had given testimonials for others previous to the advent of this before the public. It contains the requisite combination for the best, is less complicated, lighter, stronger, more durable, and more perfectly adapted to the wants and comfort of the wearer than any other leg.

It has attained a perfection in its movement which enables the wearer to walk not only with ease, but in a graceful and natural manner.

Mr. Clement has had a practical experience of twenty years in the business, and during that time has inspected every kind of leg made, and now has combined the best principles of those that had any, with new improvements of his own.

The models which received the "Great Prize Medal" at the World's Exhibition in London, and most of the others which have been exhibited before scientific bodies in this country, were made by Mr. Clement.

This limb is pronounced by many of the most eminent Surgeons as the "*best*" now made, and is endorsed by the Surgeon-General U. S. A., and adopted for the army and navy.

It is approved and recommended by the entire Faculty of Jefferson Medical College, and is the only leg before the students of this institution.

An inspection of this leg will satisfy the most fastidious of its superior merits.

The price is made to meet the circumstances of the patient.

Pamphlets, giving information and references, will be furnished, *gratis*, on application, in person or letter. ADDRESS

RICHARD CLEMENT,
929 Chestnut Street, Philadelphia.

UNITED STATES ORTHOPEDIC INSTITUTE.

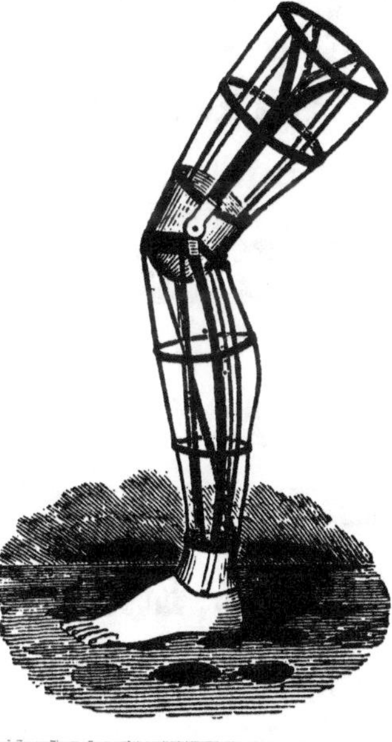

M. W. MATLACK,

Late Foreman, and Successor to JOHN F. ORD.

No. 145 North Seventh Street,

BELOW RACE, PHILADELPHIA,

MANUFACTURER OF

PATENT METALLIC
SKELETON LEG,

FOR WHICH

Premiums have been Awarded by different Scientific Associations, including

THE LONDON WORLD'S FAIR.

Improved Patent Premium Instruments for Curved S ine, Club Foot, Bow Legs, Knocked Knees, Weak and Sprained Ankles, Ununited Fractures, etc. Also, a superior article of Split Crutches.

H. A. GILDEA,

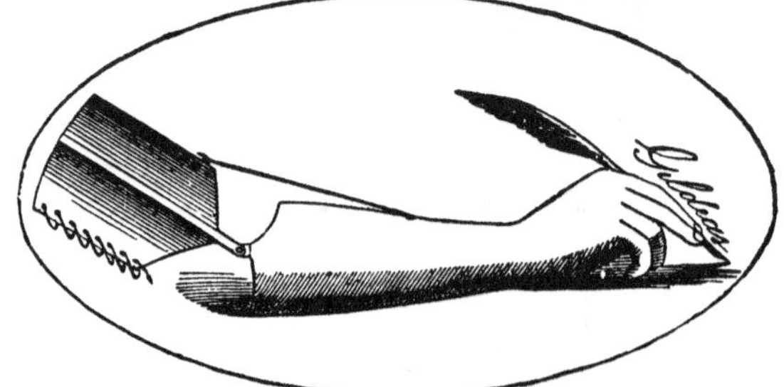

Artificial Limb Maker,

No. 312 South Fourth Street.

D. W. KOLBE,

MANUFACTURER OF

SURGICAL & ORTHOPÆDICAL

INSTRUMENTS,

ARTIFICIAL LIMBS, Etc., Etc.,

15 SOUTH NINTH STREET,

PHILADELPHIA.

HORATIO G. KERN,

MANUFACTURER OF

SURGICAL AND DENTAL

INSTRUMENTS,

SYRINGES, ELASTIC TRUSSES, Etc.

25 North Sixth Street, above Market, Philadelphia.

South is supplied from Philadelphia, on account of the cheapness and superiority of her manufactures; and greater facilities for cheap and rapid communication with Southern ports, are alone needed to secure a still larger share of the Southern trade. It is everywhere the custom of Druggists to keep more or less of Surgical Instruments on hand, which they procure from the manufacturer at a discount, and sell at his card prices. Some of the Philadelphia houses are thus engaged in supplying an extensive wholesale trade, while others confine themselves more to a retail business, or make to order.

Of the houses engaged in this manufacture, the oldest and most extensive are SNOWDEN & BROTHER, 23 South Eighth street. They are the successors of Wiegand & Snowden, who commenced business about 1821. This firm occupy for their manufacturing operations a four-story building, twenty-six by seventy feet. Their store, connecting with the manufactory, has a depth of one hundred feet, and presents an assortment of Surgical and Dental Instruments, Trusses, and similar appliances, unequalled for extent, we believe, in this country. A beautiful spiral staircase leads from the lower to the second story, which, as also the stories above, are filled with samples of their manufactures. Philadelphia is largely indebted to this firm for the pre-eminence she holds in this important branch of manufactures, and their products, which are distributed beyond the limits of the United States, have borne this reputation to foreign markets.

HORATIO G. KERN, whose Dental Instruments have been already alluded to, is probably the second largest manufacturer of Surgical Instruments in Philadelphia. His reputation for reliability, as well as mechanical skill, is well established.

Mr. J. H. GEMRIG, though not so much of a wholesale manufacturer as the others mentioned, has achieved great celebrity in this department. Nearly all the Instruments used in the principal Medical Universities of this city, and the Pennsylvania Hospital, were made by him; and the esteem in which his workmanship is held by the Professors in those institutions, and others of distinction, is evidenced by the frequent complimentary allusions to it by many of them in their lectures to their students. His Instruments for operations upon the eye, in particular, are preferred to any of European make, by one whose success, in some branches of Ophthalmic practice, is equal to that of any living Surgeon.

Mr. D. W. KOLBE, to whom allusion has been previously made as a maker of Artificial Limbs, is also a successful and respectable manufacturer of Surgical Instruments.

III.

Blacking and Ink.

Blacking consists essentially of two principal constituents: a black coloring matter, and certain substances that will acquire a gloss by friction. Each maker has, of course, proportions and methods of mixing peculiar to himself; but the chief materials used are the same in most cases—those in Day & Martin's celebrated Blacking being, it is said, Bone-black, Sugar, Molasses, Sperm Oil, Sulphuric Acid, and strong Vinegar.

Philadelphia has the largest Blacking Manufactory in the United States, and, with one exception, probably the largest in the world. We allude to that of JAMES S. MASON & Co., 138 and 140 North Front street, which was founded by the senior partner about thirty-five years ago. All the operations of manufacturing both Blacking and Ink are conducted here on a vast scale. Nearly two million sheets of tin are consumed in this one establishment, and about ten million of boxes for Blacking are manufactured annually. Two hundred hands are employed in the various departments. "Mason's Blacking" is now a standard article of commerce.

JOHN ANNEAR has for seventeen years been a prominent manufacturer of Blacking in Philadelphia. His manufactory, at 128 North Front street, is also equipped with all the necessary machinery for producing both Blacking and Ink. In the fourth story of his warehouse the Blacking Boxes are made, and he has machinery that is capable of turning out one hundred thousand boxes per day, each fastened without solder. The mixing mills are on the third floor, while the filling, labelling and packing are done in the second story. The first floor is occupied by the ware-room and offices. About fifty persons are employed in this manufactory.

Mr. Annear has been a zealous improver of the manufacture with which he is identified, and has received numerous testimonials from scientific institutes for his success.

H. A. BARTLETT & Co. are also extensive manufacturers, and besides these there are several small manufacturers of Blacking in the city.

It is a well ascertained fact that the Philadelphia Blacking stands warm climates better than the European, and is rapidly superseding the English in the South American markets.

IV.

The Brass Manufactures.

The uses and applications of Brass are so numerous, that, while its manufactures are extremely important, it is very difficult to trace them in their details. In the production of ornamental brass-work, and especially in that department distinguished as Lamps, Chandeliers, and Gas Fixtures, the manufacturers of Philadelphia are declared, by the best foreign judges, to have no superiors in the world. (See LAMPS, etc.) The same may be said of those who convert it into various *Military, Odd Fellows, Firemen's*, and *Theatrical Ornaments*, and other light and artistic forms. At one establishment, that of Samuel Croft, *Sheet Brass* is made quite extensively, though less so than we should think the demand would justify. Several foundries are chiefly devoted to making *castings* in brass, of every kind of article that may be ordered, from the largest to the smallest, either for brass-workers and finishers who finish up the foundry products, or for use in connection with other manufactures. Brass is used largely by the manufacturers of Marine Engines and Locomotives, and in Shipwork. Castings of this description are supplied by the founders to a large amount; while many of the engine and propeller builders have Brass Foundries as a part of their works.

Among the largest general manufacturers of Brass articles in Philadelphia, is the firm of MCCAMBRIDGE, FRY & Co., who employ about fifty hands. Their foundry, located at 525 and 527 Cherry street, is not only extensive, but fitted up with a variety of improved machinery for rapid production, not ordinarily found in similar establishments. In addition to their manufactory, this firm have a store at 222 North Fifth street, for the sale of their products, and connected with it is an Iron Foundry for the production of light castings.

Messrs. HOLLOWAY & REED, proprietors of the "Franklin Brass Works," 141 Race street, are also extensive manufacturers of Brass Castings, including some of the finest and rarest descriptions of work. All kinds of Brass articles required by steam and gas fitters, plumbers, and steam engine builders, are made here, together with a variety of Brass and Galvanized Tubing, Eyes, and Sockets.

Among the specialities produced at these works are Ornamental Castings in Brass for cemetery enclosures. Their new Tulip Socket and Centre Ornament is adapted to any inequality of the ground surface, and is so arranged as to admit of a free movement of the tube in any direction that may be desired, while the bottom of the socket is secured by means of a screw that holds it firmly in position. The members of this firm are practical and skillful men in their profession, and take a pride in producing good work.

Messrs. J. AUSTIN & Co., 1242 Hope street, have one of the largest Brass Foundries in Philadelphia. This firm have executed some of the heaviest castings in Brass ever made in the city, and supplied the wheels for the Government steamers Chattanooga and Ironsides, and turret plinths for Monitors, that weighed forty thousand pounds. They manufacture all the varieties of Steam, Gas, and Water Cocks, Check and Safety Valves, and their products are distributed to all parts of the United States. Messrs. Austin & Co. are one of the oldest established houses in this branch of manufactures, and now employ about twenty-five hands.

GEORGE R. KIRK, proprietor of the "Keystone Brass Works," 1030 Germantown avenue, has only been established since 1863, but by skill and close application he has built up an extensive business, which now employs twenty-five hands. He produces every description of Brass Work used by Plumbers, Coppersmiths, Steam Engine Builders, and also a variety of Brass Cocks. He is also the inventor of a Patent Compression Cock that is warranted to stand any pressure of hot or cold water, which he is now manufacturing largely.

Messrs. TAWS & HARTMAN, 1237 North Front, have been Master Brass Founders in Philadelphia for over four years, and now employ about two dozen hands. Their works are extensive, and they manufacture every description of Steam, Water and Gas Cocks, Whistles, Oil Cups and Water Gauges, and make Brass Castings to order. In addition to their Brass Work, Messrs. Taws & Hartman have recently given considerable attention to manufacturing Iron Body-Stops, Safety and Back Pressure Valves, and do a variety of machine jobbing. The members of this firm are practical and skillful workmen, and in Philadelphia, where they are best known, they are highly respected.

CHARLES PERKES, 1015 Sansom street, makes a specialty of Engineers', Plumbers' and Coppersmiths' Brass Work; and WILLIAM H. DODEMEAD, 512 Cherry street, of Brass Building Hardware, both of whom have been referred to elsewhere. Brass Stair Rods are made largely at the Philadelphia Brass Works.

The manufacture of Brass Cocks it is said originated in Philadelphia, and the firms engaged in it, of whom there are several, are probably the largest manufacturers of these useful articles in the United States.

V.

Brushes.

Few articles of manufacture admit so great a diversity of forms, sizes and qualities, or so wide a range of uses, as the production of the Brush-maker; and of none does it hold more true that the best article is the cheapest. From the delicate Pencil of the artist, to the "Whitewash," or the "Scrub," the variety in style and ornamentation is exceedingly great.

The manufacture in this city includes the usual variety of Hair, Paint, and the commoner kinds of Brushes, and employs about a dozen principal concerns, besides a large number of individuals who make to a limited extent. In this, as in other branches, our manufacturers have aimed at the production of substantial and reliable work. In the important article of a Paint Brush, particularly, some of them have successfully striven to excel; and we believe Clinton's Improved Wire Bound Paint and Varnish Brush is not surpassed by any in this country, while those of other makers are generally preferred to similar Brushes made elsewhere.*

* Edwin Clinton's Brush Manufactory

Is a representative of the best class of these establishments, not only in Philadelphia, but in the United States.

Mr. Clinton's store or wareroom, 908 Chestnut street, is probably the finest that has ever been opened in this branch of manufactures, and contains an assortment of Brushes that will meet the wants of every purchaser.

The manufacturing operations are prosecuted in a building in the rear of the sales-room, having a front on Sansom street of twenty-two feet, and a depth of one hundred feet.

The first floor is used for the dressing of the bristles, the second for the manufacture of the various kinds of Brushes and Sash Tools, and on the third is the department for nailing and tieing Whitewash and Artists' Brushes, while the fourth story is the finishing, and the fifth the storing department. The establishment employs seventy-five hands.

Mr. Clinton has a deserved reputation for the manufacture of Brushes of all kinds. His *Hair Brushes* are equal to the best made. His *Tooth Brushes* are more durable and reliable than the imported, but his marked distinction is in the manufacture of Paint Brushes. He uses in these none but the finest selected Russian Bristles, by which greater smoothness and a finer finish are attained than with any other material. In some of the Western States, Minnesota, for instance, no other than Clinton's Paint Brushes are in repute, and on several occasions the United States Government called for them by special advertisement.

Steam has not as yet been introduced to any extent in the Brush factories, and fewer Brushes are made in the Penitentiaries and Almshouses, in this State, than in some other places; but, in this city, the annual production in the House of Refuge, and Blind Asylum, is increasing. The present product, therefore, including that in Public institutions, amounts to about $225,000 annually. The Bristles are imported principally from Russia; a cold climate being indispensable, it is said, to their perfection.

Tooth Brushes, chiefly of the open-backed variety, are made here of a quality superior to the imported.

VI.

Cedar-Ware, and Wooden-Ware.

The manufacture of Cedar-Ware, though not very extensive in this city, will nevertheless well sustain the reputation of our mechanics in the minor as well as the larger branches of productive industry, by the undoubted excellence both of the material and workmanship. The chief supply of Cedar, for this business, is derived from Virginia and Carolina. There are about ten principal establishments engaged in the manufacture, besides many smaller ones.

Of these, the most extensive is that of Messrs. CLEMENT & DUNBAR, Beach and Shackamaxon streets. The factory of this firm has a front on Beach street of one hundred feet, and the premises extend to the Delaware River, a depth of seven hundred feet. In the yard are immense piles of lumber, and the Drying kilns frequently contain one hundred cords at a time. Particular attention is paid to the seasoning of the staves, and this firm, ambitious to attain a reputation for producing uniformly good articles, are extremely careful in the selection of materials.

Wooden-Ware, including all the various Wooden Housekeeping articles not made of Cedar, employs quite a number of small establishments. Only one, we believe, uses steam power in the business. Cherry Wash Boards, of a quality superior to any of the same kind from abroad, are made here, but only to a limited extent. Large quantities made of other material come here from other markets. The land of Notions and Wooden Nutmegs is still the wholesale producer of cheap articles in this branch, although each year is rendering us more independent. A larger amount of capital, we judge, might be profitably invested in this business; and we do not know why steam factories should not be sustained by our growing trade with all parts of the country.

VII.

Children's Carriages, Velocipedes, etc.

We have already alluded to the manufacture of Pleasure Carriages and Farm Wagons, but there is another class of vehicles, the construction of which is made a specialty in four establishments in Philadelphia, two of which are among the largest in the country.

In 1835 Mr. E. W. Bushnell commenced to manufacture in Philadelphia a variety of small coaches for children's gratification, and this, we believe, was the pioneer establishment in this branch in the United States. He prosecuted a constantly increasing trade until 1858, when he disposed of his business to Mr. JACOB A. YOST, who has since conducted it with remarkable success. His manufactory, 248 and 252 Girard avenue, has a front of fifty-six feet on the avenue, with a depth of one hundred and six feet, and is three stories in height. There is a great variety of ingenious machinery in the various departments of this building, and forty skillful workmen are furnished constant employment. The facilities in fact, are such, that customers are rarely kept waiting more than three days for the execution of their orders, however extensive.

Mr. Yost has established an excellent reputation for reliability and fair dealing, and using none but the best materials in his Juvenile Carriages and Velocipedes, and finishing them elegantly, his products are in demand throughout the United States. His sales-rooms are at 214 Dock street.

Mr. WILLIAM QUINN, 1005 Sansom street, has an important manufactory of miniature vehicles, and produces a variety of novel articles not usually made in similar establishments. One of his specialties is an Improved and Patented Velocipede, that is adapted alike for children and grown persons. In Paris, the use of Velocipedes by adults is becoming a fashion. He produces a variety of Bath Chairs and Invalid Carriages, that have been found by experience to combine ease and comfort with durability. His leading specialty, however, is a combined Nursery Spring Sulky and Cradle, which was patented in 1866. These are simple in construction, and can be used for the postures of sitting or reclining, which are attained by placing a chair or basket on the self-acting spring, thus saving the parent or attendant from all anxious solicitude.

Mr. Quinn has been engaged in the business since 1856, and during these eleven years, has established a reputation for using good materials, and constructing vehicles that are durable as well as elegant in design and workmanship.

VIII.

Guns and Pistols.

The manufacture of Fire Arms is not carried on very extensively in Philadelphia. The largest private Armory is that of Sharps and Hankins, near Fairmount, established by Christain Sharps, the inventor of Sharps' Rifle. There are however a number of persons who make Shot Guns, Rifles and Pistols to order, and of the very best description, including some with gold and silver mountings, and chased stocks that cost two and and three hundred dollars each.

One of the oldest and best Gun manufacturers now in business is Mr. John Krider, Second and Walnut streets, who has been identified with the pursuit since 1837. He made Pistols for Generals Henningsen and Walker that would hit a horse at nine hundred yards. Of late years he has directed a large share of his attention to the construction of Breach Loading Shot Guns, and converting Muzzle-Loading Guns into Breach Loaders. His Breach Loading Shot Guns have been repeatedly tested, and found to combine in a remarkable degree the advantages of rapidity in loading, perfect safety while loading, ease and accuracy in performing their work, and simplicity in cleaning. Mr. Krider keeps a full assortment of Guns, Rifles, Pistols and fine sporting apparatus.

Messrs. Slotter & Co., 400 Lynd street, are noted manufacturers of Rifles and Pistols. All the materials, excepting the barrels, are made under their own direction in their own workshops. They make to order Rifles of any weight or quality, from the plainest to the finest. They are also manufacturers of a variety of Pistols, including the J. Derringer pistol, and fill orders from California and the Pacific Territories. The members of this firm are practical gunsmiths, and reliable, trustworthy men.

Besides several Gun stores, some of which are large and well stocked, Philadelphia has a Bazaar of sportsmen's apparatus. We allude to the establishment of Westcott & George, successors to Philip Wilson & Co., 409 Chesnut street. Here may be found every variety of Sporting apparatus, Guns of domestic and foreign manufacture, ammunition of all kinds, Fishing Tackle of every variety, and a complete assortment of implements for Cricket, Base Ball, Archery, etc. Messrs. Westcott & George also manufacture Doubled Barrelled Shot Guns, and are about introducing a Breech Loading Gun, which they claim will be superior to any thing heretofore known.

SLOTTER & CO.,

Shot Gun, Rifle and Pistol

MANUFACTURERS,

No. 400 LYND STREET, FOURTH above GREEN,

PHILADELPHIA.

We Manufacture all kinds of Shot Guns,

From a Plain substantial Gun to the Finest Quality.

Our RIFLES are made in the most perfect manner for strength, symmetrical appearance and accuracy of shooting. The barrels we use are of REMINGTON'S make—all the other materials are made in our own factory. We make our RIFLES to suit our customers in weight and quality, from a plain Rifle to the finest, with cast-steel barrels, globe or telescope sights. Our PISTOLS are made in the neatest and most symmetrical manner, for beauty and strength. We are also the makers of the J. DERRINGER PISTOL. We make from the plain German Silver to the finest Silver and Gold mounted Pistols.

☞ ***Our work cannot be excelled, if equalled.***

J. M. MIGEOD & SON,

Firemen's Furnishing House.

FIRE HATS, BELTS, FIRE HORNS, SHIRTS, FATIGUE CAPS.

MILITARY EQUIPMENTS,

Physician's Family Medicine Chests, Pocket Cases, Gun Cases, etc.

27 SOUTH EIGHTH STREET,

Second story, entrance on Jayne Street, Philadelphia.

☞ ***Send for Catalogue.***

IX.

Hoop Skirts.

The manufacture of Hoop Skirts is comparatively a new branch in Philadelphia, but within a few years it has grown into considerable importance. There are now several establishments largely engaged in this manufacture, usually however in connection with other articles of Ladies' wear, and there are probably a hundred places in the city where Hoop Skirts are made to some extent. We shall advert to a few of those that may be called leading or representative houses.

Among the largest manufacturers of Hoop Skirts in Philadelphia, is Mr. SHADRACH HILL, whose establishment at 25 and 27 Bank street, employs one hundred hands, and a capital of fifty thousand dollars. The articles manufactured here, are of styles and qualities, favorably known to the Jobbing trade in Philadelphia and the Western cities. Mr. HILL, in addition to manufacturing Hoop Skirts, keeps in his extensive warehouses, a large stock of Elastic Cords, Braids, etc., and the celebrated Sewing-machine Needles of Jason Hill, for which he is sole agent in the United States.

JUDAH LEVY, 432 Market street, is the manufacturer and patentee of the well known improved double Spring Skirt, in which a double spring is formed by placing together two fine elastic steel springs. The advantages claimed for the use of double springs, is, that they are not braided together, as is usual in other duplex skirts, and this insures greater elasticity and comfort in carriages and crowded assemblies.

The establishment of Mr. LEVY is quite extensive, and employs sixty hands, exclusively in the manufacture of Hoop Skirts. He is careful to select the best materials, and by close attention to business has succeeded in gaining a reputation, and an extended market for his productions.

WM. T. HOPKINS, 628 Arch street, manufactures Hoop Skirts of first quality woven tape, intended principally to supply the fashionable trade. Superior Tapes, English steel springs, and Linen finished coverings are used. Improved machines are employed to secure the metallic fastenings, and prevent their becoming unclasped. Mr. HOPKINS claims for his products superior durability, cheapness, and comfort. A large city trade, and the quantity sold by merchants annually, throughout the country, attest the excellence of the articles manufactured by him.

Mr. M. A. JONES, 822 Arch street, is a manufacturer of Hoop Skirts, of the finest quality, chiefly to order, and for the most fashionable trade. Every peculiarity of form is carefully studied, and the Skirt is made to suit and grace the wearer. All the materials used are selected with the utmost care, and as none but the most experienced hands are employed, this establishment has gained a high reputation. In this Emporium, however, the manufacture of Hoop Skirts is but a branch of the general business. The Corset department is also very complete and conducted on the same principle, viz: a perfect fit and satisfaction guarantied. The business also comprises a Fashionable Dress and Cloak making department, under the supervision of one of the best and most experienced Dress-makers in the country, and, in fact, all kinds of under wear for ladies is here made to order, from a Hoop Skirt and Corset, to the finest Dress or Embroidered Opera Cloak.

Mr. Jones' salesrooms are tastefully fitted up, adorned with handsome paintings, and are among the most beautiful in the city. Many of the most fashionable of the Philadelphia ladies are indebted to his Emporium for their entire wardrobe.

There are several manufactories of Hoop Skirt Tape and Bands, in different parts of the city. The "Montgomery Mill," on Montgomery avenue above Ninth street, of which Messrs. SULLIVAN and PAULY are proprietors, is probably the most extensive of this class of factories.

X.

Lightning Rods.

Philadelphia has the largest and most celebrated manufactory of Lightning Rods in the United States, known as the "North American Lightning Rod Manufactory," 488 St. John street, of which REYBURN, HUNTER & Co., are proprietors. It was founded by the senior member of the firm in 1849, occupies a building forty-two by one hundred and twenty-five feet, and employs sixty hands. The Star Galvanized Lightning Rods, made by this firm, combine all the recent improvements in Points, Insulators, and attachments to buildings. They are made from Magnetic Iron Bars five eighths of an inch square, ground spiral, twisted and galvanized. They are warranted never to corrode, and are connected by perfectly fitting solid copper burs, which form not only a perfect and continuous connection from the points to the ground, but, by a combination of metals, form a Galvanic Battery, enabling the Rod to discharge the

628 HOOP SKIRTS. 628

WILLIAM T. HOPKINS,

MANUFACTURER OF FIRST QUALITY

EVERY LADY SHOULD TRY ONE.

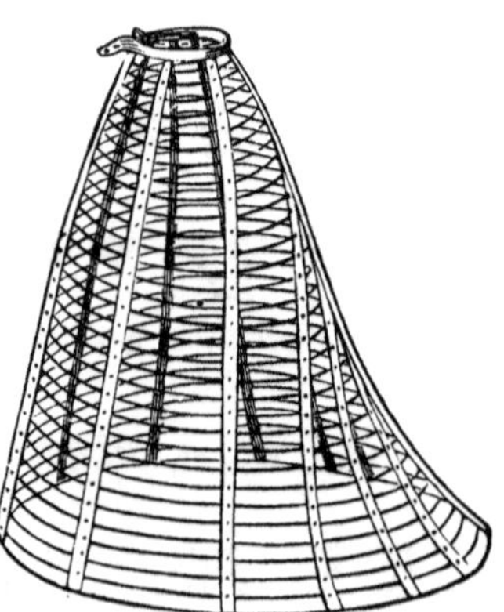

EVERY LADY SHOULD TRY ONE.

Woven Tape Hoop Skirts.

WHOLESALE AND RETAIL,

AT MANUFACTORY AND SALESROOM,

No. 628 ARCH STREET,

PHILADELPHIA.

"Our own make" of Hoop Skirts always embrace all the newest and most desirable styles, shapes and sizes of Plain and Trails for Ladies, Misses and Children, of every length and size of waist, manufactured expressly to supply the wants of first-class and most fashionable trade. Wherever introduced, they are more universally POPULAR than any others before the public. Made of very superior Tapes, English Elliptic Steel Springs, with Lined-finished Covering; the Metallic Fastenings secured by improved machinery, which entirely prevents slipping or becoming unclasped; they are lighter, more elastic, retain their shape much longer, the most *durable*, really the CHEAPEST, and afford more satisfaction and comfort to the wearer than any other *Single or Double Spring* Hoop Skirts in the American market. They are strictly what we claim for them, and will recommend themselves. *Every Lady should try one.* They are being extensively sold by merchants throughout the country. Ask for Hopkins' "own make." Buy no other, and be sure that they are stamped with our *name* and *number*.

☞ CATALOGUES sent to any address, containing styles, sizes, and prices.

Also, constantly receiving from New York and New England Manufacturers full lines of **Low Priced Hoop Skirts,** which will be found Cheaper than the same goods can be had of Jobbers.

WM. T. HOPKINS.

electricity of the most terrific thunder-storm harmlessly into the ground. For Ships and other Vessels, they manufacture a Rod composed of Galvanized Iron and Copper Wire, distinguished as the Cable Lightning Rod. About five hundred tons of Iron are annually converted by this firm into Rods, and from fifty to seventy-five tons of Malleable Iron into Hooks and Burs.

Messrs. REYBURN, HUNTER & Co., also manufacture all kinds of Lightning Rod Points, including the celebrated Kinsey, and McAllister points, to the number of at least fifty thousand annually; and of Insulators no less than four hundred thousand are made in their works annually. They also manufacture a double flange Copper Lightning Rod, made from pure copper, which is said to be the only substantial Copper Rod in use.

XI.

Lithographic Establishments.

Philadelphia has a full quota of establishments where Engraving on Stone is executed in all its styles, including *Linear and Crayon Drawings, Transfers on Stone from Steel or Copper-plate Engravings, Wood-Cuts,* or *from Lithographic Drawings themselves.*

In 1828, the second Lithographic establishment in the United States was opened in Philadelphia, and here many of the most important processes and improvements in the art have had their origin. Works have been executed here that would do credit to the artists of any city or place in the world. Several years ago, Messrs. LIPPINCOTT & Co. published an annual called the Iris, having chromo lithographic illustrations, which was so much admired in England, that the Queen ordered a dozen copies for her own household, expressing her admiration of the illustrations, and saying that it was the prettiest book she had seen from America, and reflected great credit on the city of Philadelphia.

The firm that executed this work was that of P. S. DUVAL & SON, now that of P. S. DUVAL, SON & Co., 24 South Fifth street. Mr. Duval, Senior, is now the oldest established Lithographic Printer in the city, and Mr. Duval, Jr., is also an accomplished Lithographer, having had, among other advantages, that of a three years' practice in some of the principal establishments of Paris. This firm are at present engaged in producing chromo oil portraits and landscapes, equal, in point of color and finish, to the finest oil paintings. Among their numerous productions are the oil portraits of Washington, Martha Washington, General Grant, John Chambers and Bishop Wood, and other celebrities, each printed in fifteen colors, and each color repuiring a separate stone; also a fine copy of Baker's painting of Niagara Falls, in eighteen colors. They have now in press a large Post Map of the New England States for the General Post-office Department, and are executing the illustrations for Sloan's Model Architect, a new work in two volumes, published by LIPPINCOTT & Co. They have also in hand several plates for the Surgical History of the War, published by the War Department. The works of this house are fine specimens of the art, and every kind of Lithographic work, from the cheapest label to the most elaborate show card, is executed by them.

Mr. F. BOURQUIN, 108 Hudson street, who for a long period was foreman in Mr. DUVAL'S establishment, first introduced into Philadelphia the art of transferring Plate Engravings to Stone, of so much importance in the publication of Maps, Outline Drawings, etc. To this day he has few, if any, equals in the art. His attention is chiefly devoted at present to the manufacture of Maps, of which vast amounts are made in Philadelphia for the publishers of other cities.

Mr. JAMES MCGUIGAN, South Third, corner of Dock street, is another extensive Lithographer, also engaged chiefly in the Map manufacture. He is long and favorably known as an excellent Artist, being the surviving partner of the old firm of WAGNER & MCGUIGAN. Mr. MCGUIGAN, however, is not confined to Map work, but executes orders in all the branches of Lithography.

Mr. E. KETTERLINUS, North-west corner of Fourth and Arch, first introduced colored and embossed work, for Perfumery and Mercantile purposes, into the United States, which, from the advantages of labelling, gave a great impetus to the Perfumery business. Not only were many thousands of dollars saved to the country, but a new branch of industry was created, resulting in the establishment of many Label houses with an aggregate capital of several millions, furnishing numbers of deserving Artists with employment and developing the finer tastes so necessary to advanced civilization. The establishment of Mr. KETTERLINUS is very extensive, and employs a large number of skillful persons.

Mr. THEODORE LEONHARDT, 114 South Third street, gives particular attention to the execution of Bonds, Diplomas, Certificates of Deposit, Checks, Notes, Drafts, and other mercantile work, and has no superior in this line. His productions with the Ruling machine are elegant in design, and have been considered, in many instances, equal to copper.

Messrs. BREUKER & KESSLER, South Seventh below Chestnut street, are extensively engaged in general Lithographic Engraving and Printing, producing Perfumers, and Druggists' Labels, plain or in colors, Wine and Liquor Labels, Designs and Sketches for Show Cards, in elegant style,

E. KETTERLINUS,

STEAM-POWER

Printer & Lithographer

N.-W. Cor. Arch and Fourth Streets,

PHILADELPHIA,

Recommends his old and well-known establishment for the execution of the different branches of

Plain and Ornamental Printing.

Combining the best material—selected in this Country and Europe,—with the most approved Machinery and unexceptionable workmanship;—parties may feel assured that everything emanating from his establishment to stand pre-eminent in the Art of Printing, both as regards style and execution.

A LARGE VARIETY OF

Gilt-Embossed and Illuminated Show Cards,

LIQUOR, PERFUMERY, FRUIT, DRY GOOD and FANCY

LABELS;

FANCY EMBOSSED AND ILLUMINATED

BALL, PROGRAMME AND INVITATION CARDS,

FANCY PAPERS

For Book-binders and Paper Box Manufacturers.

Just Published a New and Splendid Edition of

Latin PHARMACEUTICAL Labels

CONTAINING

Over Two Thousand Assorted Sized Labels, Edited and Revised by the Philadelphia College of Pharmacy.

JACOB HAEHNLEN'S
STEAM POWER
Lithographic & Letter Press Printing Rooms
GOLDSMITH'S HALL
J. HAEHNLEN'S STEAM POWER PRINTING ROOMS.
J. HAEHNLEN'S LITHOGRAPHIC ROOMS.
JACOB HAEHNLEN'S LITHOGRAPHIC ESTABLISHMENT
TICKETS
BILLS OF LADING
BILLS OF EXCHANGE
Checks Notes Drafts
CIRCULARS CARDS
PLANS
SHOW CARDS
BLANKS, MANIFESTS, Illustrations.
BONDS
A large Stock OF WINE, LIQUOR AND PERFUMERY LABELS always on hand
JACOB HAEHNLEN'S
LITHOGRAPHIC
AND STEAM POWER
Letter Press Printing Rooms,
GOLDSMITH'S HALL,
opposite Post Office; Library Street,
PHILADELPHIA.
LETTER HEADS, BILL HEADS.
Certificates OF STOCK AND DEPOSIT.
NEW STYLES LATIN LABELS FOR DRUGGISTS AND PHYSICIANS
BORDERS & TITLES FOR Photographic Albums.
TAGS.
RECEIPTS & STATEMENTS
Diplomas

JACOB HAEHNLEN'S
Label and Show Card
PRINTING ROOMS,
IN
GOLDSMITHS' HALL,
OPPOSITE POST OFFICE,
PHILADELPHIA.

Keep in stock the most varied and largest assortmen
OF
WINE, LIQUOR, PERFUMERY,
Druggists' and Manufacturers' Labels
In the United States.

All kinds of Lithographic and Letter-Press Printin
Executed promptly.

MANUFACTURER & IMPORTER OF

Transfer Pictures for Coach and Car Builders

Please call and examine my STOCK before purchasing or orderin elsewhere.

Philada., Oct., 1867.

etc., and are manufacturers of the Decalcomante or Transferrable Pictures. Samples of their productions illustrate this work, and it will be seen they have no superiors as practical Lithographers.

H. J. TOUDY & Co., 503 and 505 Chestnut street, have been established Lithographers in Philadelphia, since 1856, and have executed some very fine specimens of work in all departments of the art. Views of manufactories have been made by them in a style that cannot be excelled for beauty. They have shown admirable taste also in the execution of Portraits, Diplomas, Railroad Bonds, and Stock Certificates. The members of this firm are practical workmen, and their establishment has all the facilities requisite for the successful prosecution of Lithographic Drawing and Printing, especially for mercantile purposes.

M. H. TRAUBEL, 409 Chestnut street, has facilities for executing all kinds of Lithographic work, and whatever he undertakes to do is well done, whether it be a Portrait or Landscape, a Diploma or Business Card. He has been established in Philadelphia, since 1856, and while exceedingly modest in his claims, there is no doubt but that he is a master in his profession. His etchings on Stone are especially fine, and can scarcely be distinguished from Engravings on Steel.

Mr. EDWARD HERLINE, 630 Chestnut street, employs a large number of persons in Lithographic Printing, Engravings, etc., and tastefully executes all kinds of Fancy and Plain Labels, Certificates, Checks, etc. The specimens of these are exceedingly beautiful, and show to what an extent business is facilitated by the art of the Lithographer.

Besides these, there are at least twenty-five other Lithographic establishments in the city, but the most extensive in Philadelphia, and probably in the United States, is that of JACOB HAEHNLEN, 418 Library street. Mr. HAEHNLEN has purchased the extensive six-story brown stone building, known as Goldsmith's Hall, and occupies all except the lower floors, for the purposes of his business; the second floor being appropriated for the salesroom and offices, with a Machine shop and Drying room in the rear; on the third floor are the Lithographic Presses; the fourth floor is devoted principally to Card Printing; and the fifth and sixth stories to printing Pamphlets, Hand Bills, and other similar work. The Machinery is propelled by a sixty-horse-power Engine, and among the novel appliances, are several Rolling Machines to give a lustre and finish to the gold work, and two very large Embossing Presses, which are used upon Show Cards and Labels of a raised surface, enriching and greatly improving the appearance of the work. Throughout its entirety the establishment ranks among the first in the country, and is a credit to Philadelphia.

Butler & Carpenter's Engraving Establishment,

Where all the Revenue Stamps that have been required by the United States Government were made, may fairly be classed among the remarkable industrial concerns of Philadelphia. Ten years ago, when the first edition of this work was issued, we alluded to the Plate Printing Rooms of JOHN M. BUTLER, as the largest of the kind in the city. At that time he had in preparation several large and splendid line Engravings, as "Washington at Valley Forge," "Franklin before the Privy Council at Whitechapel, London," "Merry Making in the Olden Time," and the "Senate of the United States in 1850." All of these, together with several other important Engravings, as those illustrative of Dr. KANE'S Expedition, have since been issued, and constitute a valuable addition to the Fine Art Literature of America. Of late years he has discontinued the publication of large engravings, and confined his attention to executing work for which there is current demand, such as Portraits, Landscapes, and Book illustrations of various kinds. He now occupies for this purpose an immense room, forty by one hundred and fifty feet, in Jayne's Granite Building, where he employs twenty plate presses.

In 1861, Mr. JOSEPH R. CARPENTER, formerly of the firm of TOPPAN, CARPENTER & Co., well known Bank Note Engravers, became associated with Mr. BUTLER, for the Engraving and Printing of Revenue Stamps, for which they had received a contract from the United States Government. They fitted up the third and fourth stories of the building before mentioned, two rooms forty by one hundred and fifty feet, in a tasteful manner, and here they have executed all the stamps that have been issued by the Internal Revenue Department, amounting in value to many millions of dollars. The process of Engraving these stamps does not differ essentially from that of vignettes on Bank notes. The design is first engraved on a small piece of steel, which is afterwards hardened by exposure to heat, and then taken up on a roll of soft steel by means of a heavy pressure in a Bank Note transfer press. This roll is then hardened, and the picture is transferred from it, by pressure, to a large steel plate, the impressions being repeated, often as many as two hundred on a single plate. From this plate the final printing is executed on the ordinary steel plate printer's presses.

The sheets are then taken to an apartment in the fourth story, where they are gummed on the back and then dried, after which they are passed through a perforating machine, which produces the rows of holes that facilitate the separation of the Stamps. Messrs. BUTLER & CARPENTER have in operation ten presses, and employ forty persons.

In their office they have specimens of all the stamps that have been engraved by them, ranging in value from one cent to two hundred dollars, which form a most interesting collection.

It is obvious that the execution of this important contract involves great pecuniary responsibility, as well as calls for artistic skill and good workmanship, and no slight tribute of commendation is implied, in saying it is universally conceded that Messrs. BUTLER & CARPENTER have performed their part well and faithfully.

XII.

Musical Instruments.

The Musical Instruments that are made in Philadelphia comprise Organs, Piano Fortes, Melodeons, Accordeons, Violins, Flutes, Guitars, and sundry Band Instruments.

Church Organs are made by five manufacturers, J. C. B. STANDBRIDGE, JOSEPH BUFFINGTON, JOHN ROBERTS, E. C. LEDROIT, and H. KNAUFF & SON. The first named is, we believe, the oldest Organ Builder, and has constructed some of the largest instruments that are in use. Among them is the organ formerly in Concert Hall, now in St. Clement's Episcopal Church, which has four manuals and pedals, sixty registers and three thousand and fifty pipes; and that in Calvary Presbyterian Church, with three manuals and pedals, one hundred and fourteen registers and one thousand eight hundred and sixty-five pipes.

JOSEPH BUFFINGTON, of 131 South Eleventh street, is another old and distinguished Organ Builder. He commenced the business in 1842, and consequently has had an experience of a quarter of a century. No less than one hundred and fifty organs have been built by Mr. Buffington, and among them, those for the Catholic Churches, St. Philip De Neri, St. Theresa, and St. Alphonsus; the Third, Sixth, Seventh, and Penn Square Presbyterian Churches; the Western, Trinity, and Green street Methodist Churches, and the Spruce street and Tabernacle Baptist Churches. All of these are large Organs, of splendid tone and good workmanship, which have deservedly given Mr. Buffiington a high reputation.

JOHN ROBERTS, after many years' experience in Organ Building in London and in New England, located some four years since in Frankford, now a part of Philadelphia, and fitted up a shop for building Organs on an extensive scale, with large surrounding grounds for the exposure of the materials to sun and air, so essential to good workmanship. Mr. Roberts constructed some of the finest Church Organs now in Boston, where he first commenced business on his own account, and also in Chicago and other cities. He is at this time engaged in building one for the Grace Methodist Episcopal Church, in Wilmington, Del., that will contain forty-seven stops, forty musical, seven mechanical, and two thousand one hundred and twelve pipes, controlled by three full ranks of keys and one full set of pedals of two-and-a-half octaves compass, being the largest organ as yet ordered by any Methodist Episcopal Church in the United States.

Piano Fortes are made at probably a dozen establishments in Philadelphia, but the majority of the builders have so little enterprise in setting forward the advantages of their instruments, that they are completely overshadowed by the New York and New England concerns, and it is difficult even to tell who are manufacturers. Pianos of unsurpassed excellence are made in this city and sold in a quiet way, but as the makers apparently do not wish the fact to be known, we will not reveal the secrets of the trade.*

Melodeons are made in Philadelphia to a limited extent by one manufacturer; but the principal business is the sale of the instruments of this class, manufactured in the large establishments of New York and New England, by the regularly appointed agents, who charge no advance upon the factory prices.†

* We have the permission of Messrs. ALBRECHT, RIEKES & SCHMIDT, whose warerooms are at 46 North Third street, to say that they do not belong to the "Sleepy Hollow" class of Piano manufacturers. They are comparatively a young firm, who commenced business some four years since, but by energy and a combination of mercantile tact with mechanical skill, have established a flourishing trade. Their Pianos will compete in quality of tone, workmanship, and finish, with any made in this country or Europe, and they are not afraid to invite public inspection or stand competitive tests.

The manufactory of Messrs. ALBRECHT, RIEKES & SCHMIDT is at York avenue and Buttonwood streets, where they have lately largely increased their facilities for manufacturing First Class Pianos.

† For instance, Mr. E. M. BRUCE, 18 NORTH SEVENTH STREET, is agent for "Estey's Cottage Organs," which have a Patent "Tremolo," so closely resembling the human voice as to charm and surprise every one who hears it. The advantages claimed for these Organs are quickness of articulation soundness, purity and volume of tone. They are recommended by the best judges, and Bishop Simpson testifies that the Cottage Organ combines sweetness and power in an unusual degree, and is quite a favorite in his family circle. Mr. Bruce is also the agent for the sale, at both wholesale and retail, of *Taylor & Farley's Model Organs*, with the Patent Knee Swell and Sub Bass.

CHURCH & CHAPEL
ORGAN FACTORY
J. BUFFINGTON
ORGAN BUILDER
ORGANS
ADRIAN. SHARP

H. KNAUFF & SON,

649

NORTH BROAD STREET,

PHILADELPHIA,

Small Organs on hand.

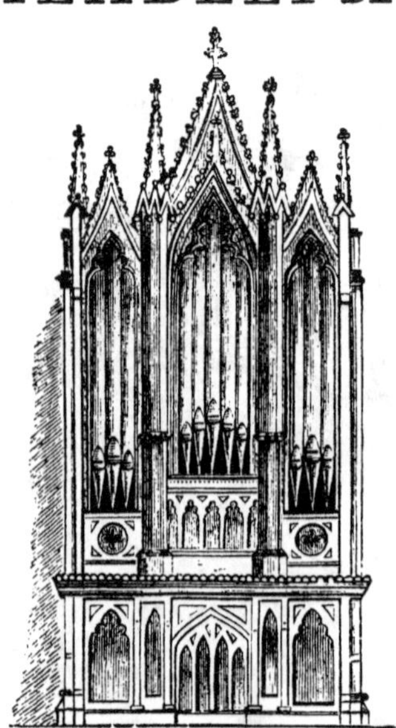

Organs repaired and tuned.

MANUFACTURERS OF

CHURCH ORGANS,

OF EVERY DESCRIPTION AND SIZE.

With a successful experience of over thirty years, the business having been established in 1834, and the constant study of new improvements, both original and European, we are able to produce Organs which are accurate in mechanical arrangement, rich in tone, and varied in combination and effect, being in every respect

FIRST CLASS INSTRUMENTS.

The use of STEAM Power enables us to manufacture on reasonable terms. We can also execute contracts of any size, having room to erect and finish, in the factory, the largest organ.

☞ Purchasers will find it to their interest to visit this establishment, it being possessed of every means to render satisfaction.

ESTEY'S
COTTAGE ORGANS

ARE STILL AHEAD OF

ALL COMPETITORS,

AND STAND UNRIVALLED IN

POWER, PURITY OF TONE AND BRILLIANCY,

And all other points which go to make a first-class Instrument.

THE "VOX HUMANA TREMOLO"

Is the greatest of all modern improvements in Reed Instruments. It CHARMS and SURPRISES all who hear it, by its wonderful resemblance to the HUMAN VOICE. Do not confound this with the common tremolo in use. It is entirely different, and far superior to any other.

ALSO, TAYLOR & FARLEY'S

MODEL ORGANS,

With two new and valuable Patents of their own Invention, viz.:

PATENT KNEE SWELL & PATENT SUB-BASS,

COMPRISING IN ALL OVER SEVENTY DIFFERENT STYLES.

Prices—from $75 to $650.

Sold Wholesale and Retail, by

E. M. BRUCE,

18 N. Seventh St., Philadelphia.

☞ ***Send for a Descriptive Circular and Price List.***

C. F. ZIMMERMANN,

No. 238 North Second Street, Philadelphia,

IMPORTER OF AND DEALER IN ALL KINDS OF

MUSICAL INSTRUMENTS,

GERMAN, FRENCH & ITALIAN GENUINE STRINGS,

WHOLESALE & RETAIL,

And manufacturer of the Lately Patented

UNION ACCORDEONS,

Patented in the United States July 10, 1866,

And January, 29, 1867, in England, France, Belgium, Russia, Saxony, etc. The Union Accordeon distinguishes itself from all others by the peculiar position of its tones and of its keys. In all Accordeons made and known, the tones obtained by drawing are different in the different octaves, while those obtained by compressing are uniform throughout the entire range of the instrument. This variety of the tones renders it very difficult to learn the Accordeon. In C. F. Zimmerman's Instruments this difficulty is obviated by arranging the tones, so that the tones both in drawing and compressing are uniform throughout the entire range of the instrument, this object being obtained by placing the sixth on a separate key.

A MORE PERFECT INSTRUMENT IS

C. F. ZIMMERMANN'S DIATONIC UNION ACCORDEON,

Which is precisely the same as that previously described in its first row of keys with the addition of a second row. The last and most complete style of Accordeons is the ***Chromatic Union Accordeon.*** This Accordeon contains all the keys described in the previous instruments, and also an additional set of keys on the treble size representing the half notes, and by these keys the Chromatic Scale can be produced the same as on a Pianoforte.

ALBRECHT, RIEKES & SCHMIDT,

MANUFACTURERS OF

FIRST CLASS

MANUFACTORY:

475 & 477 N. Fifth St., and 488 & 490 York Avenue.

ALBRECHT, RIEKES & SCHMIDT claim for their Instruments, and to which they desire to draw attention:

STRENGTH AND DURABILITY,
ELASTIC AND EVEN TOUCH,
PRECISION & PROMPTNESS OF ACTION,

A Sympathetic Tone, powerful without harshness, round in Bass, rich and singing in the middle octaves, and of clear brightness in the high treble, maintaining throughout the scales a *purity* and *evenness* unsurpassed.

ELEGANCE IN STYLE AND FINISH.

There is only well-seasoned lumber, and the best quality of all other materials used.

The different branches of the manufactory are under the personal superintendence of the several members of the firm, who are practical Piano makers, of long experience; and the employees are skillful and competent men. Besides a number of our own improvements, we have introduced all improvements which have been found of value.

Every Piano is warranted for five years.

WAREROOMS:

No. 46 NORTH THIRD STREET.

Accordeons are made in several establishments, and it is claimed that the Philadelphia instruments are in every respect superior to the French or German. Their range of notes is double that of the Foreign Accordeon, and their construction is stronger. Mr. C. M. ZIMMERMAN, who received the first premium at the World's Fair, in London, has set forth the advantages and variety of his Accordeons so fully in another place, that it is unnecessary to repeat what is there said respecting them.

Philadelphia is also the principal city in the United States in which *Violins* are made. The Philadelphia Violins enjoy the highest favor of musicians and music dealers for brilliancy of tone, and are extensively used in Orchestras.

Flutes and *Guitars* of the best quality, and German Silver Band Instruments are made to a limited extent.

XIII.

Perfumery and Fancy Soaps.

The manufacture of Perfumery is usually carried on conjointly with that of Fancy Soaps, and the combination makes the business in Philadelphia an important one. In quality the manufacturers claim, and we think justly, that the Fancy Soaps made in this city are unrivalled. In 1853 the makers of Perfumery and Soaps throughout the world exhibited specimens of their production at the Crystal Palace in New York, and a disinterested critic of that city, after examining the collection, remarked that those of Philadelphia were equal to any in the world, adding, "there is no doubt that these productions must eventually succeed in driving the French Soaps out of the American Market." The factories of X. BAZIN, the Messrs. TAYLORS and WRIGHTS, are the largest in this country, and one of them, it is believed, is the largest of the kind in the world.

One of the oldest houses in the business, is that of Messrs. GLENN & Co., 726 Chestnut street, the successors of L. W. GLENN, who commenced the manufacture in 1832. For more than thirty years GLENN's Perfumery and Toilet Soaps have been in the market, and are now regarded as standard articles, thoroughly reliable and durable as well as exquisite. All the perfumes of European celebrity are reproduced by this firm, without deterioration in quality. About sixty persons are usually employed in their manufactory, and their products are distributed to all parts of the United States.

THOMAS WORSLEY has a very extensive manufactory of Fancy Soaps at 114 Arch street. The building is twenty by one hundred and twenty five feet, four stories in height, the upper floors being used for manufacturing purposes, the cellar for storage, and the first floor for the sales-room, the boiling department being in the rear. Mr. WORSLEY has been an established manufacturer since 1846, and his Soaps are in favor in all parts of the United States, and especially with the Philadelphia trade.

Messrs. A. HAWLEY & Co., have been for many years engaged in the manufacture of Perfumery and Fancy Soaps, and the great variety as well as excellence of their preparations, has gained for them a high and well merited reputation in this country, both for the exquisiteness of their Perfumes, and the durability of their odors.

Mr. HAWLEY is a practical Chemist, and has devoted great attention to the Chemical processes involved in the manufacture of fine Perfumery and Toilet Soaps. He recently associated with him, in the Laboratory, Mr. THEODORE CHANUDET, a thoroughly practical Perfumer, who has spent fifteen years in the best establishments in Paris, to perfect himself in the art, and feels confident they are now able to produce the best preparations that are made in this country.

XIV.

Roofing.

The importance of a good Roof cannot be well over-estimated; and in a great city the selection of a material that is *Fire Proof*, as well as Water Proof, seems to be a duty which a builder owes to the public. Shingles, of course, from their combustible nature, if for no other reason, cannot be recommended. Of Metallic Roofs there are a great variety presenting claims to public attention and public confidence. *Slate* is used to some considerable extent, and *Tin* still more extensively, as a material for Roofing. Zinc has not been found well adapted to the climate; but as a protective coating for Sheet Iron, it is, as we stated, extensively used. One manufactory in this city is occupied in coating Iron for Roofing with a preparation, of which the principal ingredient is said to be India Rubber. The manufacture of Corrugated Iron for Roofing we have already alluded to. Within fifteen years, Composition Roofs have become very popular, and the manufacture of them constitutes an important business. These roofs combine many advantages, and it is to be hoped that experience may ultimately justify the large expectations that have been formed of them.

In 1852, a Composition Roofing, which had been in use for many years

in the West, was introduced into this city by Messrs. WARREN, KIRK & Co., 228 Walnut street. It is known as Warren's Improved Fire and Water Proof Roofing. This Roofing seems, in a remarkable degree, to have united the suffrages of builders and consumers of every class in its favor; and if we may judge of its merits by the degree of popularity it has rapidly attained, it must combine many points of excellence. The buildings, whether private residences, stores, warehouses, factories, depots, or public buildings, including a part of the United States Mint, which have been covered with it in this and neighboring cities, within a few years, are among the largest and best known; and the names of many of the leading business men are appended to testimonials in its favor. It has been received with equal favor in other States, and in Canada. The materials used in its construction are *Felt, Composition* and *Gravel.* The two former are said to be made of such ingredients as possess elasticity and tenacity, and are combined with the latter so as to form a Roof not only durable, with no liability to crack or decay, but one which is impervious to both fire and water—a combination never before obtained in a Composition Roof. Its fire proof qualities have been repeatedly subjected to severe tests, specially instituted for the purpose, and it seems to have passed the "ordeal by fire" with perfect impunity. The advantages upon which the manufacturers base its claims as an improvement upon all others, are, that, in addition to being *Fire* and *Water proof,* it is also cheap and durable, and requires a less pitch, and consequently less area to be covered, and less masonry upon the walls, while furnishing more facilities for light and ventilation; besides being more accessible on ordinary and extraordinary occasions than any other. They also claim that it will not expand and contract by heat like Metal Roofs, and will bear more than double as much heat without danger to the boarding beneath; that it requires only an inclination of one inch to the foot, and may be walked upon or used for drying purposes without injury; that it is a great advantage to firemen when adjoining buildings are on fire; that it is not injuriously affected by changes of temperature, or by the jarring of machinery; that it is adapted to every climate, and is easily and quickly repaired; that Gutters of the same material may be formed on the roof; and finally, the cost of it is only about one half that of Tin, and less than that of any other Fire Proof Roof now in use. If it possess these qualities, of which there seems to be no doubt, Messrs. WARREN, KIRK & Co. well deserve the success which has attended its introduction.

XV.

Scales and Balances.

Every variety of standard articles for weighing, known under the general term of Scales and Balances, is made in Philadelphia, and each branch has its representative houses, who hold the leading position. Fine Balances for weighing the precious metals are made principally by HENRY TROEMNER, the successor of F. Meyer & Co., 710 Market street. Mr. Troemner constructed all the Balances, Weights, etc., required for the U. S. Mint, Custom Houses, and Repositories, and several Scales for the Mexican Mint. Some of the Balances made for the Assay Office in New York, and for the Branch Mint of San Francisco, cost as much as one thousand dollars, and one made several years ago cost one thousand six hundred and fifty dollars. Besides Balances like these, which must turn with the thousandth part of a grain, Mr. Troemner constructs Patent Balances that will weigh twelve tons. His manufactures comprise Mint Balances, Bankers' Scales, Jewellers', Druggists', Grocers', Confectioners' Scales, etc.; in fact, any kind required for weighing purposes. Nearly all the banks in this city, New York, and other places, have his Scales in use. Mr. Troemner is the oldest manufacturer in this branch of business in the United States.

Messrs. BUCKELEW & WATERMAN, 716 Market street, manufacture the "Patent Arc Scale," which is adapted to the requirements of various trades. The Druggist Scale, made in this style, will weigh both Troy and Avoirdupois at one weighing, *without the use of weights or springs*, and is of such simple construction as to require only to be seen in order to be understood. This firm have recently opened a store at 716 Market street, as a general depot for the sale of Scales, Weights, and Measures, in addition to their manufactory and ware-room at 515 Callowhill street. Their stock is a very general one, comprising Druggists', Prescription, and Counter Scales, Jewellers' Balances, Scales for Grocers, Confectioners, Silk-dealers, Butchers, etc.; also reliable Platform Scales of various sizes, Grocers' Fixtures, Measures, etc. The Market street store of Messrs. Buckelew & Waterman is the office of the Sealer of Weights and Measures for the First District of Philadelphia.

Platform Scales are made largely by the firms of BANKS, DINMORE & Co., and BECKMAN & ENGLEMAN.

BANKS, DINMORE & Co., at Ninth and Melon streets, have the largest

BUCKELEW & WATERMAN,

MANUFACTURERS OF

STANDARD

Scales, Weights and Measures,

716 MARKET STREET, PHILADELPHIA.

Druggists' Patent Double Graduated "Arc Scale,"

Without Weights, and with Extra Balance Beam.

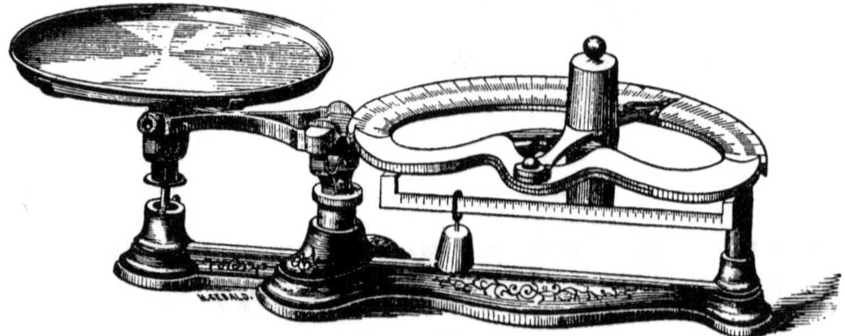

This novel invention combines with *Utility*, both *Beauty*, and *Convenience*. Its primary object is to save the annoyance and expense attendant upon the loss of weights. It works upon knife heads, as in ordinary Counter Scales, but in lieu of the plate for the reception of weights, has a graduated *Arc*, with a *Permanently Attached Weight*, through which an *Index* is passed, which latter, moving over the *Graduated Arc*, denotes with great accuracy the weight of the commodity in the opposite dish.

The scale represented by the above cut is peculiarly adapted to *Druggists'*, by having introduced upon the Arc, a scale of *Apothecaries'* weights along with the ordinary *Avoirdupois* scale, so that by using an outer and inner index, the weight in *Troy* ounces, and in commercial or *Avoirdupois* ounces and pounds is, in every instance, made to appear at a glance. It is thought that the advantage of this arrangement will be apparent to every Pharmaceutist. It obviates the necessity of keeping at hand an extra set of large officinal weights, not always readily obtained, often mislaid or lost, and always expensive. An Extra *Balance Beam* is also attached to the *Arc*—a great convenience in ascertaining the weight of bottles or other receptacles for extracts, syrups, or fluids of any kind. These scales are very compact, beautiful in their manufacture, and so pleasing to the eye, that they are an ornament to the counter of any druggist. The "Arc Scale" is also manufactured for the use of Grocers, Tea Dealers, Confectioners, and for weighing Silk.

ALSO, EVERY VARIETY OF

SCALES for DRUGGISTS', GROCERS, CONFECTIONERS, TEA DEALERS, PLATFORM SCALES, GROCERS' FIXTURES, etc., etc.

Price Lists Furnished on Application.

manufactory of Platform Scales in the State of Pennsylvania. This firm are successors of Messrs. A. B. Davis & Co., and also of Messrs. Abbott & Co., both of whom were old and well-known manufacturers of Scales. At their manufactory, established in 1846, have been made the largest as well as the most perfect of Scales and Beams, viz: several Weigh Lock Canal Scales of from two hundred to three hundred tons capacity; Railroad Track Scales, one hundred, one hundred and twelve, one hundred and twenty-five, and one hundred and thirty-five feet in length, used generally on the great Coal roads of Pennsylvania; also, an improved Crane Beam for Iron Founders' use. This beam, which is strong, durable, light, and compact, always remains in a parallel position when attached to a crane or derrick. A beam on this plan has been made for the Morgan Iron Works, New York, of which the capacity is one hundred and twenty thousand pounds. It is only ten feet ten inches in length, and requires but eight hundred and forty pounds of weights to be attached at the end, to weigh sixty tons.

Messrs. Beckman & Engelman, proprietors of the Eagle Scale Works, are extensive manufacturers of every description of Platform Scales, such as Railroad Track, Canal, Coal, Iron, Cattle, and Dormant and Portable Platform Scales, etc.; also Weighmasters' Beams, Frames, Patent Beams for weighing heavy castings, boilers', Counter and Butchers' Scales. Their recently erected factory on Ninth street, above Girard avenue, contains all the facilities for the manufacture of every thing in this branch of business. The members of this firm are both practical workmen, who give their attention closely to the business, and by guaranteeing satisfaction are winning reputation and success.

Counter Scales are made at several establishments, but probably they are more of a specialty in the manufactory of John C. Dell, 416 Vine street, than in any other. Mr. Dell is a practical and skilful workman, who possesses both the facilities and disposition to do good work. Besides Counter Scales, he manufactures Patent Balances, principally to order.

XVI.

Spices, Ground Coffee, etc.

The grinding of Spices, and the preparation of Coffee and its Essence, occupy the attention of several prominent firms in Philadelphia.

The oldest Spice Mills in the city or its vicinity are those of which Messrs. C. J. FELL and BROTHERS are proprietors, known as the "Faulkland Spice Mills." They were established more than three quarters of a century ago, by Jonathan Fell, and have since that time increased from a single-horse mill, to a large factory, possessing all the new improvements in mill machinery, and using a steam-engine and water-power equal to one hundred horses. The principal mill is located near Wilmington, and runs (speaking technically) nine pairs of stones. These are devoted to the manufacture of Mustard, all the different preparations of Cocoa, the grinding of Spices, and the making of Hominy. The last is so prepared, by a new process, that it resists the effects of any climate, and keeps sweet and good for years. The requirements in tin and wooden boxes, kegs, etc., for packing Spices, furnish employment to a large number of persons. For this purpose, the Messrs. Fell have a remarkably ingenious machine, propelled by steam, which *weighs* accurately, and packs the Spices in bundles. The motto of this house, for three generations, has been, never to sell an article otherwise than as represented, and by adhering to it they have gained the confidence of the public in their productions, and its usual concomitant, a fortune for themselves.

Another old and reliable house in this business is that of EDWARD G. MILLETT, 215 Race street, which was founded in 1835. The machinery and apparatus for grinding Spices and preparing Coffee are very complete, and Millett's Mustard and Essence of Coffee are standard articles with the Grocery trade. His products are distributed to all parts of the country, and in the Western States especially, no articles of the same class are in higher favor.

The "Oriental Spice Mills," of which WILLIAM L. GRAVER is proprietor, located at 205 and 207 North Front street, are among the most extensive of the kind in the city. Mr. Graver embarked in the business in 1854, and since 1858 has occupied his present manufactory, which is fitted up with every convenience for expeditious and economical production. The machinery is propelled by an engine of forty-horse power, and about two dozen persons are employed in the establishment. Mr. Graver's salesroom is at 208 North Front, where three stories of

ESTABLISHED IN 1835.

EDW. G. MILLET,

CENTRAL STEAM

Spice, Mustard, Coffee,

AND

Essence of Coffee

MILLS,

No. 215 RACE STREET.

Always on hand an extensive and varied assortment of the above articles, at the lowest prices known to the trade. Particular attention is called to our

EXTRA MUSTARD,

UNSURPASSED BY ANY IN THE MARKET—AND TO OUR

ESSENCE OF COFFEE,

Which for quality and price, defies all competition. Furnished wholesale by the gross or barrel.

SEE AND JUDGE FOR YOURSELVES.

a large building are filled with samples and an assortment of his products.

Besides these establishments, there are probably a dozen others, some of them large manufactories, where Spices are ground and the Essence of Coffee is prepared.

XVII.

Tin, Japanned, and Sheet-Iron Ware.

The manufacture of articles from Tin, Zinc, and Sheet-Iron, is sufficiently extensive in the aggregate to be called a leading branch, but the subject calls for no particular remark. The latest Business Directory furnishes a list of about two hundred Tin-workers in Philadelphia, but it is probable there are two hundred and fifty places in the city where Tin-ware is made. The oldest establishment in the business is that of ISAAC S. WILLIAMS, on Market street, founded by Samuel Williams and Thomas Passmore, in 1796. This house is exceedingly well provided with facilities for executing heavy orders expeditiously, and has furnished with culinary utensils some of the first-class hotels in New York, and the largest steamboats on the Western waters. Mr. Williams is said to be the most extensive manufacturer in this city of Planished-ware, of a superior quality. This ware is made by repeated hammering of the ordinary tin-plate upon highly-polished steel anvils by hammers, also highly polished. This condenses the fibre or grain of the tin, and renders it capable of a high polish, and at the same time improves its quality. Planished-ware is also made by another process, more analogous to rolling or burnishing, which, it is said, renders it nearly equal in appearance to the former, and somewhat cheaper. It is however scarcely so durable.

Within a few years a great revolution has been effected in the manufacture of Culinary and Miscellaneous Tin-ware, by the introduction of machinery. By the aid of Dies, Presses, Lathes, and other contrivances, the separate parts, or the whole, according to the degree of complexity of an article, are at once struck up into the required shape, plain or with devices, as may be desired; and the work of the Tinman is reduced to the simple act of soldering or uniting the several parts.

About two years since, Messrs. HIGGINS, MARCHAND & Co., 127 Arch street, erected a large factory, and fitted it up with machinery made after their own drawings and models, for the manufacture of *Tinned Hollow-ware*, after the French pattern, in one entire piece, without solder. By means of presses of immense power, sheet-iron in this manufactory is shaped into nearly the whole variety of Hollow-ware commonly supplied by soldered tin, cast-iron, or earthenware. After shaping, it is finished, and the utensil is then *tinned*, in a most effective and beautiful manner, leaving a surface scarcely inferior to that of silver-plated ware. The superiority claimed for this ware consists in the following points; the framework of the utensil being of one piece, is of uniform strength throughout—the plating of tin is heavier and more perfect—there are no joints to secrete fluids and promote rust. It is cheaper, since the extra quality of iron used is compensated for by the rapidity with which perfect machinery will enable the articles to be made. It is more adaptable than soldered ware, as it can be used either with or without heat.

For nearly twenty years this ware has been in general use in France and on the Continent, and as the gentlemen now engaged in its manufacture in this city are men of capital and business reputation, it is probable that this new substitute for Tin-ware will be appreciated here as in other countries. Messrs. HIGGINS, MARCHAND & Co. are now employing fifty hands.

Japanned Ware is made extensively at two Tin-ware establishments in this city; and there are also a few persons who conduct this branch separately, some of whom are not excelled for the beauty and excellence of their work.

The conversion of Sheet-iron into Stove-pipe, Coal-scuttles, etc., is carried on in several shops, and largely by GEORGE GRIFFITHS, WILLIAM EASTERBROOK, and McCOY & SNELL, to whom reference has elsewhere been made. Sheet-Iron, like Tin, is constantly being applied to new uses, and many articles, formerly made of Wrought-Iron, are now made of this material.

XVIII.

Wire Work.

WIRE WORK, of all kinds, plain and fancy, particularly *Wire Sieves, Wire Cloth, and Screens*, is made very extensively by three and to some extent by a dozen establishments. The Iron Wire is principally obtained from Easton, Pa., and from Trenton, N. J., and woven, in this city, into every article into which wire can be twisted. The Philadelphia goods are of a most substantial character, the average quality being acknowledged by all to be superior to the New England make, and command readily a higher price.

WIRE GOODS.

SELLERS BROTHERS,

No. 18 North Sixth Street, Philadelphia.

(Established 1781.)

MANUFACTURERS OF

IRON, BRASS AND COPPER

OF EVERY DESCRIPTION, FROM ONE TO ONE HUNDRED MESHES TO INCH,

Suitable for Paper, Grist and Fanning Mills. Sugar, Rice, Rosin, Turpentine, Sand, Grain, Flour, Meal, Safes, Windows, Doors, etc.

FOURDRINIER WIRES,

Made from the very best low brass stock, and of superior workmanship throughout.

DANDY ROLLS,

LAID OR WOVE, LETTERED OR PLAIN.

SIEVES,

For Flour, Sand, Grain, Drugs, Meal, Coal, etc.

SCREENS, SAFES, TRAPS, MUZZLES, ETC.

ALSO, WHOLESALE AND RETAIL DEALERS IN

IRON, BRASS AND COPPER WIRE.

The oldest house in this manufacture, and probably the oldest in the country, is that of SELLERS BROTHERS, 18 North Sixth street. It was founded by JOHN SELLERS, in 1750. His two sons continued the business, at the corner of St. James and Sixth streets, and were so employed at the close of the Revolution. The present firm is composed of the great grandsons of the original founder, and under their intelligent direction and enterprise, the business has been developed until it is now among the largest in the country. Within a few years, in fact, the sales have so increased that they are now tenfold the former amount. Messrs. SELLERS BROTHERS employ, in their manufactory, about forty hands, and have Looms, some of which weigh as much as three tons each. Their list of manufactures includes Wire-Work for buildings, Fourdrinier Cloths, Brass and Copper Wire Cloth for Paper Makers, heavy Twilled Wire-Work for Spark-Catchers, Sieves of all kinds, Circular and Standing Screens; in a word, Wire-Work, Wire-Cloth, and Sieves of every description.

Messrs. M. WALKER & SONS, 11 North Sixth street, are also an old and well established firm, which was founded about thirty years ago. It is the only firm in Philadelphia which combines the four distinct branches of Wire-Railing, Iron Bedsteads, Ornamental Wire-Work, and Wire-Weaving. In the manufacture of these articles about forty hands are employed. This firm has more recently added the manufacture of Iron Farm Fences, which until lately were imported by wealthy farmers, but now supplied by them as cheaply as the imported, and also the manufacture of Fourdrinier Wires, largely used by Paper Makers. This branch, though largely developed of late years, is probably yet only in its infancy; and the facilities possessed by this firm for its extension, are a favorable beginning for its further advancement. The productions of Messrs. WALKER & SONS are distributed throughout the United States, Canadas, South America, the West Indies, and even to China

Messrs. BAYLISS & DARBY, proprietors of the Pennsylvania Wire Works, 226 Arch street, are among the most extensive in this branch of business in Philadelphia. Both members of the firm are practical and skilled mechanics, who commenced business in 1854. Their stock of Sieves, of Silk, Hair, Brass and Copper Wire Cloth is of every mesh; and their heavy Founders' Sieves and Fancy Wire Work have been repeatedly commended. They received the Diploma at the Pennsylvania State Fair in 1854, and since then they have applied Woven Wire to Plastering purposes; the ceiling and dome of the Academy of Music being a sample of this production. Their list of manufactures includes

Wire Railing for Cemetery Lots, Piazzas, Trellis-work for vines, of beautiful patterns; also, Wire Furniture and Iron Bedsteads.

Messrs. BAYLISS & DARBY employ fifty hands, and in addition to their store and manufactory on Arch street, they have a warehouse at 123 North Second street, from which their products are distributed to all parts of the United States. Their business has doubled within a few years.

JOSIAH B. THOMPSON is largely engaged, at 29 North Twentieth street, in covering Iron and Copper Wire for Millinery purposes. Besides Bonnet Wire of all kinds, he manufactures Covered Wire for Electro Magnetic Machines, Telegraph Instruments, Chemical Apparatus, and Philosophical Instruments, in fact, every description of Covered Wires.

XIX.

Whiting.

WHITING is made from a peculiar kind of Chalk obtained from the cliffs of Dover, in England. Formerly, the raw material was brought into this country as ballast for ships, and the cost was comparatively trifling; but since the United States Government has imposed a duty of ten dollars per ton on Chalk, the articles manufactured from it are necessarily more expensive than formerly.

There are in Philadelphia three establishments for the manufacture of Whiting, of which the most extensive is that of Messrs. HASSE & PRATT, 1042 York street, Port Richmond. The buildings have a front on York street of seventy-five feet, with a depth of one hundred and fifty-three feet. A large number of hands are employed, and on the premises are all the appliances used by the best manufacturers in this particular branch. The father of Mr. Hasse was engaged in the business for many years, and his son, Mr. Charles Hasse, the senior partner of the firm, has had an experience of twenty-five years in the manufacture.

Whiting is frequently adulterated by being mixed with Terra Alba, and it is of importance to purchase from manufacturers. The products of Messrs. Hasse & Pratt are said to be strictly pure, and command the highest price in the market.

We believe we have made the fairest and most complete exhibit ever attempted of those manufactories in Philadelphia, that could be called representative in their respective branches, but we are well aware that among the six thousand different concerns there are many others worthy of mention, which we have not had time to visit. A volume, in fact, would hardly contain all that might be written upon the Miscellaneous Manufactures of Philadelphia. In by-ways and rooms concealed from the public gaze, there is at all times an army of industrious artisans busily engaged in transforming rude materials into objects of utility, or productions of taste and skill—"Inventions for delight, and sight and sound"—and aiming by superior dexterity in their handicraft operations to compensate for the lack of machinery and business facilities.

We now proceed to recapitulate, with some detail, the results of our investigations, with respect to the value of the articles annually manufactured in Philadelphia. They are given as our own conclusions, after laborious and careful examination, based partly on information furnished by manufacturers as to their own business; partly from a mean of estimates of those having some knowledge as to the business of individual manufacturers in the several branches; and partly upon information furnished from returns made to the Internal Revenue Department. We append for the sake of comparison the results of the Census of 1860, but we are convinced that what is styled the estimated product in 1866 is more approximately accurate than the returns of any Government Census that has ever been taken.

	Census of 1860. Annual Value.	Estimated Product for 1866.
Agricultural Implements and Seeds,	$103,850	$300,000
Alcohol and Camphene,	734,793	550,000
Ale, Porter, and Brown Stout,	....	2,200,000
Artificial Limbs,	41,000	150,000
Beer, Lager and other,	....	2,400,000
Blacking,	245,300	300,000
Blacksmithing,	342,568	400,000
Blank Books,	68,072	150,000
Bolts, Nuts, Rivets, etc.,	398,500	500,000
Bone Black,	190,000	250,000
Bookbinding,	909,906	1,250,000
Boots and Shoes,	5,474,587	4,237,415
Boxes, Packing,	237,750	250,000
" Paper,	188,500	250,000
Brass Founding,	361,268	400,000
Bread,	1,420,428	2,500,000
" Ship, and Crackers,	579,500	1,000,000
Brick,	1,233,416	1,500,000
Britannia Ware,	86,100	150,000
Brooms,	136,409	200,000
Brushes,	391,653	400,000
Calico Printing,	2,557,338	2,500,000
Candles, Adamantine,	551,000	608,145
Caps and Cap Fronts,	283,230	250,000
Cards, Playing,	60,000	75,000

	Census of 1860. Annual Value.	Estimated Product for 1866.
Carpentering,	1,267,120	1,500,000
Carpets, Ingrain,	2,601,325	8,100,000
" Rag, and Hemp,	84,387	1,200,000
Cars,	640,875	700,000
Car Wheels,	350,000	400,000
Carriages,	1,027,271	1,616,478
" Children's,	48,650	150,000
Chemicals,	2,412,854	3,185,098
Cigars,	1,243,342	2,500,000
Clothing, Ladies', Cloaks and Mantillas, Corsets,	680,380	750,000
" Men and boys',	11,318,564	18,994,497
Coffee and Spices, ground, Essence of Coffee, Roasted Coffee,	1,128,400	1,122,616
Combs,	56,884	75,000
Confectionery,	766,494	1,000,000
Cooperage,	415,941	500,000
Coppersmithing,	80,000	250,000
Cordage and Ropes,	237,850	2,000,000
Cotton and Mixed Goods,	6,172,437	7,426,647
Cotton Twine, Yarn, etc.,	1,038,585	1,500,000
Cutlery,	24,500	150,000
Drugs, ground,	107,500	150,000
Dyeing, etc.,	553,584	600,000
Edge Tools and Hammers,	30,050	125,000
Engraving, General Plate, Plate Printing, Wood,	322,400	350,000
Envelopes,	38,500	150,000
Fertilizers,	184,300	2,000,000
Fire Arms and Ammunition,	377,500	350,000
Fire Bricks,	72,400	150,000
Flour and Meal,	2,996,696	3,000,000
Fringe, Tassels, etc.,	1,169,845	1,500,000
Furniture and Upholstery,	1,839,416	3,000,000
Gas,	1,837,500	2,167,433
Gas Fixtures, Lamps, Chandeliers,	1,425,000	1,250,000
Gas Meters,	402,000	475,000
Glass, Fruit Jars, etc.,	1,159,300	1,500,000
Glue, Sand Paper, Curled Hair, etc.,	539,750	1,750,000
Gold and Silver Assaying and Refining, including actual expenses of United States Mint,	450,000	930,000
Gold Leaf and Foil,	228,000	250,000
Hardware, except Saws,	156,238	250,000
Hats,	1,109,842	1,250,000
Hoop Skirts,	14,930	150,000
Hosiery, Shirts, and Drawers,	2,003,665	2,500,000
Ink, Printing and Writing,	122,350	150,000
Instruments, Mathematical, Surgical, Dental, etc.,	300,307	350,000
Iron, Bar, Sheet, Railroad, and Castings,	1,777,790	11,745,515
Iron Castings, Malleable, Corrugated, Galvanized, Iron Gas Pipe, and Wrought Tubes,	845,247	932,700
Iron, Miscellaneous, Railing, Spikes, etc.,	268,520	445,000

	Census of 1860. Annual Value.	Estimated Product for 1866.
Jewelry, Gold Pens and Cases,	1,459,830	1,500,000
Lampblack,	66,792	150,000
Lamps, Coal Oil,	114,000	250,000
Lead Pipe, Sheet Lead, and Shot,	550,000	550,000
Leather, Morocco and Other,	3,729,683	3,689,134
Liquors, Distilled and Fermented,	3,406,481	8,254,397
Lithography,	386,300	400,000
Locksmithing, etc.,	109,455	150,000
Locomotive Engines,	1,420,000	1,500,000
Looking Glass and Picture Frames,	646,190	750,000
Lumber, Planed,	303,880	350,000
" Sawed,	218,080	250,000
Machinery, Tools, and Steam Engines,	3,118,976	6,500,000
Malt,	315,000	325,000
Marble and Stone Work,	982,545	1,250,000
Matches,	70,700	75,000
Medicines, Drugs, etc.,	1,015,650	1,500,000
Millinery and Straw Goods,	945,800	950,000
Mineral Waters,	237,600	250,000
Musical Instruments,	409,850	350,000
Oil Cloth,	370,000	397,365
Oils, Linseed, etc.,	1,449,836	2,500,000
" Petroleum Refined,	....	941,161
Paints, except White Lead,	1,065,574	1,250,000
Paper Bags,	121,500	150,000
Paper Collars,	....	450,000
Paper-hangings,	435,000	1,250,000
Paper, Printing, etc.,	692,000	1,215,824
Perfumery and Fancy Soaps,	646,000	850,000
Photographs and Photographic Materials,	180,200	1,000,000
Plumbing and Gas Fitting,	587,957	600,000
Pottery Ware,	90,850	125,000
Printing, etc.,	5,039,725	6,000,000
Provisions, etc.,	4,409,743	4,500,000
Roofing,	286,479	350,000
Saddlery and Harness,	929,436	1,000,000
Safes, Fire Proof,	189,500	200,000
Sash, Doors, and Blinds,	382,550	400,000
Saws,	270,599	950,000
Scales and Balances,	141,920	250,000
Ship and Boat Building,	307,829	500,000
Shovels, Spades, Forks,	299,000	350,000
Silks, Sewing, Twist, etc.,	598,000	600,000
Silver Plated Ware,	421,250	450,000
Silver Ware,	600,900	650,000
Soap and Candles,	1,480,268	1,500,000
Spices. (See Coffee.)		
Spokes, Hubs, and Felloes,	100,740	150,000
Springs, Car and Carriage,	134,082	250,000

	Census of 1860. Annual Value.	Estimated Product for 1866.
Starch,	211,275	225,000
Steel,	458,200	500,000
Stereotyping and Electrotyping,	88,800	150,000
Stoves and Heating Apparatus,	1,792,603	2,500,000
Sugar Refining,	6,356,700	21,000,000
Teeth, Porcelain, and Dental Materials,	312,100	350,000
Tin, Copper, and Sheet Iron Work,	609,583	1,250,000
Tobacco and Snuff,	72,100	1,500,000
Trunks, Valises, etc.,	215,050	250,000
Type,	308,300	450,000
Umbrellas and Parasols,	1,111,200	1,250,000
Varnishes,	402,740	500,000
Vinegar and Cider,	179,652	350,000
Wagons, Carts, etc.,	916,667	1,000,000
Watch Cases,	660,300	650,000
Whalebone, Rattan, and Steel for Umbrellas,	91,331	250,000
Whips,	58,450	75,000
White Lead,	791,100	1,500,000
Wire Work,	92,944	350,000
Woollen Goods,	4,389,701	7,178,975
Miscellaneous articles in Census, *not above specified*,	7,699,574	....
Manufactured Articles which are increased by being polished, painted, etc., on which a Revenue tax was paid in 1866,		5,887,125
Miscellaneous Manufactures, *not above specified*, on which a Revenue tax was paid in 1866,		6,788,489
Total,	**$135,979,777**	**$225,139,014**
Beyond the city limits, within a radius of one hundred miles, there are a large number of Iron Works, Dry Goods Factories, Paper Mills, Tanneries, and Wood Working Establishments, whose products, distributed from this city, amounted to not less than		$25,000,000
Total for Philadelphia and vicinity,		**$250,139,014**

In 1860, the Census returns stated the number of manufacturing establishments at six thousand two hundred and ninety-eight, the capital invested at $73,318,885, the male hands employed at sixty-eight thousand three hundred and fifty; the female hands at thirty thousand six hundred and thirty-three. According to this Census the average production of each hand was about one thousand four hundred dollars, and the capital invested was about one half the aggregate of the production. Assuming that these relative proportions were correct, and assuming that they are applicable at this time, the respective items would stand as follows: Capital invested in Manufactures in Philadelphia, $112,000,000. Hands employed, 160,500. Product, $225,139,014.

In view of this result—a result as unexpected by the Author as it probably will be surprising to the reader—a result perhaps understated, but not overstated, and of which the constituents are given with sufficient particularity to enable any one of ordinary intelligence to test its accuracy by a personal investigation, with the aid of a complete Business Directory: in view of this result then, we ask, do not the facts demonstrate the original proposition and assertion, that *Philadelphia is already a great manufacturing city—undoubtedly the greatest in the Union?* The value of the mechanical and manufacturing industry of the entire State of Massachusetts, in 1855, including Gas, an important item, was about two hundred and forty millions of dollars; we therefore may confidently say, that no other single city—not even Boston, including Lowell, and one half the State of Massachusetts, sums up so large a production of indispensable goods as are annually produced in the city of Philadelphia. We may say, moreover, that we are convinced, as the result of extended inquiry and many opportunities for comparative examination, that the goods made in Philadelphia are generally superior to the average quality of American fabrics. One reason for this superiority is, that the operations are mostly conducted in small factories, under the direct personal supervision of the owner, or in shops often illy provided with machinery for rapid production; and consequently the fabricator must give close attention to the selection of material and character of the workmanship, and master competition by the durability and intrinsic excellence of his fabrics. We hold it to be eminently safe for any consumer or merchant to infer that Philadelphia-made goods, at the same price, are invariably the cheapest. Many other considerations are suggested by the facts which we have collected, and partially submitted, and to which we would gladly invite attention, did space and circumstances admit, but we will conclude by adopting the graphic language of the late President of the Corn Association, whose prophetic vision saw what we hope we have demonstrated: "Our steam-engines are plying their iron arms in every street, in every by-way is heard the sound of the shuttle and the clink of the hammer, as the artisan contributes his mite to the vast sum of toil; whilst many a stately edifice, with its hundreds of employees and clanging machinery, sends forth a stirring music to quicken the pulse of our city life. Why then shall we not spread beyond our borders the knowledge, that in this busy hive is being made almost every article that can contribute to the wants or luxury of man? *This is the great Mart of American Manufactures, unequaled on this Continent in the extent and variety of its products. As such, let it be proclaimed!*"

APPENDIX.

REMARKABLE MANUFACTURING ESTABLISHMENTS IN PHILADELPHIA.

I.

The Bridesburg Manufacturing Co.'s Works—Barton H. Jenks, Esq., Pres't.

THESE works, of which we have already given the incidents connected with their establishment and early history, are located at Bridesburg, a flourishing town now constituting a part of the city of Philadelphia. They are built in the form of a hollow square, cover an area of one hundred and sixty thousand square feet, and consist of a Foundry one hundred and thirty by fifty feet; a Blacksmith shop, one hundred and twenty by fifty feet, having eighteen forges and four trip-hammers, for making, in addition to other things, Bolts—of which seven hundred are made and used in the machinery work daily; a building one hundred and ninety by thirty-two feet, containing an apartment used as a Brass Foundry, and also for "cleaning" the castings after they have been subjected to the process of "pickling," and a well-adapted room for storing patterns when not in use. The Machine Shop is a building two hundred and twenty-five by thirty-eight feet, and the large structure formerly used as an armory, occupying the entire southern front, is now fitted up exclusively for making Ring Spinning Frames, and one frame of two hundred and four spindles is turned out daily. A capacious Elevator is employed for raising and lowering castings and other objects between the different stories of the Machine Shop; and a Railway connects this and the Foundry. The Carpenter Shop is a building one hundred and sixty-eight by thirty feet, three stories high; and each of the various rooms and departments is supplied with tools and machinery of the most perfect construction, peculiarly adapted to the purposes for which they are designed. In the Wood-working Room are two of Daniel's Planing Machines, and one of Woodworth's, and Moulding and Sawing Machines capable of facilitating and making more perfect the wood-work required for the Carding Engines, Looms, etc. All the wood used is kept for the space of two years before being shaped by the machinery, so as to properly season it, and after it has been thus seasoned and brought to the form desired, it is placed in a commodious Drying-house, entirely fire-proof, and always kept by the heat of steam at a temperature of seventy-five degrees, for the purpose of being still more thoroughly seasoned. The tools and Machinery for performing the work in the several shops are mostly made by their own workmen; among which may be classed several Drills of new and improved construction, Boring Mills, and other self-acting machines, of the most beautiful design and perfect workmanship.

To attempt a recital of the various useful and novel Machines made in these Works, the most important of their class in the United States, would transcend our limits. We shall advert to only a few of the most important which are peculiar to the establishment, or which have received the benefit of the improvements of Mr. BARTON H. JENKS, who is one of the most ingenious of American Inventors.

1.—LOOMS.

Of Looms a large number of different styles are manufactured at these works, ranging from the single Shuttle, or ordinary Loom, through the more intricate forms of two Shuttle Looms for weaving Checks; three, and four Shuttle Looms, for weaving Ginghams and other fabrics requiring a corresponding number of colors in the weft, to the more enlarged Carpet Loom; and all of these embrace, in a greater or less degree, improvements and advantages not possessed by Looms manufactured elsewhere. The several improvements in the Looms are covered by seven distinct patents; and the main features accomplished by these inventions, so far as they relate to the two, three, and four Shuttle Looms, may be said to consist in the expeditious manner of moving the Shuttle Boxes to change the picks of weft; and, by certain new constructions, combinations, and arrangements of parts essential to this operation, and to others of an important character, by which as many picks of weft can be made by these two, three, and four Shuttle Looms, as by the single Shuttle Loom. As an exemplification of this, it may be stated, that so perfect is the arrangement of the various parts of this latter description of Looms, and the principle upon which they work, that they make one hundred and thirty picks of weft per minute, whereas the same class of ordinary Looms only make one hundred and ten. The *Muslin Loom* will run one hundred and sixty picks per minute, and produce nine pounds of muslin a yard wide, sixty-four picks to the inch, per day.

2 —JENKS' COTTON SPREADER,

Is made entirely of metal, thus insuring greater steadiness and durability. The beater, shafts, blades, and feed-rollers, are made of cast-steel; the shafts which drive the feed are braced together in such a manner that the teeth in the diagonal shaft cannot break; and, by an ingenious application of the elastic principle of air, the Machine is constructed to make the lap of uniform thickness, and of such compactness that any portion of it will sustain its own weight. Moreover, the whole machine works without producing any dust in the room.

The product of the Jenks "Spreader" is as follows: The gallows-shaft having four hundred and fifty revolutions, and the beaters eighteen hundred, produce per day two thousand four hundred and ninety-six yards. Doubled on the same machine three times, the produce will be thirty-eight laps, each twenty-two yards long, weighing each eighteen pounds; total weight, six hundred and eighty-four pounds. If two machines are used, the produce will be one hundred and thirteen and a half laps per day, each twenty-two yards long, weighing eighteen pounds; total weight, two thousand one hundred and two pounds.

JENKS' PICKER, as well as the Spreader, has an established reputation for excellence, especially in stopping action at the right point, without making the Cotton "bally and ropy." The product of this Picker is three thousand pounds per day when driven at the speed of twelve hundred revolutions a minute.

3.—THE CARDING ENGINES.

The self-stripping Cotton and Woolen Carding Engines, manufactured at these Works, are different from the Carding Machines generally used. In ordinary machines it is necessary to stop every thirty minutes, and with the laborious use of hand-cards clean out the secreted Cotton, so as to leave the teeth projecting. Jenks' Carding Engines are provided with a mechanism called a "Self stripper," which constantly picks out the Cotton from the bottom of the teeth, returns it to the Cylinder, and drops it on the points of the teeth, where the combs catch it and pass it out into a can.

One of Mr. Jenks' improvements which have given his Carding Machines the preference of the market, both in this country and in England, is his method of preparing the wood, of which the cylinders are made. The usual method of preparing it is to dry and season it by long exposure to natural or artificial heat. Wood so seasoned inevitably absorbs moisture of varying temperatures, and the cylinders of which it is made will alternately swell and shrink, and thus make imperfect carding. To prevent this, iron cylinders have been used, but there are also serious objections to their use. Mr. Jenks' plan is to hew out the wood in the rough, ready to be worked up into Cylinders, and then expose it in a vacuum chamber to extract all the moisture and juices, and also all the air and gases from its spiracles. This he accomplishes thoroughly; then, while the wood is in the vacuum, he introduces caoutchouc, paraffine, or some oleaginous fluid, which the pieces quickly and fully absorb, and the effect is to render the wood completely moisture-proof and water-proof. His Card-Cylinders and Rollers, made of wood thus treated, will not shrink or swell, and the card-teeth applied to one Cylinder will always maintain the same relative position with respect to the card-teeth of another Cylinder.

4.—JENKS' FLY FRAME,

Is well known, and used in all factories, having the most approved machinery. The productiveness of this admirable Machine is given in the following Table:

Length and Diameter of Bobbin.	Hank Roving.	Revolution per Minute of			Weight of Cotton on a full Bobbin.	Number of Sets produced per Week.	Actual Working Hours per Week.	Production per Spindle per Week.
		Spindle.	Front Roller.	Driving Shaft.				
In. In.								Lbs.
10x5	½	700	258	343	30 oz.	74.1	38.4	139
"	¾	750	233	367	"	51.2	44.1	96
"	1	800	216	392	"	38.2	47.4	71
"	1½	900	195	441	"	24.6	50.8	46
8x4	2	1100	201	360	16 oz.	33.4	47.5	33
"	2½	1200	191	392	"	26.7	49.4	26
7x3½	3	1300	184	425	10 oz.	32.4	47.2	20
"	3½	1300	167	425	"	26.2	49.1	16
"	4	1400	169	457	"	23.6	49.9	14
"	4½	1400	160	457	"	20.2	50.9	12
"	5	1400	153	457	"	17.7	51.6	11
7x3	6	1500	152	490	8 oz.	18.2	51.3	9
"	7	1500	142	490	"	15	52.5	7½
"	8	1500	133	490	"	12.4	53.2	6
"	9	1500	126	490	"	10.5	53.7	5
6x2½	10	1600	129	523	4½ oz.	16.7	51.9	4½
"	12	1600	133	523	"	14.5	52.6	4
"	14	1600	128	523	"	12.1	53.3	3⅜
5x2¼	16	1700	130	544	3 oz.	15	49.4	2¾
"	18	1700	125	544	"	13	50.4	2⅜
"	20	1700	121	544	"	11.5	51.1	2

5.—THE JENKS' PATENT SPINNING FRAMES,

Have the reputation of being the best in the United States, and on the most approved system. These machines are made with different numbers of spindles, the usual size being a machine of two hundred and four spindles. Each spindle will produce six skeins of number twenty yarn, each skein weighing one-twentieth of a pound—all weighing six-twentieths of a pound; velocity of the spindle seven thousand turns per minute, with wooden bobbin. The spindles have a self-lubricating bolster, which will run a month on six drops of oil. An important coneing motion forms the bobbins with taper ends, using paper cones on the spindles, and effecting thereby a great saving of weight, and consequently of power in carrying the wooden bobbins ordinarily in use. The reel is an important machine in a Cotton Factory where the yarn is spun to be sold in the market, and much attention has been bestowed upon this machine by the Bridesburg Manufacturing Company. The reel marks the skeins by an automatic motion, and when seven skeins are marked off at each spindle, the driving-belt is thrown off and the machine stops. By another ingenious arrangement the skeins are removable from the shaft of the reel without lifting the reel, (which is the case in all the reels excepting those made by this Company,) thus enabling smaller girls to operate the Machine.

6.—JENKS' IMPROVED CYLINDER COTTON GIN,

Is novel for the peculiarity of the construction of the cylinder, which operates in combination with a stationary straight-edge and a spirally-grooved roller called "the agitator." The straight-edge is fixed in a position parallel with, and tangential to the cylinder, with its thinner edge almost in contact with the upper side of the same and the agitator, so as to rotate rapidly at a short distance above and parallel with the straight-edge. The periphery of the cylinder consists of numerous steel-wire teeth imbedded in Babbitt metal, in positions inclined in the direction of the cylinder's motion, so that after the cylinder is "ground" or finished, each tooth presents a separate, sharp, and smooth point, tangential to the cylinder surface. When in operation, the cotton and seeds are carried by the cylinder against the straight-edge, where they are rolled over and over by the agitator, until the teeth of the cylinder have stripped off the fibre, the seeds immediately drop down, through a grating, into a receiving box. The fibre is at the same time being continually removed from the cylinder in the usual manner, by a rotating brush behind the straight-edge. The teeth of the cylinder are made of the *finest steel* needle wire, rolled into a double razor-edge section, and secured obliquely around the cylinder, with their sharper edges in directions transverse to the axis of the same; consequently, after the cylinder is "ground off" in finishing, it presents a serrulated surface, or a surface studded over with innumerable sharp and smooth tangential teeth, admirably adapted both for entering and leaving the fibres.

It will gin any Cotton, however trashy it may be, and take nothing through but the lint; it neither cuts nor naps the fibres in the least, leaving them nearly as long as when separated by hand; whilst it will clean as great a quantity in the same time as any other Gin occupying the same extent of space, and run as easy. It will also last as long, if not longer, than the Saw-gin, and cost no more for repairs.

Besides these special machines, it may safely be said that all of the machinery made at these Works is much lighter, neater, and of better iron, easier to manage, cheaper to run, and will turn off more work of equal excellence than any English machinery ever built.

Since the close of the war, and since Cotton-manufacturing has returned to its normal profits, mill-owners have commenced to throw out English cards, drawing frames, slubbers, and speeders which they had been persuaded or forced to buy, and are substituting American in their place. The whole range of English machinery is heavy and requires great power. Their cotton machinery, unimproved by the adoption of American inventions, is clumsy and weighty to the degree that might be expected in a are at their minimum value. Besides, many American improvements, for country where steam and human labor cost but little, and iron and coal which patents have been taken out in England, are inaccessible to the British machine manufacturers.

The Bridesburg Manufacturing Company has a paid-up capital of one million of dollars, and employs five hundred hands. The principal officers are Barton H. Jenks, President; Joseph G. Mitchell, formerly cashier of the Mechanics' Bank, in Philadelphia, Treasurer; Samuel O. Shouse, Secretary. The office in Philadelphia is at 65 North Front street.

II.

The Southwark Foundry—Merrick & Sons, Proprietors,

Is one of the largest and most important Machine Works in the United States. It was started in 1836, as a Foundry (for castings) only, but was soon enlarged; and now, by accessions and improvements, in buildings and tools, has become a first-class establishment for the manufacture of all kinds of heavy machinery. It occupies the entire square, bounded by Washington and Federal streets and Fourth and Fifth streets, and the principal buildings are as follows:

Building	Dimensions		Sq. Feet.
Iron Foundry	115 x 107	Area of Ground Floor	17,340
"	95 x 53		
Brass Foundry	60 x 25	"	1,500
Smith Shop	165 x 40	"	6,600
Machine Shop	162 x 54	Three Stories.	8,748
Pattern Shop and Stores	208 x 40	Two Stories.	8,320
Boiler Shop	182 x 94	"	17,108
Erecting Shop	168 x 66	"	10,758
"	130 x 34	"	4,420
Carpenter Shop	58 x 40	"	2,320
"	38 x 24	"	912
Sheds for Storage, etc.			10,000
Total area occupied by Buildings			88,026
Total area of yard-room			56,174
Total area occupied by the Works			144,200

The Pattern Shop and Stores are two stories, and entirely fire-proof.

In addition, it has a tract of land on the Delaware river, about four hundred feet front and one thousand one hundred feet deep, affording ample space for extensive iron boat yards; and on this tract there is a fine pier, sixty feet wide and two hundred and fifty feet long, with a very powerful sheers at the end, capable of lifting fifty tons.

A brief description of some of the objects of interest in this establishment will show that the arrangements, tools, and appliances in use, are on a scale proportionate to the capaciousness of the buildings.

The Foundry has two Cranes, capable of lifting fifty tons each, and three others of thirty tons lifting-power, by which any object may be transferred from one extremity to the other, or to any point on the floor. Two of Mackenzie's Cupolas are used for melting the iron, and are supplied with blast by two Mackenzie Blowers, each Blower being attached to its own Cupola, and capable of being used separately or together; capacity of the Cupolas twelve tons and six tons per hour respectively.

The ovens for drying Moulds are four in number, and of immense capacity.

In the Smith Shop the blast is obtained by an Alden Fan. There are two Nasmyth Steam Hammers, one of ten hundred and one of five hundred weight of ram, and two of Shaw & Justice's Dead Stroke Power-hammers (of which this firm are sole assignees for the State of Pennsylvania), one of one hundred pounds for general work, and one of fifty pounds, expressly adapted for making bolts. There are also, in this shop, Bolt and Rivet Machines, for the manufacture of these articles, large numbers of which are annually used. The Brass Foundry has a Cupola and four Crucible Furnaces.

The lower Machine Shop has a Boring Mill, which will bore a cylinder eleven feet diameter and fourteen feet high ; a Lathe capable of swinging seven feet diameter and thirty-two feet long.

A Planing Machine, capable of planing a piece of machinery eight feet by nineteen feet by thirty feet, and a Planer twelve feet wide, ten feet high and twenty-two feet long, besides a number of Lathes and Slotting Machines, of such power as give the firm the ability to execute orders for the heaviest description of machinery.

The upper Machine Shop is arranged in the most convenient manner, with a complete outfit of tools of improved construction ; fire-proof store-rooms for small tools ; fire-proof paint-room, etc.

The Boiler Shop is provided with a Riveting Machine capable of riveting a boiler forty feet long, and of any diameter; with a Treble Punching Machine of immense strength ; with heavy and light Shears and Punches; an Air Furnace, for heating large plates; Rolls, for bending; Cranes, etc. A large portion of the buildings described in the first edition of the present work having been destroyed by fire, new ones of the most substantial description have been erected in their stead, and they have been fitted with every convenience and mechanical appliance which long experience has proved to be necessary. All the foremen's desks are connected by magnetic telegraph with the Superintendent's and Main Office. The establishment is also in telegraphic connection with the principal lines of the country.

The firm of MERRICK & SONS is now composed of J. VAUGHAN MERRICK, WILLIAM H. MERRICK, and JOHN E. COPE, all gentlemen of experience and engineering ability.

The making of Sugar and Gas machinery forms a large proportion of the business of their establishment, which is justly considered one of the most complete of its kind in the United States. For a list of some of its most important productions, see chapter on *Iron and its Manufactures*.

III.

The Port Richmond Iron-Works—I. P. Morris, Towne & Co.,

Whose achievements in constructing heavy machinery have been previously alluded to, are also entitled to rank among the most important Machine works of the United States. They were founded by LEVI MORRIS & Co., who commenced business in 1828. At that time many of the tools which are now deemed indispensable in every machine shop, even those of the most moderate pretensions, were scarcely known. Slide Lathes and Power Drill Presses were not in general use, and the only representative of the Planing-machine in this country, it is believed, was to be found at the Allaire Works, in New York, which was originally built for fluting rollers. It was not until 1838 that a planer was purchased and fitted up in the Richmond works. In the Foundry department, the operations were also conducted with very imperfect and inefficient machinery compared with that now in use. Anthracite coal, which was introduced here about 1820, was by no means exclusively used for melting iron. The blowing machinery was of a very primitive character, with unwieldy wooden bellows and open tuyeres. The best product was not more than two thousand to three thousand pounds of iron in an hour, and in the course of the heat an average much below this. With the present improved Blowing Machinery, and improved Furnaces, eight tons have been melted in forty-six minutes, with a consumption of coal of one pound to eight pounds of iron melted.

In 1846, the works were removed from Market and Schuylkill Seventh streets to their present location, which is on the Delaware River, adjoining the Reading Railroad Coal Wharves on the south. The buildings, which are of brick, occupy a lot having a front on the Delaware River of one hundred and forty-five feet, a front on Richmond street, or Point Road, of two hundred and sixty feet, and an entire depth or length, from the Richmond side to the end of wharf, of one thousand and fifty feet.

The remarkable feature in this establishment is the extraordinary size of the tools in use, and the perfection of the machines employed in the various shops. In the *Foundry* there are three Cupola Furnaces, the largest of which will melt twelve tons of iron per hour. In the *Machine Shop*, there is a Planing Machine capable of planing castings eight feet wide, six feet high, and thirty-two feet long; a Lathe that will swing six feet clear, and turn a length of thirty-four feet; and a Boring Mill, possessing also the qualities of a horizontal lathe, which will bore out a cylinder sixteen feet in diameter and eighteen feet long. This is believed to be the largest in America or Europe. In their *Boiler Shop* they have one large Riveting Machine, and facilities for making boilers or plate-iron work, of every description that may be desired. But a few years ago, Steam Boilers, made of plate-iron, were riveted exclusively with hand-hammers; and when the City Water-Works were located at Centre Square, the steam boilers were built of wood, with cast iron furnaces. At the present time, in this, as in the best shops, circular boilers are riveted in a machine, by pressure produced by a cam operating upon a sliding mandril. In their *Smithery*, they have a Nasmyth Steam Hammer for heavy forgings; a Tilt Hammer for light work; and throughout the establishment, the minor tools, consisting of Lathes, Boring Mills, Slotting and Shaping Machines, Planing Machines, Horizontal and Vertical Drills, etc., etc., are all of the best description, and combine the latest improvements.

These works have an additional element of strength in the experienced engineering ability, and genius of the members composing the firm. There are four partners viz: Isaac R. Morris, John H. Towne, John J. Thompson, and Lewis Taws. The first-named gentleman was born in 1803, was one of the original partners in the firm of Levi Morris & Co., who commenced business in 1828, and since that period has been identified with the manufacturing interests of Philadelphia. In his business career, he has been distinguished for a discriminating intelligence, inflexible honesty, and a laudable public spirit. Mr. J. H. Towne was formerly engineering partner of the firm of Merrick & Towne, and is an engineer of unquestioned ability. Mr. Thompson, who has been connected with the establishment for many years, has under his charge the finances of the firm. Mr. Taws has been connected with the works since 1834, and until 1861, when Mr. Towne joined the firm, had exclusive control of the mechanical department of the establishment. He served his apprenticeship with Rush & Muhlenberg, the successors of Oliver Evans, and in early manhood went to New York, where he entered into the employment of the West Point Foundry Association, then under the superintendence of Adam Hall, a distinguished Scotch engineer. The present arrangement of the Port Richmond Works is the result of his experience.

The monuments of this firm's engineering abilities are found in all parts of the United States. We have already alluded to some of the great machines constructed here, and need not repeat them. See chapter on *Iron* and *its Manufactures*, article "Great Machine shops." They are now engaged in working for the Government and in building engines of various sizes, for different sections of the country, as well as a variety of work of a promiscuous character. Their list of manufactures includes every description of heavy machinery except Locomotives.

Messrs. I. P. MORRIS, TOWNE & Co., have a capital invested of over $400,000, and usually employ about four hundred hands.

IV.

The Pascal Iron-Works.—Morris, Tasker & Co., Proprietors,

Is the most extensive manufacturing establishment in the southern part of Philadelphia, and the largest of its class in the country. The buildings cover an area of about four acres, and in them are employed one thousand one hundred men, a much larger number, we think, than is at this time employed in any other iron-works in the city. The machinery is of the most approved description, much of it original with the firm, and surpasses, it is said, any that can be found in any similar establishment in England. Over forty thousand tons of anthracite coal are annually consumed in these works.

The predecessors of the present firm were the pioneers in this country in the manufacture of wrought-iron Tubes and fittings for gas, steam, and water. They commenced this business in 1836, in which year they made sixty thousand feet of tubes; the firm now manufacture nearly five millions of feet annually. Subsequently they added to these the manufacture of cast-iron Gas and Water Mains, and Lap-welded Flues for boilers.

One of the specialties of this firm's manufactures is that of Apparatus for warming public and private buildings, both by hot water and by steam. Mr. THOMAS T. TASKER, Sr., one of the original partners, is the inventor of a very popular Self-Regulating Hot Water Furnace, by which the temperature in a house can be maintained at any required degree of heat for an indefinite period of time, without further attention than an occasional supply of coal. Mr. TASKER is also the inventor of a process by which the circulation of steam is kept up through heating pipes to any extent, the condensed steam returning back to the boiler by its own gravity, thus saving the heat which was formerly lost in running off the water. Messrs. MORRIS, TASKER & MORRIS made the first public experiment to test the value of this discovery at the Pennsylvania Hospital in 1846, and since then they have warmed many houses, factories, and other buildings. The original inventor neglected to patent his discovery, and others have made fortunes by its adoption and application.

This house dates its origin back to the year 1821, when STEPHEN P. MORRIS commenced the manufacture of Stoves and Grates at the corner of Market and Schuylkill Seventh streets. In 1828, he removed to Third and Walnut streets, where he erected a foundry, and not long afterward was joined by HENRY MORRIS and THOMAS T. TASKER, establishing the firm of S. P. MORRIS & Co. Subsequently S. P. MORRIS retired and WISTAR MORRIS became a member of the firm, when the name was changed to MORRIS, TASKER & MORRIS, which continned until 1856, when the present style of MORRIS, TASKER & Co., was adopted. The partners in the firm at this time, are STEPHEN MORRIS, THOMAS T. TASKER, Jr., STEPHEN P. M. TASKER, and HENRY G. MORRIS—young men, but who have the advantage of a large capital, and of the experience of their predecessors.

V.

The McCullough Iron Company Works,

Before referred to, consist of five Rolling Mills, for the manufacture of Sheet Iron, all of which are located in Cecil County, Maryland, three of them being at North East, styled respectively the NORTH EAST, the SHANNON and the STONY CHASE; the fourth, about two miles north of Elkton, called the WEST AMWELL Mill, and the fifth, the OCTORARO, at Rowlandsville, a village about five miles North of Port Deposit.

Of these mills, four are driven by water power, and one, the largest, a bar Mill, by steam power, where the bar is manufactured and prepared for rolling into sheets. The latter operation is done by water power.

They have also "a forge," with six refinery fires for the manufacture of Charcoal Blooms, with what is termed a "run-out" fire, where all the Pig Metal is melted and "run out" with charcoal, a process of refining which the iron passes through before going to the forge. All of these are located in Cecil County, Md., with water powers yet unemployed, sufficient to drive additional works when required.

The site for the works at North East was selected over one hundred and fifty years ago, by a company composed chiefly of English capitalists, consisting of Messrs. WILLIAM and THOMAS RUSSEL, WILLIAM CHETWYND, SAMUEL and OSGOOD GEE, and WILLIAM WHITEWICK, together with AUGUSTINE WASHINGTON, the father of the illustrious GEORGE WASHINGTON. This seems to have been an old company in England, of some of the substantial and enterprising men of that age. They were succeeded by THOMAS RUSSELL, and his heirs who inherited the property, and held it until some twenty years ago, when it was purchased by the present owners.

Adjacent to the works at North East, the Company own from four to five thousand acres of land, a portion of which is laid off in fine farms, the remainder being woodland, from which they obtain the charcoal consumed in the manufacture of a superior quality of iron, which they use almost exclusively for Galvanizing purposes. The consumption of charcoal alone amounts to over one hundred thousand bushels annually.

The works known as the OCTORARO, (on the Creek of that name,) at Rowlandsville, some five miles from Port Deposit, were formerly owned by JOSEPH ROMAN, and are among the oldest Sheet Iron Works in our country. They have been the property of this Company for a number of years.

The WEST AMWELL Mill was built in 1840, and in common with the Company's other mills, enjoys a fine reputation for the quality and finish of its sheets. With the enlarged facilities which they possess, the Company is now able to produce many thousands of tons of Sheet Iron annually.

Their Galvanizing Works, as well as their Office and Warehouse, are located in Philadelphia, where the principal business of the Company is transacted. They were the first to introduce the manufacture of Galvanized Sheet Iron, Hoop Iron, Galvanized Water Pipes, etc., etc., in the United States, having in 1852 obtained skilled workmen from England for the purpose. The manufacture of these is now a very important branch of their business, and has become an important and constantly augmenting branch of the business of the country. The Galvanized Sheet Iron of the MCCULLOUGH IRON COMPANY has a wide reputation by reason of their enlarged experience and the particular attention which they have bestowed upon this specialty of their business, and is said to be superior to any other, either of domestic or foreign manufacture.

Among other advantages this Company possesses, is the exclusive right of Mr. E. A. HARVEY's valuable Patent for cleaning iron and other metals from dust, dirt or oxide, which must in time become invaluable.

The black dust from the bituminous coal used in the process of manufacturing, has heretofore remained on the sheet when finished, and was always objectionable to the workers in this article. By this patent process the sheet passes through a cleaning machine of Mr. HARVEY'S invention (consuming but a moment) when it comes out as free from dust and dirt as a sheet of the finest paper. This is a great desideratum with the worker, and must in time altogether supersede the old method of manufacture.

VI.

Henry Disston's Saw Works,

Which have been previously mentioned as probably the largest in the world, are certainly the most extensive in the United States. All the operations incidental to the manufacture of Saws of all kinds are carried on here, (including the Steel making,) on a scale of unsurpassed magnitude, and not only Saws, but all the minor constituent parts and adjuncts, from a saw screw to a saw file. It is the most complete and comprehensive establishment of the kind, and its organization attests the executive abilities and fertile genius of its originator and manager.

The buildings, 67 and 69 Laurel street, cover two hundred and fifty thousand square feet of ground, and comprise a Rolling Mill, two hundred and forty by seventy five feet; a warehouse for the reception of raw stock, one hundred and twenty by seventy feet; a Machine Shop and main Saw Factory, two hundred by one hundred feet, three stories in height; a Wood working Department, seventy-five by forty feet, four stories high; a Blacksmith's, Hardening and File Shop, and Brass Foundry, two hundred by one hundred feet, and sundry other buildings of less dimensions. In the Lumber Department, a stock of three hundred thousand feet of Beech and Apple wood for Saw Handles, is at all times in process of seasoning. On the north side of Haydock street there is another building fifty by two hundred and fifty feet, three stories high, in which Butcher Knives and Trowels and Reaping Knives, etc., are made.

These works are no less remarkable for the wonderful efficiency of their tools and machines, than for their extent. To illustrate:

To toothe five dozen Wood-Saws in an hour, is rapid work for the best mechanic in the world; Mr. Disston has machinery by which one man can toothe thirty dozen in the same time. He can toothe perfectly a sixty inch circular saw in two minutes, which by the old process would require the labor of one man two hours. The tempering process, which is patented, is most complete, and saves at least one third the labor ordinarily required, or in other words, sixty men can do as much work as one hundred formerly did. The apparatus for grinding is novel, inasmuch as it includes machinery that will grind both sides of a saw at one operation, and long as well as short saws. We believe that the machines in the Grinding Department are the only ones of the kind in the world. Mr. Disston has also a new process for stiffening saw blades, or in other words, refining the grain after tempering, by repeated blows of a steam hammer. In the Rolling Mill, there are forty melting holes and three sets of Rolls the largest being capable of turning out a saw-plate sixty-four inches in diameter. This mill gives Mr. Disston the ability to fill an order for any saw of extraordinary size in a few days that would otherwise have required months.

It will readily be perceived that this facility of production and economy of labor necessarily give the proprietor of these works great advantages, reducing the cost of manufacturing Saws to its minimum, and we are not therefore surprised to learn that, though at least three million dollars' worth of Saws are annually required in the United States, Mr. Disston supplies fully one fourth of the whole amount. His Works consume three hundred tons of coal a month, and furnish employment to four hundred men. For an account of Mr. Disston's inventions, see *Bishop's History of American Manufactures.*

VII.

Fitler, Weaver & Co.'s Cordage Works,

Are the most extensive in Philadelphia, and, with probably two exceptions, the largest in the United States. They are also among the oldest established rope manufactories in this country, having been founded by Michael Weaver in 1817. At that period, cordage spinning was carried on in the primitive way, and gangs of rigging were laid by hand, the neighbors being called in to assist. Subsequently horse power was employed for this purpose, but it was not until recently that steam power was applied, by means of which large Marline Ropes are now made with the same facility as twines. In fact, so great has the progress been, that now a gang of rigging suited for the largest vessel in the Government Marine can be finished and delivered within three days after receipt of the order. The original rope walk, with its hand spinning apparatus, is still in use by the present firm, who, however, have introduced a Corliss engine of forty horse power, and made other improvements that adapt it to finishing fine cordage. In 1847, Edwin H. Fitler became connected with the firm then carrying on the business, and possessing executive talents of a high order, soon made his influence felt in the firm's affairs. He saw that for want of a mill possessing improved machinery, the trade of the South and West was passing through Philadelphia, to the New York and Eastern manufactories, and he urged upon his partners the erection of works that would compete with any in the United States. These were begun and finished, but within ten days after their completion, were destroyed by fire. Another and larger factory was immediately erected on the former site, at Germantown road near Turner's lane, but this also was burned in July, 1866. Within four months, another mill was erected in accordance with plans drawn by Mr. Fitler, and this is now without doubt the finest and most conveniently arranged Cordage Factory in the United States.

The main building is of brick, three stories high, one hundred and eighty-four feet long, and fifty-eight feet wide. There are also attached an engine room, which contains, besides the engine, two steam fire pumps, a boiler-room, and a machine shop, twenty-four by forty-six feet. All these buildings are of brick, with gravel roofs. The factory is warmed by steam and lighted by gas, and every precaution is taken to guard against fire. There are iron water pipes with cocks and hose on every floor. On the top of the Hoisting Machine is a tank containing five thousand gallons of water, supplied from the City Reservoir, and a large well in the rear of the engine-room. The steam-boilers are enclosed by brick walls, and the tar-boiling apparatus is heated by steam. In the main building are all the machines of modern construction, that are best suited to spinning and preparing rope and cordage. The machinery is propelled by a Corliss engine of two hundred and fifty horse power, and has a capacity for performing work that would have required by the old process of hand spinning at least a thousand men. Extending at right angles from the factory proper, is a rope-walk thirteen hundred feet long, and on the premises are numerous buildings auxiliary to the business. It is probable that the cost of this factory and its machinery was not less than a quarter of a million of dollars. About three hundred men are employed in the two factories, which consume each day from sixty to seventy bales of Manilla, Russian and American Hemp, and produce about seven tons of Rope and Cordage daily, or 4,500,000 pounds per annum. The facilities for manufacturing are so complete, and the machinery so perfect, that the Cordage cannot be surpassed in quality by any made in the world.

EDWIN H. FITLER, who is now the senior partner in the firm and principal owner of the works, is a native of Philadelphia, having been born in the district of Kensington in 1825. He belongs to the energetic and progressive order of manufacturers, of whom it is often said that Philadelphia has too few. He has established a private telegraphic wire from the store on Water street to the Factory, which passes directly through his house in the city, by means of which not only orders and business reports, but private and domestic messages can be transmitted by telegraph. The system he has organized is so complete that the affairs of a vast and complicated business are managed with the minimum of trouble and labor. Every evening, an account of the various kinds of hemp on hand is taken, and the quantity of the different sizes of rope in store is made up, and thus every morning he has a complete and exact report of the state of affairs ready for his guidance during the day. Every morning also, the report of the night-watchmen, registered on their tell-tale clocks, is submitted, and their fidelity in the discharge of their duties examined. Among the minor but nevertheless novel and useful items of counting-house management, is a Diary or daily journal of events. For nearly twenty years this firm have kept, in an appropriate book, a record of each day's events, including an abstract of every important business conversation, a copy of every telegraphic despatch sent or received, and of all orders and purchases, and its utility, especially in cases where options or refusals have been given for a limited time, has been frequently demonstrated.

Messrs. Fitler, Weaver & Co., have a large warehouse at 23 North Water street, for the sale of their cordage, in connection with which is a store containing a full and complete stock of Naval supplies. Besides these, they occupy four other stores for storage purposes. The partners in the firm are EDWIN H. FITLER, MICHAEL WEAVER, and CONRAD F. CLOTHIER.

VIII.

Wilson, Childs & Co.'s Wagon Works,

Deserve a place among the remarkable Manufactories of Philadelphia, as they are probably the largest works of the kind in the United States. Their history is briefly as follows:

In 1829, D. G. WILSON, a wheelwright, and Mr. J. CHILDS, a blacksmith, formed a copartnership for making Farm Wagons, Carts, etc., and opened shops for the purpose, at the corner of St. John and Buttonwood streets. By fidelity in workmanship, and prompt attention to business, their products soon became in demand in all parts of the country, especially in the Southern States, where their plantation wagons were in the highest favor. They embarked largely in constructing Army Wagons for Government use, and the first Army Wagon made after the present improved pattern designed by General GEORGE H. CROSMAN, now Assistant Quartermaster General at Philadelphia, was built by them. Every part of these wagons is made with the same exactitude of dimensions as the Gun Carriage of a park of Artillery, and their utility was especially demonstrated in the Expedition to Utah, where they traversed the roughest roads for thousands of miles, without the breakage of any important part. In 1860 Mr. WILSON died, much regretted by his associates, leaving however, a son, WILLIAM M. WILSON, who now creditably represents his interest in the firm.

During this period, the original works were enlarged from time to time, but it was soon found that the premises, though containing two hundred and thirty feet on Buttonwood street, and one hundred and thirteen feet on St. John street, were too small to accommodate the increasing business. In 1850, they purchased a manufactory, erected by Mr. SIMONS, and additional property, comprising in all a square on both sides of Second street and Lehigh Avenue, containing two hundred and sixty by five hundred feet, or over six acres. The square on the west side of Second street, is now nearly covered with buildings. There is a main building, two hundred and fifty feet long, with a front of fifty feet, used mainly for Painting, Varnishing and storage purposes. The wheel and body shop is one hundred by forty-six feet, three stories high; the Blacksmith Shop is two hundred by thirty-five feet, the Saw Mill, Engine house and Machine shops, is eighty by forty-five feet, three stories high; the Running gear shops, one hundred by forty-five feet, and besides these there are numerous auxiliary buildings. On the east side of Second street is a Saw Mill, fifty by seventy-five feet, but the greater portion of the square, which is five hundred by two hundred and forty-eight feet, or nearly three acres, is occupied as a Lumber yard. Here is kept at all times an immense stock of lumber, amounting at times to two million feet of hard wood planks and boards, thirty thousand hubs, and five hundred thousand spokes. These, before being used, are thoroughly seasoned from one to five years, the usual allowance being one year for every inch in thickness. The hubs are made chiefly from black locust trees, sawed into suitable lengths, and before being put away to season, the bark is removed, and a hole bored in the centre to facilitate the seasoning process. In the store-rooms the firm also keep a large stock of finished work, including cart and wagon bodies, and several thousands of wheels.

These works have the capacity of turning out one hundred and fifty army wagons in a week, without interfering materially with the regular business, in Farm and Plantation wagons. The firm of WILSON, CHILDS & Co., have an established reputation for reliability, and their agregate trade amounts to hundreds of thousands of dollars per annum.

IX.

M. Landenberger & Co.'s Hosiery Manufactory,

At the intersection of Frankford Road and Wildey streets, deserves prominence from the fact that it was the first of those large mills which are destined to make Philadelphia the Leicester of America.

Twenty-five years ago Mr. LANDENBERGER commenced the manufacture of Woollen Hoisery, in Philadelphia, with a determination to compete with foreign manufacturers, and as a means to that end, erected an establishment that would rival in importance any in the chief centres of the manufacture in England. The building has a front on Frankford Road of seventy feet, and is two hundred and fifty feet deep, four stories in height. There is also a basement in which all the preliminary operations of preparing, scouring and carding the wool are performed. To follow the operations of manufacturing in their consecutive order, we must ascend to the fourth floor, which is easily effected by means of an improved Elevator. Here is a spacious room, in which a large number of persons, chiefly females, are employed in preparing yarns for the looms. In the spinning department there are eight sets of mules, with two thousand five hundred and sixty spindles. On the same floor is an apartment devoted to Stocking weaving, principally of the plainer description. The Fancy Hosiery Department is on the third floor, which is divided into two large rooms, the one for weaving, the other for finishing, in which a large number of Sewing Machines is employed. No one who has not visited a manufactory of this kind, has any conception of the immense variety of patterns and hues comprised in the term Fancy Hosiery. The second floor is principally devoted to weaving Ristori Shawls, for which this firm are celebrated, and the first floor to weaving Scarfs and other knit goods; besides the wareroom and offices.

There are no less than fifteen different kinds of looms employed in this establishment, some of them the invention of the senior partner, and as many as eight hundred different styles of goods are produced here in a single season. About five hundred hands are employed in the manufactory, and upwards of three hundred and fifty thousand pounds of American Wool are annually converted into various useful and ornamental fabrics.

Besides this extensive establishment, Messrs. LANDENBERGER & Co. have factories on White Clay Creek, near Avondale, Chester County, Pa., where the worsted is prepared for the Kensington Mills. The Avondale Blanket, so favorably known in all markets, is produced at these factories.

X.

Stephen F. Whitman's Steam Chocolate Works,

To which allusion has been previously made, were, we believe, the first to attempt the manufacture of Chocolate on a large scale by steam propelled machinery. The raw material out of which this healthy and valuable article of food is prepared, is the fruit of the Cocoa-Tree, which grows in all parts of South and Central America. The seeds are about the size and color of a sweet almond, inclosed in a pulp, but the most highly esteemed varieties are obtained from Central America. The seeds are disengaged from the pulp, dried, prepared by a peculiar process, and then shipped in large quantities to the place of manufacture.

Some years ago the best Chocolate was made in France, but now America outrivals her in the quality of the production, and Philadelphia fairly stands at the head of the manufacture in the United States.

On entering Mr. Whitman's manufactory, the first objects presented to view are bags of Cocoa seeds, piled one on the other in huge stacks, and representing large sums of money (for Cocoa seeds are costly). These seeds are first roasted, and then the bags are emptied one after the other, into a "cracker," which breaks the kernels up into bits, about the size of coffee grains. These tiny grains are then run through a fan, which cleans out the husks and shells, and throws the pure Cocoa out upon trays, where it is carefully hand-picked to remove even the slightest particle of foreign substance.

The mass is then placed in a huge mill to be ground. There are four of these, all of French manufacture, and of the most costly description. They are the largest, most delicately finished and most effective in this country. The rollers are of fine French granite, and revolve upon a granite bed, and this keeps the Cocoa of its original light brown color, without giving it that dark appearance which it acquires when ground by iron rollers. The kernels are placed in the first of these mills, and as the stones revolve they pulverize the seeds, and express the oil, leaving a rich, oily mass of a tempting appearance. From this machine the substance is removed to another mill, where it is ground even finer, and if it is desired to make sweet Chocolate, is mixed with the requisite quantity of sugar. As the substance revolves upon the rollers, steel scrapers remove it, and it falls into a receptacle, from which it is removed into moulds, settled by a curious and beautiful process, and put away to cool. From the cooling-chamber it comes out fragrant, pure and delicious Chocolate.

Mr. Whitman makes vast quantities of Chocolate of all kinds every day, and ships it to all parts of the continent. He even monopolizes a large portion of the trade in New York city, merchants there preferring an article that Mr. Whitman guarantees to be of the purest character to the adulterations that abound. Some ten or twelve different kinds are made, for all of which there is a constant and increasing demand from hotels, private families, invalids and confectioners; but the specialty of this unique establishment is the manufacture of Chocolate for table use. It is not too much to assert that in a few years the demand for this pure article will have increased, as its qualities become known, to such an extent that it will become one of the most important branches of our manufactures.

Mr. Whitman is the pioneer in the use of steam power in the manufacture in this city, and deserves the credit that he will receive, and the profits he enjoys for the skill, labor, enterprise and investment of capital in this undertaking. His fine machinery is simply wonderful in the delicacy of its movements and the elegance of its finish, and no one who has ever tasted the Chocolate produced by it will deny that it performs its work in the most superb manner.

A man who creates a new and profitable branch of manufacture is a public benefactor; but he who, in addition to this, furnishes the people with an article of food and drink which has infinitely greater nutritive qualities than tea or coffee, without any of their stimulative effects—an article which is health-giving in all its properties, and the general use of which cannot fail to elevate the health average of the people—a man who does this absolutely, deserves the gratitude of his fellow-men.

Besides his Cocoa and Chocolate Works, Mr. Whitman is one of the largest manufacturers and dealers in Fine Confectionery. See chapter on Confectionery.

XI.

Howell & Brother's Paper Hangings Manufactory.

Is believed to be the most extensive in the United States and certainly one of the largest in the world. It is located on Washington avenue between Twenty-first and Twenty-second streets and covers an area of two acres. The main building is an immense brick structure four hundred feet long, and five stories in height with a basement. Extending at right angles with the main building are two wings that have an average depth of two hundred feet. On the first floor are nine machines that have the capacity of printing twenty thousand yards of Paper in a day. Three of these machines will print twelve colors at a time. Adjacent to the printing room is a large room, sixty by one hundred feet, in which the colors are mixed. There are no less than fifty large mills employed in this department in preparing colors. On the second and third floors the clay is ground and prepared, which is used to cover the blank paper and forms a basis on which the colors rest. In the fourth story are the Bronzing and Varnishing departments and also rooms for the Designers and Engravers of blocks. Each design for a separate color, is first cut on a block of pear tree wood and the outlay for this purpose in a leading establishment like this, where new designs are constantly being made, is very large.

The facility with which Wall Paper is manufactured by the modern processes is most remarkable. Hundreds of rolls of blank paper can be printed in a variety of colors, dried, reeled, and be ready for market in a few hours. In Messrs. Howell's factory, forty tons of paper are consumed weekly, and about fifty millions of yards of Paper Hangings are produced annually, or a quantity, in miles, greater than the circumference of the globe.

This firm have been remarkably successful in producing the finer kinds of Hangings, such as velvet, velvet and gold, and satin surfaced papers. Samples of their productions can be seen at all times at their magnificent store, at the southwest corner of Ninth and Chestnut streets. This building has a front of marble thirty-three feet on Chestnut street, extends a depth of two hundred and thirty-five feet, and is generally regarded as the most elegant store in Philadelphia.

Messrs. Howell & Brothers were the only house in this manufacture who received an honorable mention at the late Exposition in Paris. Their establishment, which has grown, under their management, from a very small beginning, is the best evidence of their enterprise and energy. Its founders were the sons of John B. Howell, who came from London to the United States in 1793, and established a Paper Hanging manufactory at Albany, New York, in 1813. The kind of paper made at that time, was of common quality and unglazed, and America is indebted to the present firm more than to any other, for the introduction of those superior styles which rival the French in elegance and beauty of design.

XII.

Lockwood's Paper Collar Manufactory,

Of which the early history was recited in the article on Paper Collars, is one of the most interesting of the manufacturing establishments of Philadelphia. The buildings have a front on Third and Levant street of forty-seven feet, a depth of one hundred and eighty-seven feet, and are five stories in height. In different apartments there are about ninety machines used in the various processes, to which power is communicated by means of over six hundred feet of shafting, from a forty-horse-power engine located in the basement. The exhaust steam is utilized not only in warming the building, but in boiling water and heating glue or paste; and consequently fire is not required in any of the workshops. Many of the machines are novel, and some of them costly.

The first process, in making cloth-lined Collars, is to combine the paper and muslin, which is accomplished by means of a machine that has cost, with the various improvements necessary to overcome the unequal expansion and contraction of the two substances, nearly thirty thousand dollars. The finishing, or polishing, is effected by being passed between two highly chilled rolls, moving with different velocity, and which can be so nicely adjusted that an ordinary sheet of paper can stop an engine of forty horse-power. The Collars are cut into shape by means of a self-registering cutting apparatus, having a capacity for cutting out two hundred and forty thousand Collars per day, and requiring the attention of only two operatives. The stamping of the patent and sizes, and the imitation stitching, is effected at one operation, by steel dies, and then the Collars pass to the machine which punches out the button holes; and after that to the folding machines—which is the invention, in part, of the senior partner of the firm. All kinds and styles of ladies' and gentlemen's Collars and Cuffs, including Marseilles and fancy-printed patterns, are made; and whenever a new design of linen collar is adopted, either in this country or abroad, it is reproduced in paper in this establishment.

In connection with the Collar manufactory, there are apartments appropriated to the manufacture of Paper Boxes, of which about three thousand are made daily, or over a million in a year. The bottoms of these boxes are constructed largely of thin, circular, wooden blocks, which the firm have found to be a satisfactory and economical substitute for pasteboard. All the printing of labels, cards, etc., is executed on the premises; and so extensive is their business, that ten printing-presses, and one hundred and fifty fonts of type, are employed in this department. There is also a forge and a machine shop, where all repairs to the machinery are made. In the entire premises there are at times as many as two hundred and fifty girls and women employed, none of whom are taken without special recommendations or certificates of character. Dinner and dressing-rooms are provided for their accommodation. Elevators are used as a means of communication between the different stories of the building, and they are so arranged, by means of dogs that spring into ratchets in the sides, that any serious accident from the breakage of the wire rope is almost impossible.

A large apartment on the second floor is appropriated to the manufacture of patent direction labels, commonly called "Tags." The large amount of cuttings necessarily produced in the collar department, being as much as three thousand pounds weekly, is all, or nearly all, converted into Tags, made under a special patent, and which, from their strength, ready absorption of ink, and neatness of appearance, have been adopted as the standard tag by the Transportation Companies of the city, and are used largely by merchants of all classes. The sales of these alone amount to $25,000 per year, while of the Collars as many as three hundred thousand have been sold in a single week.

In 1862, Mr. Lockwood purchased the entire interest in the original patent, which was subsequently re-issued in four divisions, including both collars made of white paper, imitating starched linen, and collars composed of paper and muslin, or an equivalent fabric. In 1863, Mr. E. D. Lockwood became associated with his brother, establishing the firm of W. E. & E. D. Lockwood. In 1865, the firm disposed of their interests in the original patents to the Union Paper Collar Company, organized with a capital of three millions of dollars, but they still continue the business on a larger scale, working under a license from the Union Company, and paying a royalty monthly on their entire production. Recently they have fitted up machinery for manufacturing a new style of Patented Envelopes, of which they have facilities for producing a million a day. For an account of this new branch of their business, see chapter on "*Paper* and *Envelopes*."

XIII.

Wright, Brothers & Co.'s Umbrella Manufactory,

Which it is said is the largest in the world, was founded in 1820, by four brothers, John, Joseph, Edmund and Samuel Wright, natives of Oxfordshire, England, of whom only one now survives. During the first ten years of their experience as manufacturers of Umbrellas, their production did not probably at any time exceed one hundred per day; but for the long period of forty-five years, with the exception of a few months, they have kept their manufactory in operation uninterruptedly, and in the busy seasons they have produced from twenty-five hundred to three thousand Umbrellas and Parasols per day. The exception alluded to was during the early part of the late rebellion, when finding the demand for the articles of their manufacture had almost ceased, and their five hundred employees, principally females, in danger of suffering for want of their accustomed means of livelihood, they converted their establishment into a manufactory of army clothing; bnt when their contracts with the Government, amounting to a half million of dollars, had been executed and completed, they resumed their regular business, never having lost, in any degree, their perfect organization.

The manufactory and sales rooms of this firm are located at 322 and 324 Market street, in a four story building, owned by themselves, having a front on Market street of thirty-six feet, and extending in depth two hundred and six feet, widening in the rear to Hudson street, on which the front is sixty-five feet. Steam power and a great variety of machinery are employed in all the manufacturing operations, superseding, to a considerable extent, hand labor, and thereby insuring uniformity of size and strength in the various parts, and giving to the finished article a beauty and accuracy only obtainable by the use of the best mechanical means. But probably the secret of the remarkable success of this firm in their department of manufactures may be attributed to their advantages in procuring, on the most favorable terms, the raw materials that are required.

Whalebone, which was formerly a prominent article of consumption in this manufacture, is now almost entirely superseded by rattan and steel, in consequence of its great cost, caused by the singular failure of the whale fisheries. To insure a supply of rattans of the requisite quality, possessing the necessary hardness and elasticity, this firm have established connections in Hamburg, where alone the best quality, that is the Dutch East India Company's "selected," can be obtained, and their importations amount to two hundred and fifty thousand pounds annually, nearly all of which they consume in their manufactures; and that which is not suitable for their purpose, they dispose of to others, especially chair and basket makers. They have also favorable and extensive connections established with leading houses in Europe, through whom they obtain their Silks, Scotch ginghams, and various other materials direct from first hands, and of which, in a yearly production of over a half million of Umbrellas and Parasols, great quantities are necessarily consumed.

The head of this firm, Mr. SAMUEL WRIGHT, is the youngest and only surviving brother of the four who originally established it. For many years he has been the active manager of the establishment, and during a business career of a half century, has always been distinguished for honor and integrity in his dealings, and liberal and enlarged commercial views. Many of the mechanical improvements that give this firm their facilities for rapid production, are the offspring of his inventive genius. Though advanced in years, he continues an active supervision of the affairs of the firm, being aided by his three excellent sons, who are now associated with him.

XIV.

The Great Clothing Houses of Philadelphia.

Our task would not be complete without giving some account of the firms in Philadeiphia, who by the liberal use of Printer's Ink, have made their names as familiar as "household words." It is a somewhat remarkable fact, that the most illustrious examples of this too limited class, are found in a trade so standard and legitimate as the manufacture and sale of Ready Made Clothing. Yet such is the fact. No firms in any other branch of business, equal the Clothiers in the extent, and novelty of their local advertising. Every possible means to arrest public attention is availed of, and often genius of a high order is displayed in their selection of methods. The leaders and great originals in this department of the arts, are the firms of PERRY & Co., BENNETT & Co., and WANAMAKER & BROWN, each of whom has adopted a distinctive designation for the store in which they transact their business, as the "Star," "Tower Hall," and "Oak Hall."

Perry & Co.'s Star Clothing House,

Was founded by GRANVILLE STOKES, who was a pioneer in adopting the style of advertising, peculiar to these firms. He prosecuted a successful business for many years, and then disposed of his establishment to

Messrs. PERRY & Co., who are now among the largest houses in the Ready Made Clothing manufacture in Philadelphia.

The store of this firm, 609 Chestnut street, is remarkable for its spaciousness. No other house in the business, it is said, has a ground floor containing an equal superficial area. Another feature which impresses favorably, is the large amount of light which is thrown into every part of the immense room, by means of Patent Daylight Reflectors, which have been put up at great expense by the firm. Advancing from the entrance, the visitor is astonished at the immense quantity of clothing piled upon the counters, and his astonishment in this matter is not lessened, when he is informed that these piles, large as they are, need constant replenishing to supply the never ceasing demand. Advancing still further, we arrive at the Custom Department. Here we find an immense assortment of uncut goods, from which customers can make their selections, and have their goods made by measure. In this department Messrs. PERRY & Co. are at a large outlay in salaries for skillful artists, who preside at the cutting counters. This outlay, large as it is, is cheerfully made, as the firm know by experience the difficulty of obtaining first class talent, and having procured, know how to appreciate it. Still further on towards the rear entrance, are the counters for receiving goods purchased. Here is presented a scene of great activity, especially at those times when the firm are "laying in" goods to make up for the coming season. This we have been led to anticipate from the piles of Ready Made Clothing already seen. Hundreds and thousands of yards of cloths and cassimeres, of various styles and colors, both of foreign and domestic manufacture, are going through the process of measuring and examining. Every piece that is not perfect is rejected and laid aside. Those that pass this ordeal are well sponged, and then carefully rolled up and piled away for future use. On the other side of the building, opposite to the cloth department; is the place where the garments are given out to the hands to be made up. Great responsibility rests with those in charge of this department; every portion of every garment brought in has to be carefully inspected, to make sure that the most minute details have been rigidly attended to, as well as the general style and finish imparted.

Having passed this examination satisfactorily, the prices are marked on the tickets, and the garments arranged on the counters for sale. Here we are again brought to the salesroom, which we passed through almost too hastily on our entrance. Observing carefully now, we find the quality and style of the clothing for sale, far superior to what our ideas of Ready Made Clothing had led us to expect. We find that

Messrs. PERRY & Co. have been endeavoring to supply a want long felt by the community. It was formerly considered necessary to go to a Merchant Tailor, and order clothing when any thing of superior quality was desired, but here we find an abundance of garments manufactured from the very finest materials, and in every respect complete to the requirements of the most fastidious taste.

It might be supposed that with such an enormous stock as they are compelled to carry, there would be large quantities of what is known to the trade as "old Stock" on hand. This, however, is not the case. The firm take good care to prevent this, by marking down goods left over to such prices as make them a very desirable purchase for dealers from surrounding towns, and also for many clothing houses in the city, whose fashions are not required to be so far in the lead as those of the Chestnut street establishments.

Messrs. PERRY & Co. have also a store, similar to the one described, at 303 Chestnut street. When this firm embarked in the business, they resolved that their success should be founded upon the principle of giving the customer the greatest possible value for the least money, and we believe they have faithfully and constantly adhered to the rule.

Bennett & Co.'s Clothing Bazaar,

Is another of the Philadelphia establishments that has been made famous by the enterprise of its proprietors. The building is five stories in height, and surmounted with a castellated tower, the most elevated point of which is one hundred and thirty-five feet above the ground. It is one hundred and eighty feet deep, with two fronts—one on Market and the other on Minor street. The Market street front is entirely of granite, in the Norman style of architecture, with circular-headed windows deeply recessed.

Over the entire entrance to the first story is a bay window, supported by a massive bracket, composed of suits of mouldings; and on each side is a balcony, finished on each party line with octagon turrets. It has a lofty tower in two sections—the first is rectangular, about twenty feet high, and capped with battlements and turrets on the angles; the upper section, of thirty feet, is an octagon figure, capped also with battlements. A light spiral stairway leads from the fifth story up into the tower.

Upon entering one of the three main doors of the front entrance, the visitor finds himself in a vestibule of Gothic style, laid with marble flagging, and beautifully ornamented. A few additional steps lead to the main room of the building. The first thing that arrests attention, and contrasts favorably with the usual darkness of such establishments, is the abundance of light that falls through three enormous skylights. Upon either side of the main room, except where the staircase interferes, are show-cases, divided by triple columns, with rolling glass-doors.

In the semicircular tops of these are busts of distinguished statesmen and poets. Tastily-designed stands also occupy the space. The whole of this room is used as a salesroom, except the space required for the general office at the Minor-street front. The Cashier's office is partitioned off in the centre on the eastern side, and fitted up with mahogany desks and every convenience.

Immediately in the centre of the building, and running the entire height, is a spiral staircase of great width and the most elaborate workmanship. The railing is of solid mahogany, and most massive. Viewed from the top or bottom, the aspect presented is magnificent.

The archings and angles of the ceiling throughout are richly ornamented. The skylight openings through the different stories are enclosed in each with heavy mahogany railing and primitive turned balusters, while the first floor under them is laid with plate-glass so as to give light to the cellar.

In its details the second floor equals if it does not not exceed the first. There are in this apartment twenty-five cases, of a style and decoration similar to those below, containing clothing of every possible description.

On the rear of the third and fourth stories, are partitioned off private rooms. The partitions are sash, in the same general style as the building. A stairway is on the rear, for the use of Minor street, and that front has a granite basement, and all above ornamental brick-work. About eight hundred hands, on the average, are employed in the manufacture of Clothing in and out of the house; and as they are employed the year round, the proprietors always have, what steady work commands, the best; and are not obliged, like those who employ hands periodically, to take whomsoever may offer, or go without. It is urged by some, as an argument against a large establishment, that to support it the profit must be large; but the argument is void of reason. It is well known that all men in business have more or less expense in conducting it, and that expense, whatever it may be, must be met with the profits. If therefore a man's expenses are ten dollars per day, and he has but two customers, he must make five dollars on each of them in order to make himself whole; but if his neighbor's expenses are one hundred dollars per day, and he has one thousand customers, then ten cents made on each man will meet the expenses. "LARGE SALES AND SMALL PROFITS" has ever been the guiding maxim with BENNETT, and this explains why such unbounded patronage has been bestowed upon Tower Hall by a discriminating public.

The proprietors buy on the best terms, directly from the first hands, and pay no double or triple profits, while their long experience enables them to select durable fabrics and manufacture them economically, and at the same time, with an eye to their durability, knowing that well-made Clothing is always presenting a silent but impressive argument, which the wise and prudent cannot resist.

BENNETT'S TOWER HALL, from its business arrangements, may justly be considered a model Clothing House.

But the firm which has displayed an amount of enterprise, that has no precedent among the business houses of Philadelphia, is that of

Wanamaker & Brown.

On that eventful day in our national history, when Fort Sumter was evacuated, two young men, JOHN WANAMAKER and NATHAN BROWN, commenced business as Clothiers in three small rooms, at the South-east corner of Sixth and Market streets. The circumstances of the times were not favorable for business success, and the friends of the young firm were not encouraging in their predictions. It is probable, however, that even their best friends had no conception of the force, vitality and energy latent in the members of this firm. They certainly had no conception of the agencies that would be employed to direct public attention to their efforts, for these were in many respects entirely original, and without precedent. The site selected by them for their business operations, has its historical associations. It was for many years occupied by the Schuylkill Bank, an institution well known to the old citizens of Philadelphia. It was a part of a large tract of one thousand five hundred acres of ground, which William Penn sold to Robert Greenway, for thirty pounds. But all the associations of the past fade into insignificance, compared with the interest imparted to it by the firm now occupying the building. Strangers from afar come and gaze upon the lofty structure as one of the city's peculiar landmarks. Messrs. WANAMAKER & BROWN have been liberal patrons of the press, and, in turn, newspapers, magazines, and books, have furnished wings, with which the fame of their establishment has flown to every city, town and village in the Union.

When this firm commenced business, they adopted as their cardinal rule, *vigilant attention to the wants of their customers.* As a guide in ascertaining the wants of the purchasing public, they kept a register in a Special Book of every suggestion made, or objection offered, from which information could be gained. Each salesman was required to enter in a book the assigned reasons why parties left without purchasing, and all objections that it was possible to obviate were remedied with studious care. The price was marked on each article, in plain figures, and no customer was allowed to leave dissatisfied, or feeling that he had been overcharged. The result of the adoption of this careful attention to the wishes and interests of buyers was, that in the first year their sales amounted to eighty thousand dollars. In two years it was found necessary to add to the rooms occupied by them, two additional stores on Sixth street. In less than five years it was found absolutely necessary to provide additional accomodations for a business that had increased to a half million of dollars annually. It was, therefore, deemed advisable to purchase the adjoining property on Sixth street, extending to Minor, and to remodel the entire building.

In February, 1866, the work of reconstruction was commenced. The undertaking was at once difficult and expensive. The small rooms of the former erection were required to give place to those of larger dimensions and better adaptation, while an immense stock of goods on hand, and all the pressure of a large business during the work of demolition, rendered the project yet more embarrassing. But the work went on, and every appliance that could add to strength or symmetry in an architectural point of view, was called into requisition in the form of columns, girders, trusses, brackets, etc.

The erection of the new Iron Front was especially a difficult undertaking, requiring great skill on the part of the mechanical engineer. A brick wall thirty-one feet by seventy, and six stories high, was to be replaced by a massive and beautiful front of Iron. The construction of this front was entrusted to Messrs. Royer & Brothers, iron founders of this city, and so well have they performed their work, that though extremely massive, the symmetry of the line, and the spaces allowed between the arches, give the whole a light and ornate appearance, that is not probably equalled by any other Iron Front in Philadelphia.

To describe in detail all that may be seen and learned in passing through this establishment, however interesting it might be, would be a task too voluminous for our purposes. The several uses to which the various stories of the house are devoted may, however, be mentioned.

Entering from Market street through the grand arched doorway, we find ourselves on the threshold of as fine a salesroom as our metropolis affords. On our left, extending to nearly midway, is the Gentlemen's Furnishing Goods Department—a moderate business emporium of itself. On our right, arranged in perfect order, are great piles of clothing, principally Business Coats. Proceeding onward, we reach the alcove, which has been mainly appropriated for the use of the book-keepers and cashiers, and immediately adjoining this is the private office—the inner sanctum of the guiding and controlling spirits which permeate the business machinery throughout the entire building, and make the whole move steadily on with the precision of clock-work. At the rear end of this floor is the department for Boys' Clothing, well stocked with garments for boys, youths, and children. The convenience of having this department on the first floor, is highly prized by ladies who come here to fit out their sons.

We ascend to the second story by a massive walnut staircase, reminding us of the grand staircase at the Capitol at Washington, and find ourselves in a spacious salesroom, elegantly fitted up, and used solely for the *Custom Department.* In glancing around over the maze of fabrics and busy life which here greet us, we fancy that if two ordinary cloth stores and a dozen old fashioned tailor shops were thrown together into one, the effect produced would be very similar.

The central and north portions of this floor present a stately gallery-like effect, from which the first floor is seen in its entire extent, and around which are arranged pants and vests in endless profusion. Every style, quality, and price of vest has its special gallery and bracket, by which even the very walls of the room are brought into requisition for storing purposes. Neat and convenient dressing-rooms, with all the etceteras, are here provided for the use of customers, and every facility afforded them to be well fitted.

As the *light* of this house is a very striking and most valuable feature to buyers, we will notice it here. The Sixth street side of the building is alone perforated with more than one hundred windows, through which the light has free ingress at all hours of the day. To purchasers the advantages of a corner store in this respect can hardly be over-estimated.

The third floor is wholly devoted to overcoats and fine dress coats. Forty-nine counters are here arranged, all loaded with garments—coats enough apparently to dress an empire—made up in every desirable mode and style, to suit people of all sizes, ages, tastes, and occupations.

The fourth story brings us to the Wholesale and Manufacturing Rooms, where a large portion of the garments are cut and sent out to be made up. On this floor, also, the *inspection of the work* is conducted, where every garment that enters the house is subjected to the closest scrutiny. Nothing that does not come up to the highest standard is accepted by the inspector, who justly insists upon the *best workmanship* in consideration of paying the fullest prices for making. We may notice here as an interesting fact, that the sewing hands of this house are paid regularly on the *delivery of their work.* We doubt not that the general adoption of this rule by clothiers would add greatly to the comfort and convenience of their hands. The result of this policy has been to insure to them at all times the very cream and choice of workmen.

In this room we encounter one of the most interesting phases of this branch of our manufactures. Scattered around are a large number of professional cutters. Here a batch of youths' garments are being cut out ready to be sent away; at the next counter, men are cutting dress vests; and next to that, handsome beaver walking coats are being got under way; there heavy chinchilla overcoats are receiving attention, and on the next table, men are cutting out suits to match; directly opposite are stacks of black pants waiting for the trimmers, and further on a set of coats for a Lancaster fire company are being marked out; by the side of the latter, orders are being filled for Washington; a report has just come up from the salesroom that the Melton suits are sold out and ordering the stock to be duplicated immediately, and at once preparations are set on foot to expedite the order. Every thing here moves with the exactitude and regularity of a machine; the thorough system which prevails throughout prevents collision, jar, confusion, or delay in the entire department. While the cutting above referred to is going on, a score of hands are waiting to receive the work, and carry it to their homes to be finished and placed in the salesroom.

The fifth story of the building is the Cloth, Cassimere and Trimming Room. In this room all the piece goods are examined with the greatest care, every facility being secured in order to detect the slightest imperfections, either in the face or texture of the articles, or in their strength. All imperfect goods are studiously rejected. After inspection, each piece of cloth goods is placed under the sponging rollers, when it is ready for the cutters. *Every piece of woollen goods that enters this house is sponged* before it leaves this room, and all linen fabrics are boiled, thus rendering subsequent shrinkage impossible.

This department of the house employs a special corps, whose whole time is occupied with the care of materials, a large proportion of which is specially made or imported for this firm. The amount of these fabrics now stored here is enormous. They are piled around like woollen mountains, and in them in incipiency repose the personal comfort and adornment of thousands.

The sixth story of the building is used only for storage purposes.

We would state, in concluding this part of our description, that in all the details the evidences of thoughtfulness for the business requirements of the firm are strikingly apparent. The hands who are employed in the upper part of the building enter it from Minor street by means of a bridge constructed along the eastern outside, that answers ingeniously the object intended.

We have already alluded to some of the principles in accordance with which the business of this house is conducted, but we may enumerate a few additional peculiarities. None but well known makes of goods are bought, either in fabrics or trimmings, and every customer has the assurance that what he gets is *reliable.*

The great extent and character of the stock is constantly before the proprietor's eye, recorded in a running stock-book, so that no part of the stock can become depleted without detection.

One gentleman is employed at a high salary, whose sole duty it is to receive and direct customers to the respective departments they wish to visit.

The clerk who sells, never delivers the parcel to his customer. He simply makes his ticket of sale, and another, whose business it is, packs and delivers.

The house is provided throughout with speaking-tubes and dumb-waiters, by which means all parts are brought into direct communication with the office, without incurring labor, or losing time.

Every garment in the house, from the lowest cost to the most expensive, is numbered, and has its history and constitutional make-up recorded for future reference. This sometimes becomes practically a most important feature.

The number of persons to whom this establishment now gives steady employment, by actual count, is over six hundred—men and women. Of these ten are clerks and book-keepers; forty-one salesmen and assistants; twenty one cutters; and the balance are hands making up clothing. The great majority of these persons have families which are supported from their earnings, so that it is no extravagant estimate to suppose that this establishment affords constant support to no less than twenty-five hundred people—enough to populate a county town. Hands are seldom changed, dismissals being rarely made, and never without indubitable cause. In this way the establishment has attained a degree of homogeneousness alike advantageous to the employers and the employed.

No one can predict with confidence the future of any great commercial house, but Messrs. WANAMAKER & BROWN certainly deserve continued success, and are entitled to the good wishes of all the citizens of Philadelphia, who appreciate enterprise, conjoined with a high order of mer cantile integrity.

TRANSPORTATION FACILITIES.

[Philadelphia, it will be seen, has recently increased her facilities for transportation, especially to the Southern ports. The Lines referred to in annexed pages are all reliable, cheap, and expeditious. The "SOUTHERN MAIL STEAMSHIP COMPANY" will soon have NINE FIRST-CLASS STEAMSHIPS running to the leading Southern ports.]

For New York.

SWIFTSURE TRANSPORTATION

COMPANY,

VIA DELAWARE AND RARITAN CANAL.

Leaving Daily at 12 M. and 5 o'clock, P. M.

THIS LINE is composed of EIGHTEEN FIRST-CLASS PROPELLERS, making their trips in about TWENTY-FOUR hours.

GOODS forwarded, free of all commissions, to any points North or East of New York.

APPLY TO

WM. M. BAIRD & CO., Agents,
132 South Delaware Avenue, Philadelphia.

WM. KIRKPATRICK,
Pier 10 North River, New York.

J. & N. BRIGGS,
40 South Street, New York.

FOR HARTFORD, CONN.

VIA DELAWARE AND RARITAN CANAL.

STEAM PROPELLERS ARE NOW RUNNING

EVERY WEEK TO THE ABOVE PORT.

FOR FREIGHT, APPLY TO

WM. M. BAIRD & CO.,
132 South Delaware Avenue.

PRINCIPAL HOTELS.

ASHLAND HOUSE,
Frederick Belcher, 709 Arch Street.

AMERICAN,
Samuel W. Hewlings, 515 Chestnut Street.

BINGHAM HOUSE,
Curlis Davis, S. E. corner Market and Eleventh Streets.

CONTINENTAL,
J. E. Kingsley & Co., S. E. corner Chestnut and Ninth Sts.

GIRARD HOUSE,
Henry W. Kanagy, N. E. corner Chestnut and Ninth Streets.

LA PIERRE HOUSE,
Baker & Farley, South Broad Street, below Chestnut.

MERCHANTS',
J. & W. C. McKibbin, 42 North Fourth Street.

MOUNT VERNON HOUSE,
M. & T. P. Watson, 119 North Second Street.

RIDGEWAY HOUSE,
J. B. Butterworth, 1 Market Street.

ST. CHARLES,
(European Plan), 64 North Third Street.

ST. LAWRENCE,
John H. Wiemann, 1018 Chesnut Street.

STATES UNION,
T. & B. Sanders, 606 Market Street,

UNION,
Christ & Weber, 317 Arch Street,

WALNUT ST. HOUSE,
J. B. Bloodgood, 2 Walnut Street.

WASHINGTON HOUSE,
J. B. De Haven, 709 Chesnut Street.

ARCH STREET,

ABOVE SEVENTH,

PHILADELPHIA.

THE ABOVE HOUSE HAS BEEN THOROUGHLY

RENOVATED

AND

REFURNISHED,

WITH EVERY CONVENIENCE.

The LOCATION is the most desirable, being but a short distance from all places of AMUSEMENT and BUSINESS portions of the city.

The more remote places of INTEREST are of easy access by numerous lines of PASSENGER CARS which pass the DOOR.

A. F. BELCHER,

PROPRIETOR.

INDEX.

www.ingramcontent.com/pod-product-compliance
Lightning Source LLC
LaVergne TN
LVHW021058110826
845150LV00001B/112

* 9 7 8 1 4 2 5 5 6 6 2 9 6 *